Metal Vapour Ion Lasers

Metal Vapour Ion Lasers:
Kinetic Processes and Gas Discharges

I. G. IVANOV
E. L. LATUSH
M. F. SEM
Rostov State University, Russia

Edited by
C. E. LITTLE
University of St Andrews, St Andrews, Scotland

Translated by
D. N. ASTADJOV
Institute of Solid State Physics, Sofia, Bulgaria
and
C. E. LITTLE
University of St Andrews, St Andrews, Scotland

John Wiley & Sons

Chichester · New York · Brisbane · Toronto · Singapore

National 01243 779777
International (+44) 1243 779777

e-mail(for orders and customer service enquiries): cs-books@wiley.co.uk

Visit our Home page on http://www.wiley.co.uk
or http://www.wiley.com

Other Wiley Editorial Offices

John Wiley & Sons, Inc., 605 Third Avenue,
New York, NY 10158-0012, USA

Jacaranda Wiley Ltd, 33 Park Road, Milton,
Queensland 4064, Australia

John Wiley & Sons (Canada) Ltd, 22 Worcester Road,
Rexdale, Ontario M9W 1L1, Canada

John Wiley & Sons (Asia) Pte Ltd, 2 Clementi Loop #02-01,
Jin Xing Distripark, Singapore 0512

Library of Congress Cataloging-in-Publication Data

Ivanov, I. G. (Igor 'Grigor'evich)
(Ionnye lazery na parakh metallov. English)
Metal vapour ion lasers : kinetic processes and gas discharges /
I.G. Ivanov, E.L. Latush, M.F. Sem / edited by C.E. Little ;
translated by D.N. Astadjov.
p. cm.
Includes bibliographical references and index.
ISBN 0-471-95563-9 (cloth : alk. paper)
1. Metal vapor lasers. I. Latush, E. L. (Evgeniĭ Leonidovich
II. Sėm, M. F. (Miroslav Frant͡sevich) III. Little, C. E.
TA1695.I8813 1996
621.36'63—dc20 96-3190
CIP

British Library Cataloguing in Publication Data

A catalogue record for this book is available from the British Library

ISBN 0 471 95563 9

Typeset in 10/12pt Times by Mackreth Media Services, Hemel Hempstead
Printed and bound in Great Britain by Biddles Ltd, Guildford, Surrey
This book is printed on acid-free paper responsibly manufactured from sustainable forestation, for which at least two trees are planted for each one used for paper production.

Contents

Preface **ix**

Editorial Preface **xiii**

Chapter 1. Pumping Mechanisms and Methods of Establishing Population Inversion **1**

1.1 The active media of metal vapour ion lasers 1
1.2 Pumping and the formation of population inversions in ions by electron collision 4
1.3 Excitation of excited ion states by heavy-body collisions 7
1.4 Pumping and inversion in recombining plasmas 13
1.5 Optical pumping and radiative decay 21
1.6 The influence of isotope shifts and hyperfine splitting of laser levels on population inversion and laser oscillation 22
References 24

Chapter 2. Gas Discharges Used in Metal Vapour Ion Lasers **26**

2.1 Types of discharge 26
2.2 Steady-state positive-column discharges 28
2.3 Pulsed positive-column discharges 46
2.4 Steady-state hollow-cathode and transverse-RF discharges 53
2.5 Pulsed hollow-cathode discharges 68
2.6 Efficiency of excitation of metal vapour ion levels in positive columns and negative glows 71
References 75

Chapter 3. Pulsed Metal Vapour Ion Lasers with Positive-Column Discharge Excitation **79**

3.1 Laser engineering 79
3.2 Lasers excited by electron impact 82
3.3 Lasers excited by heavy-body collisional energy transfer 87
References 113

Chapter 4. Metal Vapour Ion Lasers using Recombining Plasmas **116**

4.1 Laser oscillation in mercury, cadmium and zinc ions 116
4.2 Laser oscillation on ion transitions of Sr and Ca 119
4.3 Laser oscillation on transitions of other ions 133
4.4 Alternative mechanisms of ion-laser pumping in relaxing plasmas 141
4.5 Engineering of strontium and calcium ion recombination lasers 143
4.6 Pump mechanisms in segmented-plasma-excited recombination lasers 158
4.7 Laser-induced-plasma recombination lasers 160
References 162

Chapter 5. Positive-Column Excitation of CW Metal Vapour Ion Lasers **167**

5.1 Tube construction 167
5.2 Energy characteristics and pumping mechanisms 174
5.3 Optimum conditions for laser oscillation 180
5.4 Laser power and efficiency 188
5.5 Laser radiation noise 196
5.6 Technology and applications of cataphoretic lasers 192
References 194

Chapter 6. Metal Vapour Ion Lasers Excited by Fast Electrons **197**

6.1 Metal vapour ion lasers with hollow-cathode discharges 197
6.2 Metal vapour ion lasers with transverse-RF excitation 218
6.3 The efficiency of transversely excited metal vapour ion lasers 221
6.4 Electron-beam-excited lasers 223
6.5 The technology and applications of transversely excited metal vapour ion lasers 225
References 228

Chapter 7. Multi-Colour CW Metal Vapour Ion Lasers **233**

7.1 Multi-colour laser oscillation in metal vapours 233
7.2 Metal vapour ion lasers with combined active media 236
References 242

Appendices

A.1 Excitation potentials of metastable levels and first and second ionization potentials of atoms 244

A.2 The ranges of temperatures for which the vapour pressures of the elements lie between 0.13 and 133 Pa 247
A.3 Cross-sections and rate constants for de-excitation, recombination and heavy-body reactions in MVIL discharges 248
A.4 Laser transitions of metal vapour ion lasers 253
References 273
Index 281

Preface

Metal vapour lasers (MVLs) oscillate on the transitions of monatomic vapours of elements that do not exist in the gaseous phase under normal conditions. This class of lasers is taken to incorporate all the metals of the periodic table (mercury as well), but can also include elements such as selenium and tellurium which are not ordinarily classified as metals. Thus the term 'metal vapour laser' is used to embrace all lasers where special measures must be taken to produce the necessary vapour pressure of lasant atoms, due to the low saturated vapour pressure of the element at room temperature and atmospheric pressure. Another criterion for inclusion in the class of MVLs is that the lasant particles should be atoms or ions, but not molecules.

In MVLs, laser oscillation is obtained on the transitions of atoms or ions, and the properties of laser oscillation such as monochromaticity, coherence and divergence are determined by the same factors as for all gaseous lasers.

Laser oscillation was first achieved in metal vapours in caesium by optical pumping in 1961, and in atomic and ionic mercury by gas-discharge pumping in 1963 and 1964, respectively [1]. Activity in the search for new laser lines and the development of high-efficiency gas lasers gained momentum with the appearance of metal vapours as active media. To date, more than 700 lines have been obtained on the atomic and ionic transitions of 42 elements, which are not normally gases under normal conditions. With MVLs based on the transitions of neutral atoms, the best results have been achieved with self-terminating transitions from resonance to metastable levels in metals such as copper and gold — efficiencies of around 1% are available with high peak (100's kW) and average powers (100's W) in the visible region.

An important sub-division of the class of MVLs incorporates devices that oscillate on ion transitions. The present book deals with the physical processes in metal vapour ion lasers (MVILs), and is a survey of the major issues in the field.

The energy level structures of ions are often disposed to favour continuous-wave (CW) generation in the visible and UV spectral regions. Such MVILs offer a unique combination of high monochromaticity, coherence, low beam divergence and low noise; output powers from the milliwatt range through to around 1 W have been demonstrated (including the UV spectral region). Some devices can be made to oscillate on more than one transition in various regions of the spectrum simultaneously, including the blue, green and red, which makes possible 'white' light lasers for applications in colour holography, reprographics and image processing.

Pulsed MVILs have more recently been demonstrated to produce outputs in the visible and near-UV with pulse recurrence frequencies (PRF) of tens of kilohertz, and average output powers of several watts. Simultaneous oscillation on several lines is possible, as is discrete tuning.

It is worth noting that some of the proposed schemes for vacuum ultraviolet and X-ray lasers include transitions of multiply charged metal ions [2].

In the majority of MVILs the laser active medium is a weakly ionized 'non-equilibrium' plasma, produced by an electric discharge through a mixture of a noble gas and the metal vapour, or more rarely in the metal vapour alone. In these plasmas the levels of metal ions are populated in processes that do not ordinarily occur in neutral-atom MVLs and noble-gas ion lasers, processes such as charge transfer (Duffendack reactions), Penning ionization, and other collisions between atoms and ions of the metal and of the noble gas. These processes are energetically allowed since the first ionization potential of most elements is less than the potentials of excitation and ionization in noble gas atoms, whereas the second ionization potential of most elements is higher. Hence, such collision processes tend to populate the energy levels of singly ionized metal atoms. By careful selection of a metal and noble-gas atom pair, high rates of reaction can be obtained for population of the chosen ion levels and they can be strongly pumped. Because of the large number of ways to pump metal ion levels, most of the lines obtained in metal vapours (>500) are ion transitions.

Research on laser oscillation in metal ions has been carried out for over 25 years, and many experimental and theoretical data have been gathered, although they wait to be systemized. The few survey papers that have been published have not considered many of the experimental studies. In many respects they are out of date, and much of the work carried out in the USSR has been overlooked. The field of MVILs is not well served by the literature. Other books and surveys published on the subject have been devoted to either atomic and molecular gas lasers [3], or self-terminating lasers based on neutral metal atoms [4,5]. The purpose of this book is to fill the gap in the available general knowledge of MVILs. The present time is apt because MVILs are now making the transition from lab devices to commercial products.

Laser oscillation on metal ions has largely been obtained and studied with gaseous discharge excitation, and most of our attention will be focused on ion laser transitions in various types of gaseous discharge. The two newer techniques of pumping by electron beam and by laser-produced plasma shall also be examined.

Since the search for new laser transitions in ions is a continuing one, we list for the convenience of the researcher the laser lines obtained from metal ions to date. In an appendix, we also present cross-section data for important processes in MVILs, temperatures required to attain the necessary working pressures for different elements, their ionization potentials and the potentials of excitation of the metastable states of atoms.

Compared with the Russian edition (1990), this English version of the book has been greatly updated with the latest material on MVILs, as well as having been corrected.

The authors are deeply grateful to all the colleagues at the Physics Department and the Physics Research Institute of Rostov State University who at various times took part in discussions which aided the book so much. The authors are very grateful to the editor of this book, Dr C. E. Little, first for initiating this English edition and, second for invaluable help in the course of work on it.

Likewise, we are thankful to Dr D. N. Astadjiv together with Dr C. E. Little for their excellent work in translating this book from troublesome Russian.

This English edition of the book gives a much better opportunity for interested readers worldwide to become acquainted with the material discussed, and we hope this acquaintance will be of some value to them.

The Authors
Rostov State University

REFERENCES

[1] *Laser Handbook*, ed. A. M. Prokhorova, Moscow: Sovyetskaya Radio (1978) (in Russian).
[2] L. I. Gudzenko and S. I. Yakovlenko, *Plasma Lasers,* Moscow: Atomizdat (1978) (in Russian).
[3] A. V. Yeletski and B. M. Smirnov, *Physical Processes in Gas Lasers,* Moscow: Energoatomizdat (1985) (in Russian).
[4] G. G. Petrash, *Sov. Phys. - Usp.* **14**(6), 747–65 (1972).
[5] A. N. Soldatov and V. I. Solomonov, *Gas-Discharge Lasers Utilizing Self-Terminating Transitions in Metal Vapours,* Novosobirsk: Nauka (1985) (in Russian).

Editorial Preface

The Russian language original of this book was the first comprehensive review of the research and development of metal vapour lasers that oscillate on transitions of free metal ions. Like most texts published in Russia, and in the old Soviet Union, many excellent scientific books, written by some of the leading practitioners in their respective fields, have never been made available in any English language edition. This translation is the first text in the English language to be published on pulsed metal vapour ion lasers. The field of metal vapour lasers has been an active one in both the East and the West for over 30 years now, and the excellent properties of these lasers for many important applications in industry, medicine and science continue to stimulate the development of new devices as well as deeper investigations of their fundamental properties. This book should be of interest to many researchers and developers of metal vapour lasers and their applications from around the world.

The editor has enjoyed his task immensely, not only because the book is on one of his fields of interest, but also because it allowed him to learn and appreciate some of the intricacies of the Russian language.

Chris E. Little
St Andrews

1

Pumping Mechanisms and Methods of Establishing Population Inversion

1.1 THE ACTIVE MEDIA OF METAL VAPOUR ION LASERS

The working medium of a metal vapour ion laser (MVIL) is usually a mixture of the metal vapour and a noble gas, the latter referred to as the 'buffer' gas. The proper selection of this mixture can allow effective pumping of metal ion laser levels due to collisional transfer from excited buffer-gas atoms or ions. However, even in those cases when buffer-gas species do not participate directly in the kinetics of the laser levels, the presence of a buffer is often necessary. First, the buffer gas permits a discharge to be initiated before the metal vapour is present; second, with a buffer gas, there is a reduction in the loss of the metal atoms by diffusion to the ends of the laser tube; third, in some cases the buffer gas permits transport of metal atoms along the tube by cataphoresis; fourth, buffer-gas atoms may also facilitate introduction of metal atoms to the discharge by cathode sputtering. The electron temperature T_e in the discharge is a strong function of the buffer-gas pressure, the variation of which permits some control of T_e. In particular, high pressures of a light buffer gas are used to promote strong cooling of electrons after removal of excitation, which is of importance for the operation of recombination pumped lasers. In a few cases it may be feasible to obtain ion laser oscillation using the pure metal vapour; however there are many serious difficulties that arise in practice, which may be solved simply by the addition of a buffer gas.

The ionization potentials of metal atoms are about two or more times lower than those of the noble gases (helium, neon etc.). For excitation of the electronic levels of metal ions, particles (electrons, excited buffer-gas atoms or ions) with high (15–25 eV) energies are required, and the vapour pressure of easily ionized species must be much lower than the buffer gas pressure to maintain a high T_e. Therefore, in MVILs the partial pressure of the metal is usually small (0.1–10 Pa).

Various techniques are used to produce the required metal vapour pressure. In the majority of MVILs, the metal vapour is introduced to the laser tube where a discharge through a noble gas is already present. The active medium containing the metal vapour lies between the electrodes, and the buffer gas prevents diffusion of the metal atoms to the window regions of the tube, which are at room temperature. The most obvious method of introducing metal vapour is to heat the discharge tube, along the inside of which are distributed pieces of the metal, to provide the necessary vapour pressure; alternatively, the metal is placed in a heated side extension to the main tube. In the latter case, the temperature of the main discharge tube is usually maintained slightly above that of the side-arm, so that the vapour pressure in the active region may be controlled by the side-arm temperature. Without a buffer gas, the situation is considerably complicated by the need to heat the entire discharge chamber, including the window regions.

The introduction of metal vapour from a reservoir is extremely convenient, since it permits independent control over discharge current and vapour pressure. This is particularly important for investigations where a search is being made for laser oscillation in metal atoms, or for the optimization of laser characteristics. A drawback of the method is the requirement for one or more bulky ovens (especially for high temperatures).

To heat metal that is located in the discharge channel (or close to it), it is possible to use 'waste' power from the discharge, in which case no external heater is necessary. If high temperatures are required, the active region may be insulated to reduce the rate of heat loss. Lasers utilizing this technique are referred to as 'self-heated'.

Self-terminating lasers based on the vapours of copper, gold and other metals employ self-heating to provide temperatures of up to 1700 °C in the active region [1, 2]. A shortcoming of self-heating, however, is that the vapour pressure of the metal depends on the average power delivered to the discharge tube. Since both metal vapour pressure and input pulse energy strongly affect the laser output energy, the optimum temperature for pulsed lasers may (to a large extent) be obtained by selection of the pulse recurrence frequency, whilst maintaining the input pulse energy close to that optimum for laser output energy. In the case of lasers with continuous excitation, other means of thermal management must be employed to secure optimum input power simultaneously with optimum operating temperature for laser oscillation.

To reduce the temperature necessary for laser operation (which is particularly important for elements requiring high temperatures to provide a sufficient vapour pressure) it is possible to replace the elemental metal with a more volatile molecular donor of the metal. In this case, metal atoms are produced in the active region by dissociation of the metal-bearing molecules in the discharge. The method is widely applied to metal vapour lasers utilizing atomic transitions [1, 2]; however it has not been applied successfully to metal vapour ion lasers. Experiments [3, 4] to investigate the application of metal halides to MVILs demonstrate laser output powers that are lower than those obtainable from lasers using the pure metal. This may be attributed to the reduction in T_e by collisions of electrons with the halide dissociation products in the discharge; ion lasers are considerably more sensitive to the parameter T_e than atomic lasers.

In continuous-wave (CW) MVILs, it is possible to obtain a uniform concentration of metal atoms along the length of the discharge by making use of cataphoresis [5–8]. With this method metal vapour is introduced to the discharge from a small heated reservoir located close to the anode. Since the metal atoms have ionization potentials that are considerably lower than those of the noble-gas buffer atoms, the metal is ionized preferentially and transported along the discharge towards the cathode as ions. In this manner the active region is seeded with metal vapour. In a relatively cool region near the cathode, the metal condenses out of the discharge. Discharge heating prevents condensation of the metal on the tube wall in the active zone (see section 2.2). The use of cataphoresis has led to the commercial realization of CW MVILs for the visible and near-UV spectral regions with laser tubes little more complicated than those employed for He–Ne lasers.

The small concentrations of metal atoms typically required in MVILs can also be established by the sputtering of a metal cathode in hollow-cathode discharges or in transverse RF discharges [9]. The technique is especially useful for non-volatile metals that sputter well. Its use has permitted laser generation, notably in the UV, on transitions of copper, silver and gold ions [10–12]. There are, however, complications in the application of the method due to the interdependence of metal atom density and discharge current, which can lead to unstable discharges at large currents, and other effects (see Chapters 2 and 6).

In metal vapour lasers, the laser medium (a mixture of the metal vapour and a noble gas) becomes active when the gas discharge produces a 'non-equilibrium' plasma, in which a population inversion is established between two levels, 2 (upper) and 1 (lower), i.e.

$$n_2 > (g_2/g_1)n_1 , \tag{1.1}$$

where $n_{1,2}$ represent the population densities, and $g_{1,2}$ are the statistical weights of levels 1 and 2. In a steady-state non-equilibrium plasma, when the excitation of levels is characterized by the pump rates W_1 and W_2, equation (1.1) leads to the following well-known relation:

$$W_2\tau_2 > W_1\tau_1 \cdot g_2/g_1 , \tag{1.2}$$

where τ_1 and τ_2 are the lifetimes of the levels 1 and 2.

In the pulsed regime, condition (1.1) can be realized, even though the condition (1.2) may not be satisfied in the steady state. This is a result of transient processes in the rapid creation of a plasma (ionization regime), or in the relaxation of a plasma following rapid removal of its excitation source (the recombination regime[1]). Thus, laser oscillation can occur on a larger number of transitions in the pulsed regime than can be obtained with continuous excitation. Moreover, in the pulsed regime it is easy to realize excitation conditions of high input power which, in the continuous regime, it would be very difficult to attain, and consequently the peak laser output power can be considerably higher than that available using continuous excitation. It should also be noted that investigations in the pulsed regime permit observations of the transient dependences of excited-state populations and laser oscillation, and often facilitate the identification of pumping mechanisms that lead to population inversion.

In other aspects, CW generation can have definite advantages (highly stable power and frequency of laser output, high monochromaticity and coherence) of importance in a number of practical laser applications.

The distribution of populations in excited states, and therefore the possibility of producing population inversion, is determined by the excitation (pumping) and de-excitation processes. Pumping processes to be considered in the population of excited states in ion lasers include the following: inelastic collisions with electrons (either direct or stepwise), energy transfer by collision from excited atoms or ions, recombination into excited ion states, cascades of transitions from higher-lying levels (by radiation or collision), and absorption of photons. Deactivation of excited states can be realized by radiation (spontaneous emission), or by collisions (with electrons or heavy-body species). We shall now consider a few details of the above pumping and deactivation processes in relation to ion energy levels.

1.2 PUMPING AND THE FORMATION OF POPULATION INVERSIONS IN IONS BY ELECTRON COLLISION

1.2.1 Excitation by Electron Impact

Excitation of ion levels by collision with electrons can be accomplished from the atomic ground state (direct excitation)

$$A + e \longrightarrow A^{+*} + 2e \ , \tag{1.3}$$

or from the ion ground state (stepwise excitation)

$$A^{+} + e \longrightarrow A^{+*} + e \ , \tag{1.4}$$

where A is a lasant atom in its ground state, A^+ is a ground-state ion, A^{+*} is an excited ion, and e is an electron. The rate of pumping, W, in process (1.3) is

$$W = nn_e\langle q'v\rangle \propto n_e\,, \tag{1.5}$$

and in process (1.4) is

$$W = n^+n_e\langle q''v\rangle \qquad \sim n_e^2 \text{ (since } n^+ \sim n_e) \tag{1.6}$$

where $\langle q'v\rangle$ and $\langle q''v\rangle$ are averaged products of excitation cross-section q and electron velocity v, often referred to as rate constants; n_e, n^+, n are concentrations of electrons, ground-state ions and ground-state atoms, respectively. In the first case, the pumping rate is approximately proportional to current, and in the second, varies quadratically with discharge current. Often the dependence of the pumping rate (and output power) on discharge current allows processes (1.3) and (1.4) to be distinguished.

The cross-section q for excitation by electron impact, which depends on the electron energy distribution function, is well known to be characterized by a threshold electron energy below which excitation is not possible, a maximum, and a subsequent decrease in magnitude. The peak value and form of $q(v)$ depends on the type of excited level (or transition) involved.

In the Born approximation, the cross-section is proportional to the square of the matrix element of the transition. The cross-section is greater for levels that decay quickly by radiation, thus often making production of a population inversion (especially steady state) by electron collisional pumping difficult.

Experimental data on the cross-sections for excitation of ion levels by electron collision are sparse in the literature. Therefore, in order to calculate the pump rate, one is usually forced to use empirical formulæ and approximations [13].

The frequency of electron excitation, $\nu = n_e\langle qv\rangle$, depends on the types of excited levels involved (which determine, as noted above, the form and magnitude of $q(v)$ at the peak), and the type of discharge and its conditions (which determine n_e and the electron energy distribution function).

Since the energies of the excited ions produced in reactions (1.3) and (1.4) are in general high, only the small fraction of electrons that belongs to the high-energy region of the electron energy distribution function can perform the excitation. This contributes to the overall low efficiencies of gas-discharge ion lasers in general, and metal vapour ion lasers specifically.

In order to increase the effectiveness of excitation of ion levels by electron collision, it is necessary to increase the electron temperature in the discharge. This is usually achieved by lowering the gas pressure p or by raising the electric field E between the electrodes, so that the ratio E/p, which governs the electron temperature in the discharge, is made as high as possible.

Other ways of increasing the effectiveness of excitation include the use of discharges, such as hollow-cathode (HC) discharges or transverse RF discharges, in which there exist groups of high-energy electrons that can effectively excite ion levels, or alternatively a beam of accelerated electrons may be used for excitation.

It should be noted that steps taken to increase pumping by increasing the concentrations of the metal atoms (see 1.5) are limited by the fall in electron temperature which inevitably accompanies such a move. Using this method, the best results are attained in types of discharge where the increase in metal vapour density, n, does not strongly reduce the number of fast electrons capable of exciting the ion levels (see Chapter 2).

For excitation of ions from the ion ground state, the largest cross-sections ($\sim 10^{16}$ cm^2) correspond to excitation of the resonance levels. These cross-sections exceed those for direct excitation to ion levels from the atomic ground state by typically two orders of magnitude, whereas the concentration of metal ions can be one tenth or more of the ground-state population of metal atoms. In a similar manner to atomic resonance levels, ion resonance levels are most effectively pumped by electron collision from the ion ground state, leading as a result to the highest laser powers being attained in the pulsed regime on transitions from resonance levels to metastable levels [1, 2]. The selectivity of pumping by electron collision depends of course on the natures of the upper and lower laser levels. The absence of an allowed decay channel for the lower level leads to the self-terminating character of these laser transitions.

1.2.2 Deactivation of Ion Levels by Electron Collision

As well as excitation of ion levels by electron collision, it is possible to induce de-excitation of ion levels. Using the Bethe approximation [13] for cross-sections, it is possible to show that, for allowed transitions, the rate constant for electron de-excitation from an upper level 2 to a lower level 1 is

$$k_{12} = <q_{12}v> \approx 3.2 \times 10^{-6} f_{12} \frac{g_1}{g_2} \frac{1}{\Delta\varepsilon_{21}\sqrt{T_e}}, \tag{1.7}$$

while for the reverse process—electron excitation—the Principle of Detailed Balance can be applied to write

$$k_{12} = <q_{12}v> \approx 3.2 \times 10^{-6} f_{12} \frac{1}{\Delta\varepsilon_{21}\sqrt{T_e}} \exp\left(-\frac{\Delta\varepsilon_{21}}{T_e}\right), \tag{1.8}$$

where $\Delta\varepsilon_{21}$ is the energy difference between levels in electron-volts; f_{12} is the oscillator strength; T_e is the electron temperature in electron-volts. From

(1.7) and (1.8), it is obvious that the probability of an inelastic collision with an electron is greater the smaller $\Delta\varepsilon_{21}$ is, and for low T_e the probability of electron de-excitation can significantly exceed the probability of excitation. Note that the probability for electron de-excitation in the (low energy) threshold region of the cross-section is considerably higher for ions than for atoms, since $T_e < \Delta\varepsilon_{21}$, and

$$\frac{k_{21}(\text{ion})}{k_{21}(\text{atom})} \sim \left(\frac{\Delta\varepsilon_{21}}{T_e} + 1\right)^{1/2}. \tag{1.9}$$

The rate constant for de-excitation by electron collision in the threshold region is 10^{-4}–10^{-5} cm^3 s^{-1} for $T_e \approx \Delta\varepsilon_{21} \sim 0.2$–1.0 eV. Therefore, for electron concentrations of 10^{11}–10^{14} cm^{-3}, which are normally produced in gaseous lasers, the probability of transitions to lower levels by electron collision is often of the same order of magnitude as the probability of radiative decay, and for certain transitions (for which $\Delta\varepsilon_{21}$ is low and the probability of radiative decay is low), transitions to lower levels are governed primarily by electron collisions. As a result, de-excitation by electron collision can have an important influence on laser level populations.

The role of electron-impact de-excitation in the formation of population inversions in ion lasers was first established in [14], where it was shown that without including this process it is not possible to explain experimental observations of pulsed cadmium and zinc ion laser characteristics (see Chapter 3). Electron de-excitation appears as an important mechanism in steady-state discharges, e.g. in HC discharges [15], where there can be large populations of slow electrons.

The exploitation of electron de-excitation is favoured by particular configurations of energy levels in ion lasers. Such cases occur where de-excitation channels pump the upper laser level from high-lying levels, and where electron collisional de-excitation augments the clearance of lower levels by radiation, as is the case for certain transitions in ions of strontium and calcium (see Chapter 4). If the disposition of energy levels is unfavourable from the point of view of electron collisional kinetics, electron de-excitation can be the cause of rapid saturation of laser output power with current.

1.3 EXCITATION OF EXCITED ION STATES BY HEAVY-BODY COLLISIONS

1.3.1 Types of Collisions

In discharges through mixtures of gases, or a metal vapour and a buffer gas, it is possible to populate ion levels by the transfer of energy from excited atoms (including metastables) or ions of one species to atoms of another in the mixture in processes called collisions of the second kind [16]. The ionization potentials of the majority of elements lie below the energies of excitation of the noble gases, which are most often employed as buffer gases, and therefore collisions of the second kind with excited noble-gas atoms lead to excited ions of the metal. In this case, the pump rate is determined by the relationship

$$W = n_A N_B \overline{QV}, \tag{1.10}$$

where n_A and N_B are the concentrations of species A and B; Q is the collision cross-section; V is the relative speed of the particles.

The processes are used to greatest effect when the concentrations of the colliding particles are large, which occurs when the participating atoms are in the ground and metastable levels of atoms and ions, as in charge transfer (Duffendack reactions), for example:

$$A + B^+ \longrightarrow A^{+*} + B + \Delta\varepsilon \ , \tag{1.11}$$

or in so-called Penning ionization (or Penning collisions)

$$A + B(m) \longrightarrow A^{+*} + B + e\,. \tag{1.12}$$

In these and subsequent reactions, A and B are lasant and buffer-gas atoms in their ground states; $A(m)$, $B(m)$ are atoms in metastable levels; A^+, B^+ are ground-state ions; A^{+*} is an excited ion; and $\Delta\varepsilon$ is the energy defect of the reaction.

There are also other types of energy-transfer collisions which need to be taken into account when discussing pumping and inversion in ion lasers. For example, when the energy of double ionization of metal atoms is less than the first ionization potential of the buffer-gas atoms, process (1.11) can assume the form

$$A + B^+ \longrightarrow A^{2+} + B + e\,, \tag{1.13}$$

where single-step heavy-body collisions lead to the production of doubly ionized metal atoms. The existence of reaction (1.13) was first experimentally established for helium-magnesium mixtures [17], and subsequently for other mixtures of alkaline-earth metals with helium and neon [18]. When He is used as the buffer gas, the process is energetically allowed for 30 chemical elements. Reaction (1.13) leads to the production in discharge afterglows of doubly charged ions, which can undergo recombination with electrons to populate excited levels in singly charged ions, thus producing a population inversion in them (see Chapter 4).

Reaction (1.12) can also lead to the formation of doubly charged ions, provided that the potential for double ionization is less than the energy of the metastable buffer-gas atoms:

$$A + B(m) \longrightarrow A^{2+} + B + 2e\,. \qquad (1.14)$$

Excitation of ion levels may also be obtained by other processes, referred to as ‘stepwise’ processes, in which metastables or ions of the lasant species participate, e.g.

$$A(m) + B(m) \longrightarrow A^{+*} + B + e\,; \qquad (1.15)$$

$$A(m) + B^{+} \longrightarrow A^{+*} + B + \Delta\varepsilon\,; \qquad (1.16)$$

$$A^{+} + B(m) \longrightarrow A^{+*} + B + \Delta\varepsilon\,. \qquad (1.17)$$

Their roles in pumping laser levels are not as important as those of processes (1.11) and (1.12), since the concentrations of excited and ionized lasant atoms are usually lower than the ground-state concentration. Process (1.15) was proposed in [19] as an explanation for the selective pumping of the upper level of the 615 nm mercury ion transition. Process (1.16), according to the data of [15, 20], is responsible for charge transfer to excited levels of arsenic and copper ions, where the excited levels lie above the limit due to charge-transfer excitation from the ground states of those ions.

If the energy of metastables B(m) is sufficient to produce doubly charged ions from singly charged ones as a result of collisions of the type

$$A^{+} + B(m) \longrightarrow A^{2+} + B + e\,, \qquad (1.18)$$

then ion transitions will appear in the afterglow following recombination of the doubly charged ions. This is observed, for example, in mixtures of mercury and helium [21].

With regard to the cross-sections for reactions (1.11–1.18), we may make the following general remark: larger cross-sections exist for those reactions in which the smallest number of bound electrons change their states. For example, it is well known that resonant charge transfer between atoms of the same species, with electron transfer without alteration of the states of the other electrons bound to the colliding particles, has a large cross-section ($\sim 10^{-14}$ cm^2) [16]. With charge-transfer excitation (process (1.11)), two electrons change state—one changes state in the noble gas ion, which is neutralized, and the other is excited to the metal ion state. Typically, the cross-section for charge transfer with excitation of the resonance level is smaller[2]: $\sim 10^{-15}$ cm^2. When a Penning collision results in the formation of an excited ion (process 1.12)), the states of as many as three electrons are altered (de-excited helium, ionized and excited metal), and consequently for this case we expect small cross-sections. The cross-sections can be large, however, as one might expect from a process similar to (1.11), when the metal ion is formed in the ground state and the process again involves two electrons.

Two-electron changing processes appear in reactions of charge transfer with ionization (1.13), Penning ionization of ions (1.18), and reactions (1.16) and (1.17). For those processes large cross-sections can be expected. Reactions (1.14) and (1.15) involve three electrons; therefore their cross-sections should be much smaller, which is confirmed by experimental data (see Appendix 3). We shall now discuss in detail the characteristics of pumping excited states via the heavy-body assisted processes of Penning ionization and charge transfer.

1.3.2 Penning Ionization

Penning ionization is not a resonant process, since the energy defect can be transferred to the liberated electron. Therefore, using this process it is possible to excite all ion levels that lie at energies below those of the metastable levels of the buffer gas. According to [22–24], the cross-sections for Penning ionization are greater, the larger the energy difference between the metastable levels of the buffer gases and the excited levels of ions, and maximize for ionization into ion ground states (see Appendix 3). It should be borne in mind that attempts to increase the concentration of metastable atoms in the discharge by increasing the discharge current are limited by the occurrence of stepwise ionization, which quickly leads to saturation of the density of metastables with increasing current, an effect that has been noticed in helium–neon lasers [25].

The non–selective nature of Penning collisions requires favourable relative lifetimes of the laser levels for the production of inversion, which fact limits its application in establishing inversions in ion lasers.

However, for the ions of some metals, Penning ionization can provide preferential excitation of upper laser levels. For example, this situation occurs for cadmium and zinc (electron configurations of the atomic ground states are $nd^{10}(n+1)s^2$; n equal to 3 for zinc and 4 for cadmium). It is possible to ionize these elements either by removal of an outer s electron, forming the ion in its ground state $nd^{10}(n+1)s$, or by removal of an inner d electron, forming the ion in the 'ground' state $nd^9(n+1)s^2$ – this term is $^2D^*$. Between these states in cadmium and zinc lie the lowest (resonance) terms $^2P_{1/2,3/2}$ of the $nd^{10}(n+1)p$ configuration, which are optically connected to the states $nd^{10}(n+1)s$ and $nd^9(n+1)s^2$ (Fig. 1.1).

All three states may be excited by Penning ionization involving helium

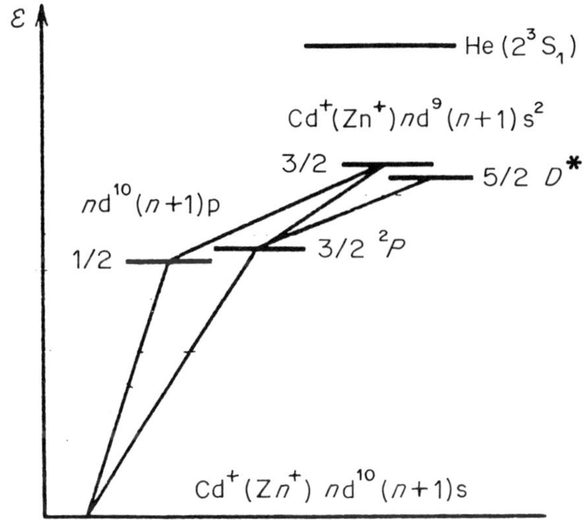

Figure 1.1 Partial energy level diagram of Cd^+.

metastables. However, since excitation to 'ground' configurations (two-electron transitions) is more intense than to excited ion states $^2P_{1/2,3/2}$ (three-electron transitions), excitation of the states $^2D^*$ (upper laser levels) by Penning collisions in these elements is greater than excitation of the ^2P states (lower laser levels). According to [22], the relative rates of excitation to these two terms is 5:2. When we also take into account the longer lifetimes of $^2D^*$ compared with $^2P_{1/2,3/2}$, it becomes clear why in mixtures of cadmium or zinc with helium strong laser action on these transitions is obtained by Penning collisions.

1.3.3 Pumping and the Production of Inversion due to Charge Transfer

In contrast to Penning ionization, charge transfer is a resonant process. The successful application of charge transfer to pumping levels in gas lasers has encouraged theoretical and experimental studies of charge transfer between particles colliding with thermal velocities, i.e. for conditions pertaining to gas laser discharges [26–28]. This work shows that the greatest effective cross-section for charge transfer is attained for values of the energy defect, $\Delta\varepsilon$ of 0.2–0.5 eV. The largest cross-sections are of magnitude ~10^{-15} cm^2 (see Appendix 3).

The selectivity of the charge transfer process can be utilized by choosing the appropriate combination of buffer gas and lasant species; usually only the chosen excited ion state (upper laser level) is strongly pumped. By altering the buffer–lasant pair, a great number of upper levels with large pumping cross-sections can be obtained.

The selectivity of charge transfer also makes the requirements of relaxation lifetimes of the laser levels less stringent than would otherwise be the case (e.g. with Penning ionization), and a larger number of potential laser transitions is found. Furthermore, in some cases lasing occurs from levels that are not excited directly by charge transfer, but via transitions (radiative or collisionally induced) from higher-lying levels that are excited directly, and for which conditions of efficient charge-transfer pumping exist. This is the reason for the large number of laser transitions pumped by charge transfer in the ion spectra of many elements (see Chapter 3 and Appendix 4). It is important to note that the pump rates for charge transfer are often greater than those for collisions with metastable atoms, since the ion concentration increases with increasing current and can greatly exceed the concentrations of the metastables.

To ensure the effective creation of inversion by charge transfer, the laser levels should satisfy the following criteria:

(i) The upper laser level should have a large cross-section for excitation due to charge transfer, and low probabilities of deactivation by other processes aside from the emission of laser radiation (the probability of radiative decay should not be too high);
(ii) Pumping of the lower laser level other than by stimulated emission should be minimal, and the level should be effectively cleared by collision or by emission of radiation.

A large pump rate is attained if $\Delta\varepsilon$ for the upper laser level corresponds to the maximum cross-section for excitation, and there are no other levels that can be effectively pumped by charge transfer. In all cases, it is desirable that the number of levels be minimal—otherwise the pumping will be distributed between the various levels, and their individual populations become too small for inversion.

Minimal pumping of the lower laser level is realized if there are no strong radiative or collisional transitions to it from higher-lying levels, and there is no probability of it being excited by electron impact or by radiation trapping; the latter condition requires that the lower laser level is not an ion resonance level. Of course, how closely the laser levels of any type of ion approach the ideal form in reality determines the attainable laser output parameters in each case.

1.3.4 Clearance of Lower Laser Levels by Atomic Collisions

In order to clear the lower laser level in an MVIL, it is possible to use collisions with gas (vapour) additives. The additives must be selected so that as a result of the collision, the additive atom C deactivates the excited ion from the lower laser level, A^{+*}. An example is charge transfer:

$$A^{+*} + C \longrightarrow A + C^{+} + \Delta\varepsilon\,, \tag{1.19}$$

which occurs with a frequency

$$\nu = N_c \overline{QV}. \tag{1.20}$$

The correct selection of gas atoms C can enable quasi-resonant charge transfer with a cross-section $\sim 10^{-15}$ cm^2, yielding a highly selective clearing process.

Nevertheless, because upper levels of ion laser transitions normally have large Einstein A coefficients of 10^7–10^8 s^{-1}, lower level clearing is only effective for values of ν of the same order of magnitude. This requires a concentration of the additive N_C (assuming $\overline{QV} = 10^{-10}$ cm^3 s^{-1}) of around 10^{18} cm^{-3}, which usually poses a serious problem for gas discharge lasers in that the additive may affect the discharge characteristics and reduce the pumping rate to the upper laser level.

1.4 PUMPING AND INVERSION IN RECOMBINING PLASMAS

1.4.1 Characteristics of Recombination Pumping

For the pumping of ion laser transitions it is most appropriate to use the high rates of collisional or collisional–radiative recombination found in 'dense' plasmas. When molecular ions are present, it is possible to make use of dissociative recombination.

The first mention of the possibility of using recombination to pump laser levels and sustain inversion was made in [29]. Theoretical studies of plasma recombination lasers are summarized in [30] (see also [31]). Despite the stimulation of experimental work provided by the attainment of lasing on strontium and calcium ion transitions with recombination pumping in pulsed discharge afterglows [32], the number of theoretical studies still outnumbers the reports of experiments.

Recombination pumping has a number of important advantages. First, in the recombination regime the flux of excited atoms is from high-lying levels downwards. This guarantees a high effective pump rate to the upper laser level, and eases the restrictions on energy-level structure required for an inversion compared with excitation in the ionization regime, where the flux of excitation is from lower levels upwards. As a result, in the recombination regime, stray pumping of the lower laser level is found to be smaller. Second, the potential energy of ionized atoms is effectively utilized; this energy can be stored to a significant degree during the course of a current pulse. Third, as will be shown below, in the recombination regime, collisional transfers between levels are established due to impacts with slow electrons. These transfers considerably increase the range of possible laser transitions, and widen the regime that permits significant population inversions to be attained. Because absorption of radiation does not affect the lasing process, it is possible to volume-scale the active medium to increase the available output power.

In three-body recombination with two participating electrons,

$$A^{Z+} + e + e \longrightarrow A^{(Z-1)+} + e\,, \tag{1.21}$$

the rate of decay of the concentration of multiply charged ions after terminating the current pulse (neglecting diffusion losses) is given by the equation

$$dn^{Z+} / dt = -\alpha n^{Z+} n_e^2\,, \tag{1.22}$$

where α is the recombination coefficient, leading to[3]

$$n^{Z+} = n_0^{Z+} \exp(-t / \tau_r);\ \tau_r = 1 / \alpha n_e^2\,. \tag{1.23}$$

The coefficient of three-body recombination depends on the charge of the ion, z, and the electron temperature, T_e:

$$\alpha = 9.21 \times 10^{-27}\, Z^3 L_C T_e^{-9/2} = c(Z) T_e^{-9/2}, \tag{1.24}$$

where $L_C = \ln(Z^2 + 1)^{1/2}$ is the Coulomb logarithm, T_e is in electron-volts and α is in units of $cm^6\ s^{-1}$. It is clear from (1.24) that the rate of recombination, and therefore the pump rate of levels, greatly increases with reducing T_e and with increasing n_e and n^{Z+}, and also with the charge of the recombining ions.

Even in the discharges of pulsed gas lasers it is difficult to obtain significant concentrations of ions with $Z > 2$, and the rate of decay of singly charged ions due to recombination is small, so that the creation of population inversion in the recombination regime pumping favours transitions in singly charged ions. In this case the initially high concentration of doubly charged ions required can be realized with currents typical of pulsed discharges, and fast collisional recombination can readily be established.

In order to obtain intense recombination pumping of levels it is necessary to create a dense plasma with a high concentration of ions. It is therefore desirable for the lasant atoms to have low single and double ionization potentials. The ionization potential of the buffer gas, however, should be as high as possible.

The plasma electrons should be cooled as quickly and as deeply as possible after the current pulse. For this it is possible to use a high pressure of a light buffer gas, e.g. helium or hydrogen, so that intense cooling of the electron gas can occur via elastic collisions between the electrons and these atoms and ions.

Recombination of doubly ionized atoms is most suitable, because the necessary values of n_e and T_e for their production during the current pulse, and for intense recombination in the afterglow, are easiest to attain.

Thus, the following scheme for the recombination pumping of laser levels in a gas discharge is advisable. Large concentrations of metal ions should be produced by the action of a high-density current pulse through a mixture of the vapour of an easily ionized element and a light buffer gas; helium is the most practical buffer gas for this purpose. Because of its high ionization potential, helium permits a high T_e to be established during the current pulse, and does not hinder the production of large numbers of ions of the other atomic species present. After termination of the current pulse, intense cooling of the electrons occurs via elastic collisions with atoms and ions of helium, resulting in rapid recombination and pumping of the excited levels of singly ionized metal atoms. It is important that the current pulse decay time is not too long, otherwise it will reduce the recombination rate.

1.4.2 Energy Level Configurations Favourable to the Creation of Population Inversions in Recombining Plasmas

During the stage of recombination non-equilibrium, which is characterized by large values of n_e and low T_e, additional means of establishing a population inversion are possible. This is connected with the controlling influence on excited state populations of inelastic collisions with slow electrons (primarily by electron de-excitation (see section 1.2)).

In the majority of atoms and ions, energy levels are disposed more or less in groups, the spacings between which can considerable exceed the spacings between individual levels within the groups (Fig. 1.2). For sufficiently large concentrations of electrons, and for spacings between levels within groups comparable to or less than the average electron energy, the probability of collisionally induced transitions between levels, both up and down, will considerably exceed the probabilities of radiative transitions within groups and to lower-lying groups. In that case, the populations of levels within groups will correspond to a Boltzmann distribution of temperature T_e.

The probabilities of electron collisional de-excitation and, even more so, of excitation between levels of different groups are considerably lower than between levels within groups, provided that $\Delta\varepsilon$ between groups is much greater than between levels. Therefore, between groups that are widely

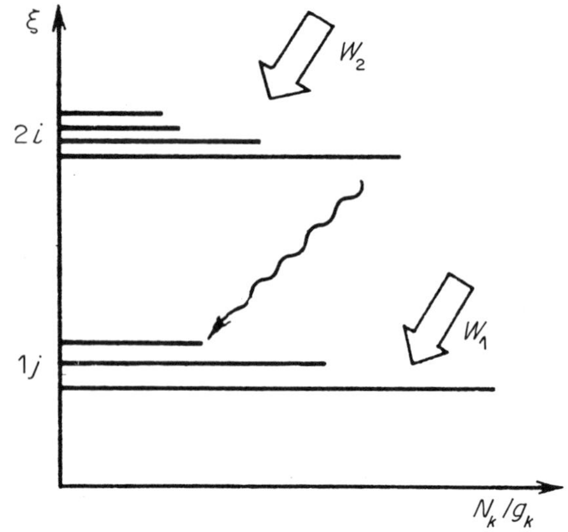

Figure 1.2 Configurations of excited-state populations in groups of closely spaced levels.

separated, a Boltzmann distribution cannot be established even though n_e may be significant[4]. Then, it is possible to create a population inversion between the lower levels of one group and the upper levels of the group below it, even if the total population of the lower group exceeds that of the upper group.

In order to establish the criteria for producing an inversion in the case of such a configuration of energy levels, we consider transitions between groups which can occur radiatively with probability A_{ij}, and by electron collision. For the population of the ith level in the upper group, we may write

$$\frac{N_i}{\sum\limits_i N_i} = \frac{g_i \exp_i}{\sum\limits_i g_i \exp_i}, \tag{1.25}$$

where $\exp_i = \exp(-\Delta\varepsilon_i / T_e)$, and $\Delta\varepsilon_i$ is the energy difference between the ith level and the lowest of the same group. For the case of quasi-steady-state pumping to the upper group, the pump rate W_2 is equal to the total decay rate of the population in that group:

$$W_2 = \sum_i N_i C_i \quad , \tag{1.26}$$

where $C_i = A_{i1} + k_{i1}n_e$ is the total probability for decay of the ith level due to radiative and collisionally induced transitions to lower groups.

From (1.25) and (1.26) follows

$$W_2 = \frac{\left(\sum_i N_i\right)\left(\sum_i C_i g_i \exp_i\right)}{\sum_i g_i \exp_i}, \tag{1.27}$$

whence is obtained

$$N_i = \frac{W_2 g_i \exp_i}{\sum_i C_i g_i \exp_i}. \tag{1.28}$$

Similarly, for the lower group it may be shown that

$$N_j = \frac{W_1 g_j \exp_j}{\sum_j C_j g_j \exp_j}. \tag{1.29}$$

Here, W_1 includes pumping due to transitions from group 2, as well as other higher-lying groups. Therefore, in general, $W_1 > W_2$. For the existence of a population inversion, it is necessary that

$$\frac{g_j N_i}{g_i N_j} = \frac{W_2}{W_1} \frac{\sum_j C_j g_j \exp_j}{\sum_i C_i g_i \exp_i} \exp \frac{\Delta\varepsilon_j - \Delta\varepsilon_i}{T_e} > 1\,. \tag{1.30}$$

The quantity F, defined as

$$F = \frac{\sum_j C_j g_j \exp_j}{\sum_i C_i g_i \exp_i}, \tag{1.31}$$

compares the probability of decay of the lower group with the probability of decay from the upper group.

Using (1.31), the criterion for existence of a population inversion is

$$F > \frac{W_1}{W_2} \exp\left(-\frac{\Delta\varepsilon_j - \Delta\varepsilon_i}{T_e}\right). \tag{1.32}$$

By taking the lowest level of the upper group ($\Delta\varepsilon_i = 0$) and the highest level of the lower group, it follows that in expression (1.32) the exponential, $\exp(-\Delta\varepsilon_j / T_e)$, is much less than unity provided that $\Delta\varepsilon_j > T_e$. If $F \approx 1$, then an inversion will be maintained between the two levels in question even if $W_1 \gg W_2$. If $W_1 \approx W_2$, which is usually the case in the recombination regime, where the recombination flux occurs down through successive groups, then the inversion is maintained for $F \ll 1$, i.e. the probability of decay from the lower group is lower than that from the upper group.

The above conclusion is of particular import, since the case $F \ll 1$ occurs quite frequently. This is linked, first, with the trend for the number of levels in a lower group to be smaller than in a higher group. Therefore the pump rate to the lower group, W_1, is distributed between fewer levels and the population of each level is larger compared with a higher group. Second, the rate of decay of levels within groups is determined primarily by the probability of decay of the lowest levels, because they are usually the most populated; if the lowest level is metastable, the populations of levels within the group will consequently be relatively high, and F will be low. Therefore, generally speaking, a metastable lower level in the lower group is undesirable, even though in this case, for the appropriate value of $\Delta\varepsilon_j$ and a low T_e, the criterion (1.32) for inversion may still be met.

It may in some cases be difficult to isolate groups of sufficiently closely spaced levels. In that case, the existence of inversion must be ascertained by solution of the complete system of rate equations. It is possible to approximate the case of a small number of levels by applying the criterion to the case where a group consists of one level only. Then, from (1.32) for the maintenance of inversion, we can derive

$$\frac{g_1(A_1 + k_1 n_e)}{g_2(A_2 + k_2 n_e)} > \frac{W_1}{W_2}. \tag{1.33}$$

Here it is interesting to examine the cases of high and low n_e. In the case of a low electron concentration (1.33) becomes

$$g_1 A_1 / g_2 A_2 > W_1 / W_2, \tag{1.34}$$

i.e. the inversion is maintained by radiative transitions. For this it is usually necessary that the spacing between levels 2 and 1 is smaller than that between levels 1 and 0 (Fig. 1.3a), because this distribution of levels yields the required relationship between the radiative transition probabilities. However, with increasing n_e, and hence with a faster rate of pumping due to recombination, inversion in this case disappears as the terms $k_1 n_e$ and $k_2 n_e$ come to prevail (the values of these terms, in accordance with (1.7) and (1.8), are larger the smaller the spacing between levels). Thus, such a form for the distribution of levels in the recombination regime is not conducive to the production of large gain and high laser output powers.

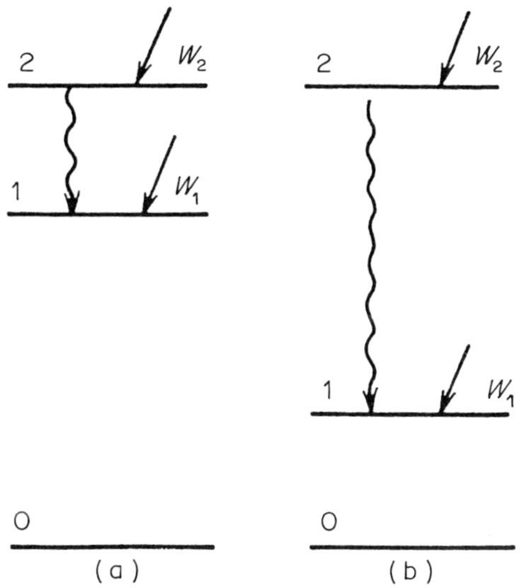

Figure 1.3 The two cases of energy-level distribution considered.

In the case of high electron concentration, inversion is attained if

$$g_1 k_1 n_e / g_2 k_2 n_e > W_1 / W_2 \,. \tag{1.35}$$

For this case, the favoured distribution of levels is presented in Fig. 1.3(b). Inversion due to favourable radiative probabilities is not possible in this case, since $\Delta\varepsilon_{21} > \Delta\varepsilon_{10}$ leads to $A_{21} > A_{10}$. However, an inversion based on favourable relative probabilities for electron de-excitation becomes possible, because now $k_1 n_e > k_2 n_e$. Inversion on such a transition will exist in highly dense plasmas with rapid recombination pumping, and the laser will have large output powers and high gain.

Nevertheless, in the scheme illustrated in Fig. 1.3(b), it is desirable that level 1 not be situated too close to the ground state, which in practice translates to $\Delta\varepsilon_{10} > T_e$. If this inequality is not obeyed then the excitation of level 1 by single-step electron collision will be an important process. Hence, a necessary condition for laser oscillation in the second case, besides the need for a high n_e, is to have as low a value of T_e as possible. This requirement should also be realized for effective recombination pumping. Therefore, this second scheme of levels is conducive to creating an inversion by recombination pumping.

From the point of view of attaining high laser output powers and efficiencies, together with short-wavelength output, it is desirable that the laser levels be disposed as low as possible in the ion energy-level structure. In this case, the pumping will be concentrated on the lowest levels, where the spacing between levels is relatively large. The positions of the laser levels are particularly important when level 0 is the ground state of the singly charged ion, which is long lived (10's μs) and highly populated. We can derive several details concerning such a scheme by assuming that only level 2 is pumped, at a rate W, and the decay due to radiation and collision falls only onto levels 1 and 0 (Fig. 1.3b). Here, the duration of pumping is considerably less than the lifetime of the ground state, whose change in population during the pumping is negligible. Then, for the populations of the laser levels, we have

$$N_2 = W / (A_{21} + k_{21}n_e)\,; \tag{1.36}$$

$$N_1 = \frac{W + N_0 k_{01} n_e}{A_{10} + k_{10} n_e}\,. \tag{1.37}$$

Using the relationship $k_{01}/k_{10} = g_1/g_0 \exp(-\Delta\varepsilon_{10}/Te)$ and assuming that for large n_e that $A_{10} \ll k_{10}\, n_e$, which easily holds when we allow for self-absorption, the condition for the existence of a population inversion is found to be

$$(A_{21} + k_{21}n_e)[g_2 / (g_1 k_{10} n_e) + \exp(-\Delta\varepsilon_{10} / T_e) g_2 N_0 / g_0 W] < 1\,. \tag{1.38}$$

If, however, n_e is such that $k_{21}n_e > A_{21}$, it follows that the condition is

$$\frac{g_2}{g_1}\frac{k_{21}}{k_{10}} + \frac{g_2}{g_0} k_{21} n_e \frac{N_0}{W} \exp\left(-\frac{\Delta\varepsilon_{10}}{T_e}\right) < 1\,. \tag{1.39}$$

This form, compared with the criterion for inversion (1.35), which is correct only when N_0 is low, has a second term that increases as N_0 / W and T_e increase. Inserting into (1.39) expressions for W and k (relations (1.7), (1.8), (1.22) and (1.24)), and assuming $N_0 \approx n^+ \sim n_e$, we find that

$$a + b\frac{T_e^4}{n_e n^{2+}} \exp\left(-\frac{\Delta\varepsilon_{10}}{T_e}\right) < 1\,, \tag{1.40}$$

where a and b are constants, and n^{2+} is the concentration of doubly charged ions.

From the last relation we see that, for this distribution of levels, the inversion depends critically on T_e, and quickly disappears as T_e increases. An increase in n^{2+}, on the one hand, or W, on the other hand, assists the maintenance of inversion for such energy-level configurations in singly charged ions. Thus, the main process terminating inversion when the lower level is close to the ion ground state is the excitation of the lower level by electron impact from the ion ground state.

In summary of the above discussion, the general requirements for establishment of inversion in the collisional–recombination regime are:

i. the upper laser level should be one of the lowest in an upper group of closely spaced levels;
ii. the lower laser level should be one of the highest in a lower group of closely spaced energy levels;
iii. the transitions between levels should be allowed;
iv. the electron concentration should be sufficiently high that the probability of collisionally induced transitions within groups exceeds the probabilities for radiative transitions; and
v. the electron temperature should be as low as possible.

With these criteria, electron de-excitation causes pumping to be channelled to the upper laser level, and effectively clears the lower laser level. These requirements are also valid for pumping in discharge afterglows by second-kind collisions such as charge transfer and Penning ionization.

This formulation of the general requirements for discharge conditions and energy-level arrangement serves as a starting point for the analysis of existing recombination lasers, and as a guide in the search for new recombination laser transitions.

1.5 OPTICAL PUMPING AND RADIATIVE DECAY

1.5.1 Optical Pumping

The major advantage of optical pumping is its high degree of selectivity. The excitation of ion laser levels by optical pumping with photoionization is associated not with the detachment of outer-shell electrons, but with the removal of inner-shell electrons [33]. Furthermore, in contrast to electron-impact ionization, for which the peak of the cross-section lies at energies much greater than the threshold energy, photoionization has a maximum cross-section near the threshold [34]. Thus, with the appropriate pump source, photoionization can be made to proceed mainly via the loss of inner-shell electrons, and it may be possible to use the excited ion state so produced as an upper laser level. Using this technique, transitions in the UV region can be accessed. Alternatively, the excited ion can be either optically re-pumped (by a tunable dye laser, for example) to a nearby state of the same ion, or used to collisionally excite another atom to a more suitable upper laser level [33].

The authors of [33] suggested that photoionization of d electrons in Cd and Zn could be used to achieve inversion on the $D^* \longrightarrow P$ transitions (see section 1.3.2). Between the ground state and the D^* states created by photoionization lie the ^2P levels of the $nd^{10}(n+1)$p configuration, which can be used as lower laser levels by virtue of their short radiative lifetimes (allowed radiative transitions to the ground state). Laser oscillation is observed on the $D^*_{5/2} \longrightarrow {}^2P_{3/2}$ transition of Cd when cadmium vapour is pumped by the radiation produced at a tungsten target which is illuminated by a powerful laser pulse [35]. Laser oscillation has also been obtained on other transitions, when a dye laser was used to pump higher excited levels from the D^* states. The principal difficulty with the use of photoionization of inner-shell electrons is the development of a practical high-energy photon pump source, and the isolation of the source from the active medium.

Another technique that can be employed is the combined use of gas discharge and optical pumping. In the first step, long-lived ion states (metastables) are excited under gas-discharge conditions. The energies stored in the metastable populations can then be transferred to potential upper laser levels by optical pumping.

1.5.2 The Radiative Decay of Excited States

In weakly ionized plasmas where the electron density is not too high, the main decay mechanism of excited states is by spontaneous emission. According to the expression (1.2), a population inversion can be created if the decay of the lower level is faster than the decay of the upper level. To avoid radiative trapping and stepwise electron excitation, it is desirable that the lower laser level should not be the resonance level of an ion.

A favourable ratio of lifetimes can be obtained if the lower laser level is connected to the resonance level by an optically allowed transition of a wavelength much shorter than that of the laser transition, since the probability of spontaneous emission is proportional to $\Delta\varepsilon_{ij}^3$. The use of such configurations is responsible for the largest number of CW laser transitions in both atoms and ions. For ions, the laser wavelengths lie mainly in the visible, whereas in the case of atoms, whose energy levels are more compact, they lie in the infrared. At high electron densities, electron collisional transitions must be considered also, as discussed in section 1.2.

1.6 THE INFLUENCE OF ISOTOPE SHIFTS AND HYPERFINE SPLITTING OF LASER LEVELS ON POPULATION INVERSION AND LASER OSCILLATION

When isotopic or hyperfine splittings are larger than the Doppler width, the use of even isotopes instead of the natural isotopic mixture can increase the available gain and output power [36]. The gain coefficient, g, is proportional to Δf^{-1}, where Δf is the linewidth of the laser transition. Isotopic and hyperfine splitting can be treated as additional broadening of the line. Thus, the use of a single isotope in place of the naturally occurring mixture must increase the gain coefficient. If isotopic splitting is less than the Doppler width, i.e. the components of isotopic splitting greatly overlap, the use of single isotopes will not enhance the gain coefficient. The use of odd isotopes is to be avoided (even if a large fraction of the isotopes is even) because their splittings (hyperfine structure) often exceed the Doppler width. Thus, in order to increase the gain, even isotopes should be used in place of the natural isotopic mixture, but only if the isotopic splittings exceed the Doppler broadening.

The nuclear mass does not influence the cross-sections of excitation, and pump rates to the levels of different isotopes are equal. Hence, the number densities of ions of each isotope excited to the laser levels are proportional to X_i, the fraction of the ith isotope in the natural isotopic mixture:

$$N_{2i} = aX_i \, , \; N_{1i} = b\,X_i \, , \tag{1.41}$$

where N_{2i} and N_{1i} are the densities of ions of the ith isotope in the upper and lower laser levels; a and b are factors accounting for the rates of excitation and de-excitation of the levels. The population difference between the levels of the ith isotope is

$$\Delta N_i \;=\; (a - b)\, X_i \, . \tag{1.42}$$

If isotope splitting is larger than the Doppler width, laser oscillation will begin on the components of the isotopic structure having small-signal gains higher than the cavity losses. The maximum small-signal gain coefficient belongs to the isotope with the largest concentration in the mixture:

$$g_{\max} \sim (a - b)\, X_{\max} \, . \tag{1.43}$$

The use of a single, even isotope means $X_{\max} = 1$. Thus $g \sim (a - b)$ and the ratio of the small-signal gain coefficient for the even isotope to the maximum one for the natural mixture is

$$g \,/\, g_{\max} \;=\; 1 \,/\, X_{\max} \, , \tag{1.44}$$

i.e. the use of only one isotope increases the gain coefficient $X_{\max}^{-1}$ times. The increase in gain coefficient with the use of a single, even isotope should increase the laser power as well [37].

REFERENCES

[1] G. G. Petrash, *Sov. Phys. - Usp.*, **14**(6), 747–65 (1972).
[2] A. N. Soldatov and V. I. Solomonov, *Gas-Discharge Lasers Utilizing Self-Terminating Transitions in Metal Vapours,* Novosobirsk: Nauka (1985) (in Russian).
[3] G. J. Collins, H. Kuno, S. Hattori, K. Tokutome, M. Ishikawa and N. Kamide *IEEE J. Quantum Electron.*, **8**(7), 679–80 (1972).
[4] J. A. Piper, *Opt. Commun.*, **19**(2), 189–92 (1976).
[5] J. P. Goldsborough, *Appl. Phys. Lett.*, **15**(6), 159–61 (1969).
[6] T. P. Sosnowski, *J. Appl. Phys.*, **40**(13), 5138–44 (1969).
[7] J. R. Fendley Jr., I. Gorog, K. G. Hernqvist and C Sun, *RCA Rev.*, **30**(3) 422–8 (1969).
[8] M. F. Sem, V.. F Keidan and V. C. Mikhalevskii, *Elektron. Tekh., Ser. 1 Elektronika SVCh,* No. 1, 69–76 (1971) (in Russian).
[9] É. K. Karabut, V. C. Michalevskii, V. F. Papakin and M F Sém, *Sov. Phys. - Tech. Phys.*, **14**(10), 1447–8 (1970).
[10] J. R. McNeil, G. J. Collins, K. B. Persson and D. L. Franzen, *Appl. Phys. Lett.*, **28**(4), 207–9 (1976).
[11] J. R. McNeil, W. L. Johnson, G. J. Collins and K. B. Persson, *Appl. Phys. Lett.*, **29**(3), 172–4 (1976).
[12] R. D. Reid, J. R. McNeil and G. J. Collins, *Appl. Phys. Lett.*, **29**(10), 666–8 (1976).
[13] L. A. Vainshteyn, I. I. Sobelman and E. D. Yukov, *Excitation of Atoms and Spectral Line Broadening,* Moscow: Nauka (1979) (in Russian); *Cross-sections for Electronic Excitation of Atoms and Ions,* Moscow: Nauka (1973) (in Russian).
[14] E. L. Latush, V. C. Mikhalevskii and M. F. Sem, *Opt. Spectrosc.*, **34**(2), 120–3 (1973).
[15] J. M. Green and C. E. Webb, *J. Phys. B.*, **8**(9), 1484–500 (1975).
[16] B. M. Smirnov, *Atomic Collisions and Elementary Processes in Plasmas,* Moscow: Atomizdat (1968) (in Russian).
[17] E. L. Latush and M. F. Sem, *JETP Lett.*, **15**(11), 457–9 (1972).
[18] V. V. Zhukov, E. L. Latush and M. F. Sem, *Sov. Phys. J.*, **20**(7), 957–9 (1977).
[19] N. Suzuki, *Jpn. J. Appl. Phys.*, **4**(6), 452–7 (1965).
[20] R. Solanki, E. L. Latush, W. M. Fairbank Jr. and G. J. Collins, *Appl. Phys. Lett.*, **34**(9), 568–70 (1979).
[21] E. L. Latush, *Izv. Ser.-Kav. Nauch Tsentra Vissh. Shkoli. Ser. Yestyestv. Nauk.*, No. 2, 101–2 (1973) (in Russian).
[22] V. Cermak, *Collect. Czech. Chem. Commun.*, **36**(2), 948–50 (1971); **33** 2739 (1968).
[23] S. Inaba, T. Goto and S. Hattori, *J. Phys. B.*, **14,** 507–12 (1981).
[24] H. Hotop and A. Niehaus, *Z. Phys.*, **228**(1), 68–88 (1969).
[25] A. V. Yeletski and B. M. Smirnov, *Physical Processes in Gas Lasers,* Moscow: Energoatomizdat (1985) (in Russian).
[26] V. S. Aleinikov, *Opt. Spectrosc.*, **28,** 15–17 (1970).
[27] A. R. Turner-Smith, J. M. Green and C. E. Webb, *J. Phys. B.*, **6**(1), 114–30 (1973).
[28] V. I. Bylkin, *Opt. Spectrosc.*, **29,** 552–5 (1970).
[29] L. I. Gudzenko and L A Shelepin, *Sov. Phys. -JETP,* **18**(4),. 998–1000 (1964)

[30] L. I. Gudzenko and S. I. Yakovlenko, *Plasma Lasers,* Moscow: Atomizdat (1978) (in Russian).
[31] V. A. Boiko, F. V. Bunkin, V I Derzhiev and S I Yakovlenko, *Izv. Akad. Nauk SSSR Ser. Fiz.,* **47**(10), 1880–97 (1983) (in Russian); **48**(8), 1626–38 (1984) (in Russian).
[32] E. L. Latush and M F Sem, *Sov. Phys. - JETP,* **37**(6), 1017–18 (1973).
[33] V. B. Rozanov, *Kvantovaya Elektron.,* No. 3, 54–60 (1971) (in Russian).
[34] E. W. McDaniel, *Collision Phenomena in Ionized Gases,* New York: John Wiley (1964).
[35] W. T. Silfvast, O. R. Wood II, H. Lundberg and J. J. Macklin, *Opt. Lett.,* **10**(3) 122–4 (1985).
[36] V. F. Papakin and M. F. Sém, *Sov. Phys. J.*, **13**(2), 230–1 (1970).
[37] V. G. Il'yushko, V. F. Papakin and M. F. Sém, *Sov. Phys. J.*, **16**(3), 416–17 (1974).
[38] B. M. Smirnov, *Excited Atoms,* Moscow: Energoatomizdat (1982) (in Russian).

Notes

1. Lasers operating in this regime are generally referred to in the text as recombination lasers.
2. An exception is energy transfer in collisions between atoms of the same type, one of which is in a resonance level. The cross-section for this process can exceed the cross-section for resonant charge transfer, even though it involves two electrons. This is the result of a strong dipole-dipole interaction with the associated exchange of resonance photons [38].
3. Equations (1.22) and (1.23) are not completely correct for the recombination of singly charged ions, because in a weakly ionized plasma the concentration of singly charged ions n^+ is proportional to n_e and during recombination the recombination time increases due to the decrease in n_e.
4. Naturally, for very large concentrations of electrons, a Boltzmann distribution may be established between the groups of excited levels and the ground state. This is the principal factor which limits the specific output power of recombination lasers.

2

Gas Discharges used in Metal Vapour Ion Lasers

2.1 TYPES OF DISCHARGE

For the active media of MVILs, longitudinally and transversely excited electric discharges are employed in the DC, pulsed and RF regimes. In a longitudinal discharge, the field is applied along the length of the active medium, and use is made of the uniform quasi-neutral positive column (PC) component of the gas discharge. Each element of the PC is independent in the sense that electrons receive energy from the local field. Thus, the length of the PC—the length of the active medium—can be varied over a large range without any changes in the local properties of the PC.

Transverse discharges are usually applied between two closely opposed electrodes, as in hollow cathode (HC) and RF discharges. In this case, the active medium is not the PC, which here is very small or even non-existent, but the negative glow (NG) region of the discharge. The electron population in the NG is derived mainly from electrons emitted by the cathode with energies of the order of 100 eV, created as a result of the potential across the cathode dark space (CDS) lying between the cathode and the NG region. The source of ionization in the NG lies in the CDS, i.e. it is external to the NG. The electron energy distribution function (EEDF) is shaped by the energy losses of these fast, injected electrons due to excitation and ionization of the gas. In the PC discharge, however, electrons are accelerated from effectively zero energy to the level of the ionization energy of the gas, and frequent collisions with gas molecules prevent large numbers of electrons with higher energies from being established. Due to the different mechanisms of gain and loss of electron energy in PC and NG discharges, their EEDFs are quite different.

Longitudinal RF discharges can be of the inductive (H) or capacitive (E) type. In the former, the discharge is produced by an alternating magnetic field, whereas in the latter an alternating electric field is applied across one or more pairs of electrodes mounted along the outside of the discharge tube. Inductive discharges were first used to excite pulsed lasers based on the vapours of Br, Se and As [1], which will corrode internal electrodes. They have also been used to excite a CW He–Cd laser, by applying a magnetic field via an inductor wound on the silica laser tube [2]. The plasma of a longitudinal RF discharge is similar to that of a direct current PC discharge, because, just as in the latter, each part of an RF discharge is excited independently of the rest of the discharge, and the frequency of the applied field (usually 10–100 MHz) is, as a rule, higher than the relaxation rate of particles which transfer energy to the laser levels.

The capacitively coupled RF discharge of greatest use for the excitation of MVILs is the transverse RF (TRF) discharge, operated in the so-called γ mode [3]. The plasma of a TRF discharge is similar to that of a HC discharge, since the space charge near the wall (the sheath region) is characterized by a stationary field [4], similar to that of a CDS.

As for all gas lasers, MVILs can also be excited using longitudinal and transverse discharges in the pulsed regime. Here, conditions of input powers and energies can be utilized that would require complex technical solutions to be realized, if they were possible at all, under DC and RF stationary conditions.

Below, we shall deal with the external (electrical) and internal (plasma) characteristics of each of the above-mentioned types of longitudinal PC and transverse NG discharges. Knowledge of the external parameters is necessary for the design of laser power supplies. The internal parameters are used for the optimization of discharge conditions for particular pump mechanisms in order to achieve maximum output power and efficiency, and to predict the range of conditions for laser operation.

The development of discharge lamps, mercury rectifiers and other gas-discharge devices stimulated many investigations of discharges in metal vapours and in mixtures of inert gases with metal vapours. The key results have been reviewed in a number of publications, e.g. [5–7]. However, the discharges examined in those investigations have parameters that differ from those required for the excitation of MVILs. Metal vapour ion lasers are operated in discharges where the gas species have widely disparate ionization potentials; the discharge channel often has a small diameter (~1 mm); a wide range of pressures (1–10^5 Pa) of gas mixtures is employed; high discharge current densities are common (1–10^4 A cm^{-2} in PC discharges, and up to 10 A cm^{-2} at the electrode surface of a transverse discharge); ionization or recombination non-equilibrium conditions of the active medium occur with microsecond or sub-microsecond pulse excitation. The investigation of MVIL gas discharges has parallelled the search for new active media and methods of pumping. This has fostered new experimental techniques (e.g. utilizing the laser emission of the active medium for self-pumping, etc.), engendered ideas which have enriched the more fundamental 'pure' studies of gas discharges, and generally extended the boundaries of the field.

2.2 STEADY-STATE POSITIVE-COLUMN DISCHARGES

Among the most advanced MVLs today are the so-called 'cataphoretic' ion lasers which use the vapours of Cd, Se, etc., and are excited in stationary PC discharges [8–11]. The discharges of these MVILs are supported in medium pressures (0.1–4 kPa) of buffer gas (usually helium) with current densities of up to 15 A cm^{-2}. Since the PC of a glow discharge is a self-sustaining uniform region of plasma, which does not depend on processes at the electrodes, the discussion below may be applied to longitudinal RF discharges as well.

The initiation of a discharge in a mixture of two gases leads to a variety of energy-transfer (see Chapter 1) and momentum-transfer processes between plasma species. The great difference between the ionization potentials of the constituent atoms (helium and metal atoms) causes significantly different rates of ionization of those species and leads to axial and radial separation of the gas mixture (axial and radial cataphoresis). The mechanism responsible is the drift of easily ionized metal ions (typically 9–10 eV ionization potential) in axial and radial electric fields towards the cathode and tube wall. Electrophoresis—the drift of buffer-gas atoms in the opposite direction, towards the anode, as a result of electron-atom momentum transfer—can be much stronger in a gas mixture than in a gas of one species.

2.2.1 Electron Temperature and Axial Field Intensity

At medium gas pressures, the stationary PC discharge is diffusion-limited. Here, the free paths of ions are less than the radius of the PC discharge, and ambipolar diffusion of charged particles to the wall determines the rate of neutralization of charged particles there. At low current densities, the EEDF (usually specified by the electron temperature T_e) may be characterized by a depletion of electrons at energies above the gas excitation potential, when compared to a Maxwellian EEDF with the same T_e [12]. The condition to be satisfied to prevent electron depletion zones from appearing in the EEDF is that the electron–electron collision frequency, ν_{ee}, be higher than the collision frequencies of electrons with gas atoms in elastic and inelastic interactions, i.e. $\nu_{ee} > \nu_{el}$, and $\nu_{ee} > \nu_{inel}$. We can estimate these frequencies using the formulæ of [11] and [13]:

$$\nu_{ee} = \langle q_{ee} v \rangle n_e = 3 \times 10^{-5} T_e^{-3/2} n_e \, ; \tag{2.1}$$

$$\nu_{el} = mv / (M\lambda_{el}) \, ; \tag{2.2}$$

$$\nu_{inel} = v / \lambda_{inel} \, , \tag{2.3}$$

where λ_{el} and λ_{inel} are the mean free paths of electrons for elastic and inelastic collisions respectively, and v is the electron velocity. Consideration of these equations shows that the condition, $\nu_{ee} > \nu_{el}, \nu_{inel}$, is approximately satisfied (the higher n_e, the closer the approximation) for electron energies up to the first excitation potential of helium (19.8 eV), while at higher energies, depletion of electrons occurs. This leads to two-step ionization of helium atoms, i.e. ionization from the lowest excited states (the metastable levels, $2^{1,3}S$). Since only the number of electrons near the excitation threshold is important (as we shall see below), the use of an EEDF of Maxwellian form instead of the true EEDF allows good agreement with experiment. It should be noted that a high-frequency ν_{ee} results in a high thermal conductivity of the electron gas, and T_e becomes radially uniform.

In a one-component gas, the distribution of charged and excited particles along a radius in a PC discharge is assumed to be described to a first approximation by a Bessel function of zero order, $J_0(r)$ [6]. In a discharge through a mixture of gases with greatly disparate ionization potentials, the radial field in the PC is $E_r = 5T_e/(er_0)$ (where r_0 is the radius of the PC), and is usually higher than the axial field E_z. This causes strong radial cataphoresis and a sharp increase in the metal vapour density in a thin layer near the wall (~0.1 r_0). Beyond this layer, however, the radial distribution of charged particles and metastable He atoms is approximated by $J_0(r)$.

In the steady state, the ion densities of each component are determined by the balance between rates of formation and loss by diffusion to the wall (charge transfer of He ions in a PC discharge is negligible compared with diffusion losses), i.e.

for metal ions: $$z_M^+ n_e = -D_{aM}\nabla_r^2 n^+ ; \qquad (2.4)$$

for buffer-gas ions: $$z_B^+ n_e = -D_{aB}\nabla_r^2 N^+ , \qquad (2.5)$$

where n^+ and N^+ are the metal (M) and buffer (B) ion densities, z_M^+, z_B^+ are the frequencies per electron for ionization of the metal and buffer atoms, and $D_{aM,B}$ are the respective ambipolar diffusion coefficients. Assuming a Bessel profile for the electron density, the solution of equations (2.4) and (2.5) is

$$z_M^+ / D_{aM} + z_B^+ / D_{aB} = 5.76 / r_0^2. \qquad (2.6)$$

Helium is ionized in a stepwise fashion via its low metastable states (ionization rate $k_{mi} = \langle q_{mi} v \rangle$), which are populated by electron impact from the atomic ground state (rate constant $k_{0m} = \langle q_{0m} v \rangle$), while depopulation of the states is due primarily to the processes listed in Table 2.1. Hence, we can write

$$z_B^+ = n_e N k_{0m} k_{mi} / (\nu_d + n_e k_{mi} + K_P n) . \qquad (2.7)$$

Table 2.1 Evaluated rates of excitation and deactivation of metastable helium atoms under conditions appropriate to a cataphoretic MVIL (tube diameter 3 mm, $n_{Cd} = 5 \times 10^{13}$ cm^{-3}, $i = 0.12$ A).

Process	Rate	Value of rate (10^{17} cm^3 s^{-1})	Process rate contant (cm^{-3} s^{-1})
Production of He atoms in metastable states:			
$He_0 + e \longrightarrow He_m + e$	$N_0 n_e k_{0m}$	6.5	$k_{0m} = 9 \times 10^{-12}$
De-excitation processes:			
Diffusion to the tube wall	$N_{He_m} \nu_d$	0.5	$\nu_d = 4 \times 10^3/p$ (p in kPa)
Electron impact			
$He_m + e \longrightarrow He^+ + 2e$	$N_{He_m} n_e k_{mi}$	3	$k_{mi} = 6 \times 10^{-8}$
$He_m + e \longrightarrow He_0 + e$			
Penning collisions			
$He_m + Cd \longrightarrow He + Cd^{+*} + e$	$N_{He_m} n K_P$	2.5	$K_P = 10^{-9}$
Associative ionization (Biondi process)			
$He_m + He_m \longrightarrow He + He^+ + e$	$N^2_{He_m} K_i$	0.5	$K_i = 2 \times 10^{-9}$
$He_m + He_m \longrightarrow He_2^+ + e$			

Ionization of metal atoms (see Chapter 1) occurs by electron impact—either directly or stepwise, i.e. via metastable states—and by collisions of the second kind (charge transfer and Penning ionization). However, it has been established that, for $n_e \geq 10^{12}$ cm^{-3}, the influences of second-kind collisions (see section 2.2) and stepwise ionization can be neglected compared with direct impact ionization (the cross-sections for direct ionization and indirect ionization via metastable states of metal atoms are of the same order [14], and the metastable density is $n_m \approx 0.01n$). Therefore we have

$$z_M^+ = nk_{0i}^M . \tag{2.8}$$

Substituting in (2.6) the ionization frequencies given by (2.7) and (2.8), and assuming for D_a the forms

$$D_{aM} = \mu_M^+ T_e / e; \quad D_{aB} = \mu_B^+ T_e / e , \tag{2.9}$$

(where μ_M^+, μ_B^+ are the ion mobilities in the mixture), we obtain an expression similar to the Schottky equation, which relates electron temperature T_e (both explicitly and implicitly through the rate constant k) to key discharge parameters, i.e. the number densities of heavy bodies and electrons, and radius of the discharge tube:

$$\left(\frac{Nk_{0m}k_{mi}}{\nu_d + n_e k_{mi} + K_P n}\,\frac{n_e}{\mu_B^+} + \frac{nk_{0i}^M}{\mu_M^+}\right)\Big/ T_e = 5.76/r_0^2\,. \qquad (2.10)$$

Although equation (2.10) has been obtained after some simplification, it accurately describes experimental observations (Figure 2.1). With an increase in metal vapour pressure, T_e at first falls slowly, after which there is an abrupt decrease caused by the fall in the effective ionization potential of the gas mixture and the enhanced inelastic interactions of electrons with metal atoms. When the metal atom density is low, inelastic collisions have a weak effect on T_e, and the intensities of atomic and ionic lines of the metal are enhanced with increasing atom density. The accelerating fall in T_e reduces the ionization rate of helium (and the rates of excitation of its levels—in Fig. 2.1 the intensity of the He 471.3 nm line is shown) and leads to saturation and subsequent reductions in the intensities of the lines of the metal ions. For example, in a He–Cd mixture at optimum cadmium vapour density (5×10^{13} cm^{-3}) for laser oscillation, the original straw-coloured glow of pure

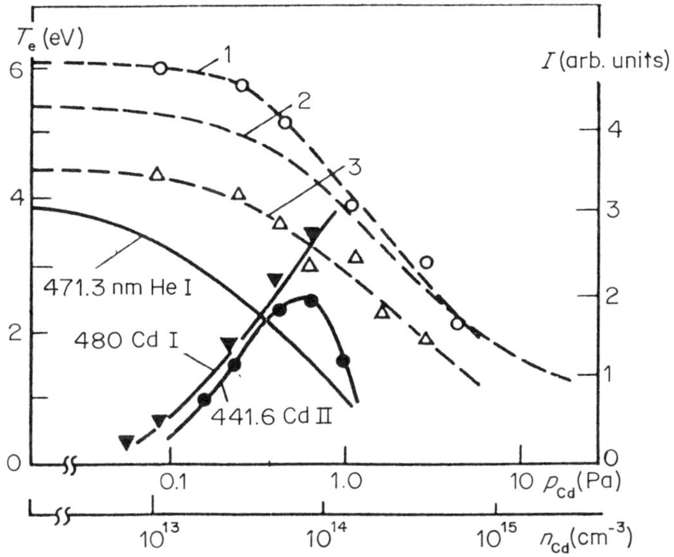

Figure 2.1 The dependence of T_e (broken lines) and spectral line intensities (solid lines) in a He–Cd mixture on the density and pressure of cadmium vapour: 1, 3 – measurement [20] by the double-probe method [18] for d = 3 mm, j = 1.5 A cm^{-2}, $p_{He}d$ = 0.12 kPa cm (1) and 0.25 kPa cm (3); 2 – calculation from equation (2.10) for $p_{He}d$ = 0.12 kPa cm.

helium is supplanted by a white-blue glow, and when $n > 2 \times 10^{14}$ cm^{-3}, the glow becomes bright blue (because only the atomic spectrum of cadmium is excited). The decrease in T_e is as sharp as the fall in the effective ionization potential of the gas mixture. For this reason, among others, the density of other 'impurity' species should be kept to a minimum. If another element is admitted to the tube (for instance, after dissociation of a chemical compound), it is desirable that it have a high ionization potential.

It should be noted that as a result of the saturation of the density of He metastable atoms (see section 2.2), T_e depends weakly on current (Fig. 2.2).

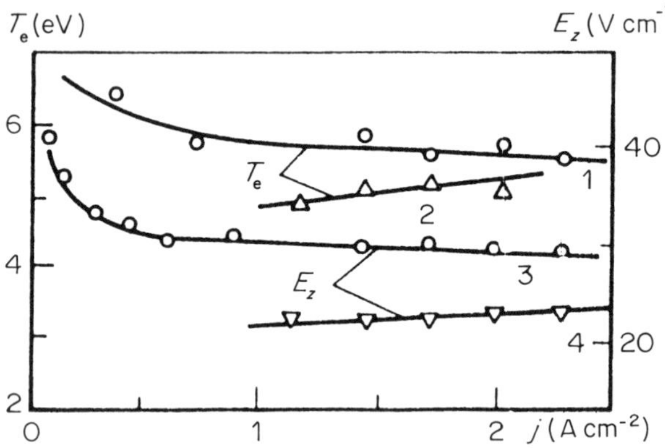

Figure 2.2 The dependences of T_e (1,2) and E_z (3,4) on j with d = 3 mm in helium, with pd = 0.12 kPa cm (1,3), and in a mixture with cadmium vapour with $n_{Cd} = 5 \times 10^{13}$ cm^{-3}, p_{Cd} = 0.4 Pa (2,4).

It is well known that the longitudinal electric field, E_z, in a self-sustained PC discharge is related to T_e and the electron mean free path, λ_e, by [6]

$$eE_z = 1.5\sqrt{\delta^*}\, T_e / \lambda_e \,, \tag{2.11}$$

where δ^* is the average total fraction of electron energy lost in collisions. (In a gas with one kind of atom of mass M, $\delta^* \approx 2m_e / M$, where m_e is the electron mass.) Physically, this means that the value of E_z established in the discharge determines the rate of ionization (i.e. the electron energy) necessary for a current to flow in a given gas composition (i.e. for a given $\sqrt{\delta^*}\, \lambda_e$). Because $\lambda_e \sim 1/p$, we can obtain from (2.11) one of the similarity laws for gas discharges:

$$E_z / p \sim T_e \sqrt{\delta^*} \,. \tag{2.12}$$

In gas mixtures,

$$\lambda_e = \lambda_e^{(B)} \left[1 + q_M n / (q_B N)\right]^{-1} , \tag{2.13}$$

where q_M and q_B are, respectively, the total cross-sections for electron impact with metal and buffer atoms; $\lambda_e^{(B)} = 1 / q_B N$ is the electron mean free path in the buffer gas. It is evident that for low metal vapour densities, when $n/N \ll q_B/q_M$, T_e and E_z exhibit similar behaviour. When the vapour density increases, λ_e is strongly reduced (since q_M is much higher than the cross-section q_B for light He atoms) and E_z begins to rise in order to sustain the electron energy necessary for ionization (the two lower curves in Figure 2.3a).

There are two points to be made here. First, for vapour pressures up to 1.5 Pa (covering the entire range of MVL performance), there is a simple relationship between E_z and the vapour pressure, i.e. E_z is a measure of the vapour pressure in the discharge, and can be employed in stabilization of this vapour pressure during laser-tube operation. Second, with an increase in

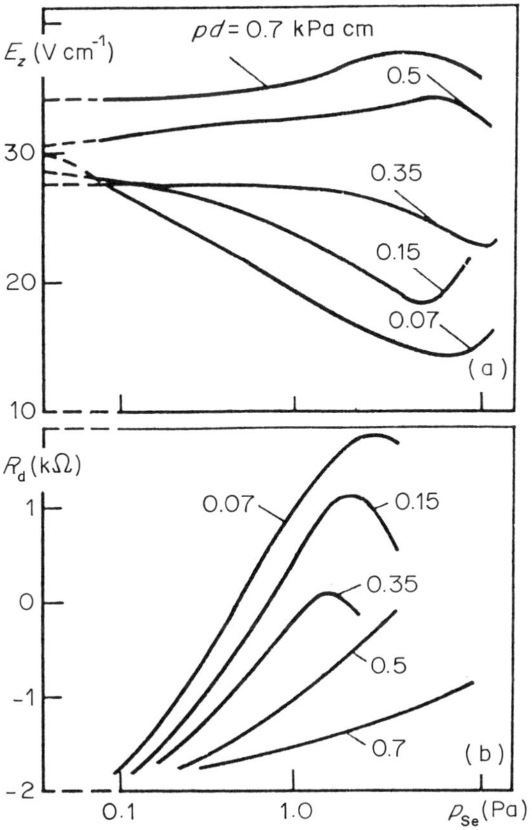

Figure 2.3 The dependences of (a) E_z and (b) the dynamic resistance R_d on the selenium vapour pressure in a He–Se mixture.

buffer-gas pressure the dependence becomes weaker; in the regime of *pd* (i.e. gas pressure x discharge tube inner diameter) ≈ 0.5 kPa cm, E_z virtually ceases to depend on vapour pressure; for lasers with *pd* in this regime, E_z cannot be used as a measure of the metal vapour pressure. Circumvention of this difficulty for lasers with high values of *pd* is possible if the vapour density is controlled by the value of E_z in an additional short capillary section of inside diameter $d_{add} < d$, for which, therefore, $pd_{add} < pd$ [15]. In order to avoid aperturing of the laser beam, this capillary section should be located at the side of the main one (not along the optical axis), for example, between the metal vapour source and the main capillary section in a tube of the kind depicted in Figure 5.1(h). At higher *pd* values, under conditions of non-lasing, E_z increases with the increase of vapour pressure from the very beginning (Figure 2.3a), due mainly to the lower T_e in 'pure' helium.

In Figure 2.2, it can be seen that, in the absence of metal vapour, the current–voltage characteristic of the discharge is one of negative slope due to thermal processes, and also because the mechanism of ionization changes from single-step electron impact to stepwise impact via the metastable states. When metal vapour is present, the slope is positive, i.e. the dynamic impedance of the discharge [5] changes polarity. This effect is due mainly to radial cataphoresis which increases with rising current, leading to a deficit of the easily ionized metal atoms on the axis of the discharge tube. In Figure 2.3(b), it is seen that for all values of *pd*, the dynamic impedance is more sensitive to vapour pressure than E_z is, and thus is a more convenient measure of the metal vapour density in the discharge than E_z. Note that at high rates of current change the dynamic impedance becomes complex [16, 17].

2.2.2 Charged Particle Densities

In a two-component gas mixture, when only atomic ions are present,

$$n_e = N^+ + n^+ . \tag{2.14}$$

If the density of charged particles is measured by the probe method [18], the total ion current to a negative cylindrical probe inserted in the discharge will be the sum of the ion currents of both gas components, i.e.

$$I_i = T_e s (R_c \ln(l / R_c)^{-1} (\mu_M^+ n^+ + \mu_B^+ N^+) , \tag{2.15}$$

where l and s are, respectively, the length and area of the probe, and R_c is the radius of the charge layer. Hence, by measuring I_i (for example, by a double-probe method [18]) and n_e, the densities of ions of each gas component may be found from (2.14) and (2.15).

In discharge tubes of small diameter, the measurement of electron density is quite difficult. The most convenient methods are microwave techniques, which only weakly perturb the plasma, and so are more reliable [19]. Moreover, the form of the EEDF is not of importance for microwave techniques. Most suitable is the resonator method, where the laser tube passes through a microwave resonator and the modes E_{0n0} are examined, which gives the critical electron density, n_e^* (e.g. at a frequency of 5.9 GHz [20], $n_e^* = m_e\omega^2/4\pi e^2 = 4.1 \times 10^{11}$ cm^{-3} [19]). If the discharge tube is axially disposed, there is no plasma polarization, because the electric field of the standing wave inside the resonator is perpendicular to the electron density gradient. If the field frequency, ω, exceeds ν_{ee} (which is usual for $\omega > 1$ GHz), perturbation theory can be applied to measurements when n_e is an order of magnitude higher than the critical n_e in small-diameter tubes. This is confirmed by the small change in resonator Q factor when the discharge is initiated.

In Figure 2.4, it is seen that the electron density rises with increasing discharge current and vapour density according to the expression for charge transfer with a Bessel current density profile:

$$j = 0.43 n_e^{(0)} e \mu_e E_z \,. \tag{2.16}$$

With an increase in vapour pressure, E_z and μ_e fall (with the reduction in λ_e),

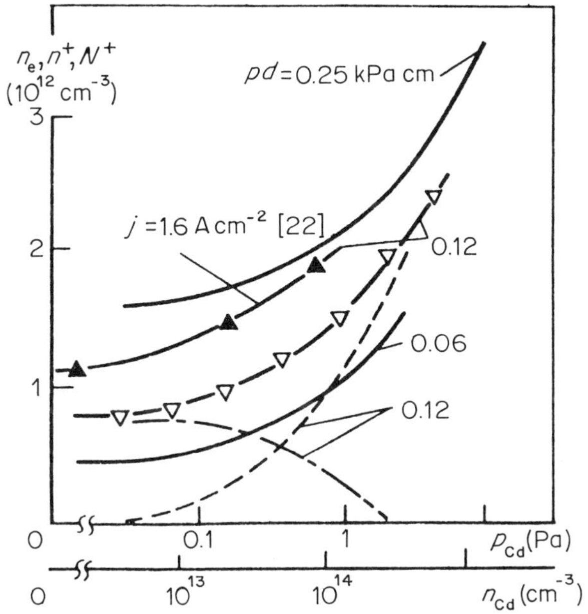

Figure 2.4 The electron density n_e (solid lines), the cadmium ion density (even-dash lines) and the helium ion density (long-dash lines) with j = 1.5 A cm^{-2}.

and n_e increases nearly linearly at low pressures. Meanwhile, T_e decreases, and He ions are quickly replaced by metal ions.

The electron density n_e can also be measured by the simpler double-probe method [21], but less accurately. First, we find $n_e^{(0)}$ in a 'pure' buffer gas from the saturated ion current, and $E_z^{(0)}$ in 'pure' gas and E_z in the metal–buffer mixture at the same discharge current. By applying (2.16) to both the 'pure' gas and to the mixture, we find that

$$n_e = n_e^{(0)} E_z^{(0)} \mu_e^{(0)} / (E_z \mu_e) , \tag{2.17}$$

where μ_e in a mixture of given composition can be evaluated using Blanc's law. Using electron densities measured in [20–23], we can approximate n_e under 'lasing' conditions for capillary tubes of diameter 2–4 mm, with pd = 0.06–0.4 kPa cm, in the region of falling E_z (Figure 2.3), as

$$n_e = 4.3\text{–}5.0 \times 10^{13}\, j\, p_{\mathrm{He}} / E_z , \tag{2.18}$$

where n_e is in cm^{-3}, j is in A cm^{-2}, p_{He} is in kPa and E_z is in V cm^{-1}. Using Figure 2.4, we can estimate the degree of ionization of helium atoms in the discharge as ~10^{-5}, and of metal atoms as ~10^{-2}; the degree of ionization decreases as the metal vapour density increases.

2.2.3 The Density of Metastable Buffer-Gas Atoms

The density of metastable buffer-gas atoms is an important parameter because, being long-lived particles and having high densities, metastables participate in the ionization of both helium atoms and metal atoms, and also in the pumping of the most intense laser transitions of zinc and cadmium.

In Table 2.1 are given the constants and rates for population of the He metastables by electron impact, and their depopulation under optimum laser conditions for a He–Cd gas mixture, i.e. for $n_e = 1.2 \times 10^{12}$ cm^{-3}, $N_{\mathrm{He_m}} = 5 \times 10^{12}$ cm^{-3}, and $n_{\mathrm{Cd}} = 5 \times 10^{13}$ cm^{-3}. It can be seen that He metastables are chiefly deactivated by ionizing collisions with electrons and by Penning collisions.

Most of the methods used to determine the densities of excited atomic states are based on the measurement of the absorption coefficient κ_0 at the centre of a line which terminates on the investigated state [12]. For the metastable He levels 2 ^1S and 2 ^3S, and the low excited levels 2 ^1P and 2 ^3P, which are optically connected to the metastable states, absorption is usually measured on the following lines: 501.6 nm (3 $^1P \longrightarrow$ 2 ^1S transition), 388.9 nm (3 $^3P \longrightarrow$ 2 ^3S), 492.2 nm (4 $^1D \longrightarrow$ 2 ^1P) and 447.1 nm (4 $^3D \longrightarrow$ 2 ^3P).

Measurements [21, 22, 24] have shown that in 'pure' helium for pd = 0.12 kPa cm, the metastable states 2 ^1S and 2 ^3S become saturated at low

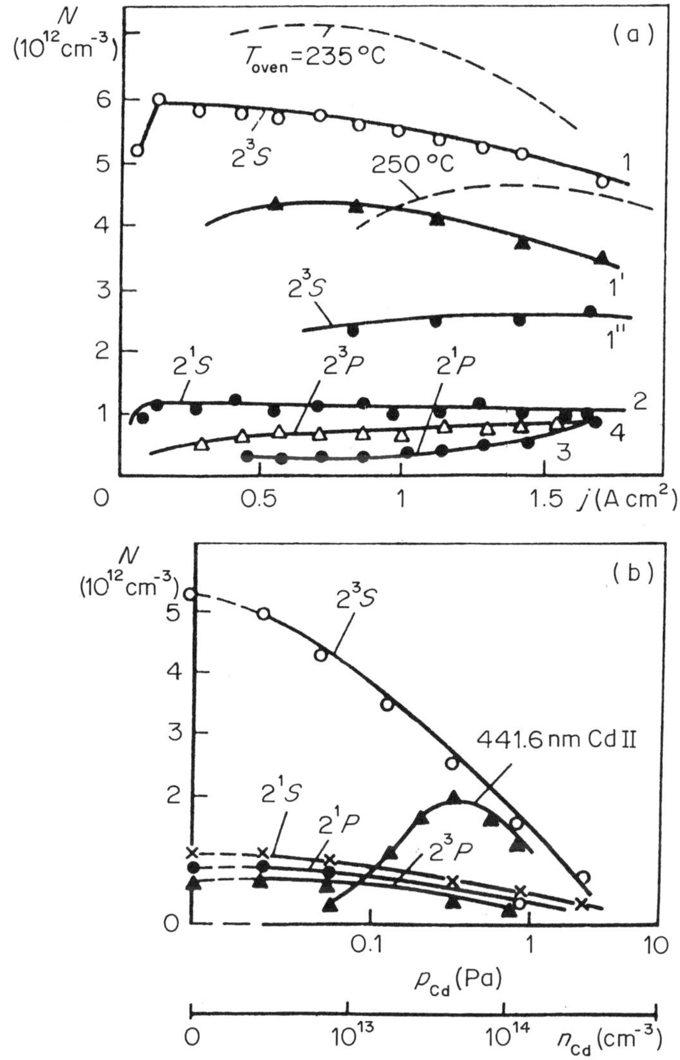

Figure 2.5 The concentrations of excited helium atoms measured by absorption in a He–Cd mixture, with $p_{He}d = 0.12$ kPa cm and $d = 3$ mm, as functions of (a) j and (b) n_{Cd}. (a) $p_{Cd} = 0$ (1–4), 0.1 Pa (1′) and 0.4 Pa (1″); the broken lines are for $2\,^3S$ [24] with $pd = 0.2$ kPa cm and two different oven temperatures, T_{oven}. (b) $j = 1.5$ A cm^{-2}.

currents, while the resonance levels $2\,^1P$ and $2\,^3P$ are not saturated (Figure 2.5a). At higher cadmium densities, saturation of $n(2\,^1S, 2\,^3S)$ sets in at higher currents, which can be seen from the kinetic equation for N_{He_m}

$$N_{He_m} \sim n_e k_{0m}/(\nu_d + n_e k_{mi} + nK_P)\,. \tag{2.19}$$

The saturation of metastable state populations at low currents means that they are mainly depopulated by stepwise ionization, and hence, for higher vapour pressures, saturation appears at higher n_e. The subsequent decrease in metastable state density with the increase in current is caused by a reduction in k_{0m} due to falls in T_e and E_z (Figure 2.2).

When Cd vapour is admitted to the discharge and its density is increased, the populations of all four He levels decrease (Figure 2.5b). This is due to the accompanying fall in T_e, which reduces the pump rate constant k_{0m}, and to the increased rate of depopulation via Penning collisions, nK_P. The dependence of N_{He_m} on n is typical of the various He levels.

Taking into account the data of different authors, by analogy with [24], the saturated density of metastables is approximated as

$$(N_{He_m})_{sat} \approx 0.5\text{–}1 \times 10^{12}(E_z - 30p)\,, \qquad (2.20)$$

where n is in cm^{-3}, E_z is in V cm^{-1} and p is in kPa.

2.2.4 Ionization of Atoms in the Gas Mixture

It has sometimes been suggested that metal-ion laser levels can be excited by charge transfer between molecular helium ions and metal atoms. However, molecular helium ions are not readily formed in typical PC laser discharges. It is most likely that He_2^+ ions are produced by associative ionization of He metastable atoms (see Table 2.1) and via the ion conversion reaction $He^+ + 2He \longrightarrow He_2^+ + He$ (see Appendix 3). Under 'optimum' laser conditions, the associative ionization rate is one to two orders of magnitude higher than the ion conversion rate, and the overall rate of production of He_2^+ is approximately two orders of magnitude less than the production rate of He^+. Although the rate of ion conversion increases with pressure, for $p < 4$ kPa, the density of He_2^+ in the discharge is very small. At higher pressures (>8 kPa), atomic ions are depleted via the ion conversion route rather than diffusion to the wall.

It is important to clarify the contribution of Penning ionization to the overall rate of metal atom ionization, because it is a factor leading to fluctuations and noise in laser emission. Its important role in the production of Cd^+ has been verified, but under conditions different from those in laser discharges [25, 26].

The total rate of metal (cadmium) atom ionization is the sum of the total rate of Penning ionization, the total rate of charge transfer and the total rate of electron-impact excitation of all the levels in the Cd II spectrum. Due to a sparsity of metastable states in Cd II, the total rate of ionization is equal to the formation rate of the ion ground state. cross-sections for all these processes have been measured in [27–33].

Using the formulæ (2.18) and (2.20), it can be found that the greatest contribution of the Penning process occurs at 0.3 kPa helium pressure, and decreases with increasing helium pressure. By considering the dependence of metastable density on discharge current (Figure 2.5a), we find that Penning ionization of the 5 $^2S_{1/2}$ state is stronger than electron excitation (and charge transfer) up to $n_e \approx 2.5 \times 10^{11}$ cm^{-3}, which corresponds to a current density of 0.3 A.cm^{-2}. Under these circumstances (for $j \approx 13$ mA cm^{-2}) the measurements of [25] were conducted. At higher current densities, j ($n_e > 0.05\ N_{He_m}$), the situation reverses, and for n_e corresponding to laser conditions (~10^{12} cm^{-3} and higher), electron excitation dominates over Penning ionization and becomes the major mechanism of metal atom ionization (Figure 2.6).

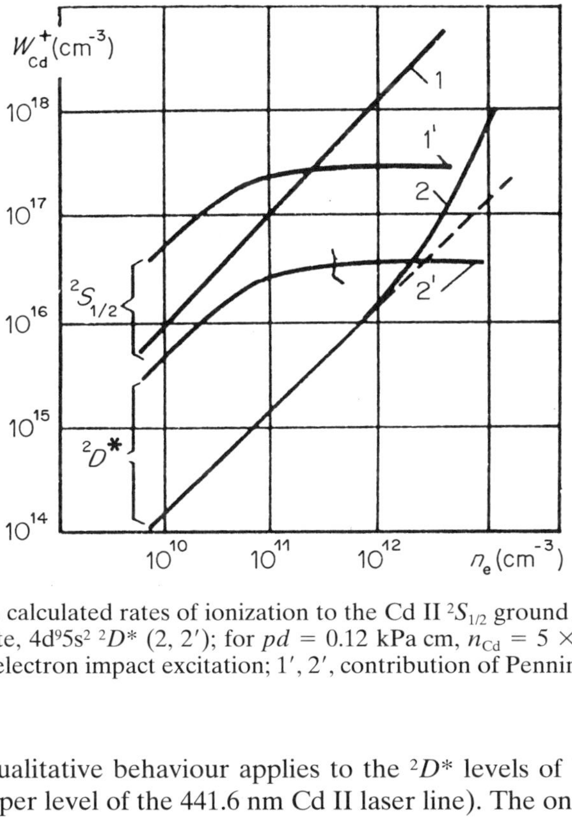

Figure 2.6 The calculated rates of ionization to the Cd II $^2S_{1/2}$ ground state (1, 1′) and to the Cd II state, 4d^{9}5s^2 $^2D^*$ (2, 2′); for pd = 0.12 kPa cm, n_{Cd} = 5 × 10^{13} cm^{-3}: 1, 2, contribution of electron impact excitation; 1′, 2′, contribution of Penning ionization.

The same qualitative behaviour applies to the $^2D^*$ levels of Beutler's term system (the upper level of the 441.6 nm Cd II laser line). The only difference is that the rate of electron excitation does not exceed the Penning ionization rate until much higher currents, i.e. for $n_e \approx 2 \times 10^{12}$ cm^{-3}. This means that under optimum He–Cd laser conditions ($n_e \approx 1.2 \times 10^{12}$ cm^{-3}), excitation of the $^2D^*$ levels of Cd II is primarily by Penning ionization (Figure 2.6). Under the same conditions, the total rate of ionization to the 5 $^2S_{1/2}$ ion ground state is about an order of magnitude higher than the rate of ionization to the $^2D^*$ levels.

The dependences plotted in Figure 2.6 are in many respects typical, and have a character that is qualitative due to the low accuracy of the known cross-sections and, hence, of the rate constants k. Therefore, for analyses of laser-level excitation mechanisms (including that for the $^2D^*$ Cd II levels), it is advisable to use techniques that do not depend on rate constant calculations, such as those described in [21, 25, 26] (for more detail, see section 5.3).

2.2.5 Cataphoresis

It is well known that a binary gas mixture in a PC discharge can to some extent be separated due to cataphoresis, and for that reason the process is used in gas purification. It relies on the selective transport (along the longitudinal z axis) towards the cathode of the low ionization component, whose flow rate, Γ_i, given by

$$\Gamma_i = 0.43\theta n \mu_M^+ E_z \tag{2.21}$$

(where θ is the radially averaged ionization coefficient of the metal atoms), is larger than the flow rate due to diffusion of neutral particles in the opposite direction. As was shown above for a He–Cd discharge, metal atom ionization takes place from the atomic ground state, and the metal ion initially has no directed velocity. The ion is then accelerated between collisions in the electric field of the PC, and fractions of its momentum, $\delta_1(mv_z)$ and $\delta_2(mv_z)$, are transferred in collisions with, respectively, buffer-gas atoms and metal atoms. Momentum can be found from

$$m\dot{v}_z = eE_z - (\delta_1 N\langle q_1 v\rangle + \delta_2 n\langle q_2 v\rangle)mv_z\,. \tag{2.22}$$

Integrating (2.22), we obtain the following expression [34] for the directed ion velocity, v_z (with $v_z = 0$ at $t = 0$):

$$v_z = v_{z0}[1 - \exp(-t/\tau)]\,, \tag{2.23}$$

where $\tau = 1/(\delta_1 K_{el1} N + \delta_1 K_{el2} n)$ is the time taken for the directed ion velocity to be established (K_{el1} and K_{el2} are the rate constants for elastic collisions of metal ions with buffer-gas and metal atoms, respectively); v_{z_0} is the final directed velocity.

The ion is also accelerated in the radial electric field towards the negatively charged tube wall. The radial electric field is larger than the axial electric field, and the radial ion velocity is an order of magnitude higher than the axial ion velocity ($v_z = 1.3-2.1 \times 10^4$ cm s^{-1}; $v_r = 1.3 \times 10^5$ cm s^{-1} [35, 36]). The drift time of an ion with such values of v_r to the tube wall is 0.1–1 μs, and this process may determine the lifetime of ions in long-lived states.

As mentioned above, the degree of ionization of the metal atoms (10^{-2}) is much higher than for the buffer-gas atoms (10^{-5}), due to the much higher ionization probability for the low ionization potential metal atoms. The relatively high degree of metal ionization and the high ion drift velocities lead to cataphoresis in the axial direction (towards the cathode) and in the radial direction (to the tube wall) which are not compensated for by diffusion.

In the simplest design of the discharge tube [9], where the vapour source is placed near the anode, and the metal vapour does not condense along the tube (except at a special condenser near the cathode), three distinct zones of vapour density and spontaneous emission are observed along the tube axis [9, 21] (Figure 2.7). Assuming that the rates of ionization are not functions of z,

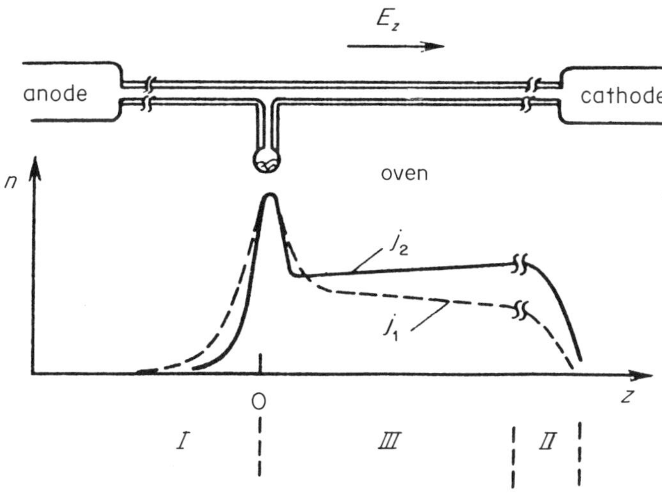

Figure 2.7 Typical distribution of metal atoms along a discharge tube due to cataphoresis; $j_1 < j_2$.

and that ambipolar diffusion to the tube wall is the main process of depletion of the ions, we obtain [9]

$$\frac{\partial^2 n}{\partial z^2} + \frac{\beta}{l}\frac{\partial n}{\partial z} = 0 \quad \left(\beta = \frac{\Theta E_z l}{T_g}\right), \tag{2.24}$$

where l is the distance from the Cd source oven to the condenser, and T_g is the gas temperature. The solution of equation (2.24) is

$$n(z) = n(0)[1 - \exp\beta(1 - z/l)]/(1 - \exp\beta) , \tag{2.25}$$

where $n(0)$ is the vapour density in the oven. In early studies [6] of gas separation by cataphoresis, only zone I, from the oven to the anode, was studied (Figure 2.7). There are two types of ions in the discharge, and in that zone the atomic metal density falls exponentially towards the cathode:

$$n(z) = n(0)\exp(\Theta\mu_{\mathrm{M}}^{+}E_z z\,/\,D)\ (z<0)\,. \tag{2.26}$$

Mass transport to the anode through zone I of length l_1 is

$$\Gamma(l_1) = 0.43n(0)\Theta\mu^+E_z\exp(-\Theta\mu^+E_z l_1\,/\,D)\,, \tag{2.27}$$

from which can be calculated the distance from the oven towards the anode, over which mass transport cannot be supported, for a given ionization degree and value of electric field E_z, i.e. when vapour 'blockage' occurs where the metal vapour cannot reach the anode. Note that at relatively high vapour densities ($>5\times10^{14}$ cm^{-3}), when $n_e = n_{Cd+}$ (i.e. only the metal is ionized), instead of (2.26) there is a linear decrease of vapour density

$$n(z) = n(0) + \mu^+jz\,/\,(\mu_e D)\,. \tag{2.28}$$

The distribution of the vapour near the condenser (zone II) is also non-uniform, with a length that diminishes with increasing discharge current. The distribution of vapour is most uniform in zone III, for which optimum conditions for excitation of the metal can be maintained over a sufficient distance for laser operation. Mass transport of metal atoms towards the cathode along zone III will be equal to $m\Gamma_i$, and the metal drift velocity will be

$$V_{dr} = \overline{n^+}\mu^+E_z\,/\,\bar{n}\,, \tag{2.29}$$

where $\bar{n}$ and $\overline{n^+}$ are radially averaged densities. The drift velocity can be measured visually by the directed velocity of a 'pulse' of metal vapour along zone III (Figure 2.8). For a constant oven temperature, with an increase in j, the radially averaged metal vapour density remains constant, but the increase in n^+ is slow due to the increasing radial separation of metal and buffer. If the radial flow of metal atoms is compensated for by increasing the oven temperature, while E_z is maintained constant, then V_{dr} will gradually increase due to the increase in n^+ (Figure 2.8).

From Figure 2.8(b), it is apparent that for constant current and increasing n, despite a decrease in E_z (see Figure 2.3) and in μ^+, V_{dr} increases quickly due to the rise in the ratio $\overline{n^+}\,/\,\bar{n}$ which is caused mainly by the increase in n^+ on the tube axis. From (2.29) and Figure 2.8, we can determine the radially averaged metal density, $\bar{n}$. The radial distribution of metal atoms can be approximated by an exponential function, similar to (2.26). If we take the radial average, then we find that the vapour density at the tube wall is 3 to 4 times higher than the average density $\bar{n}$ and two orders of

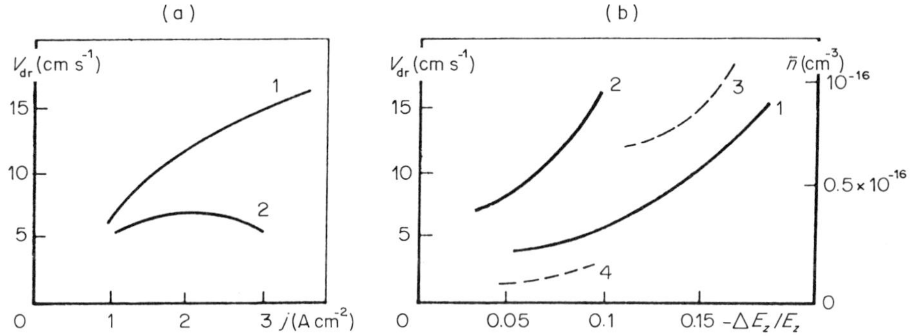

Figure 2.8 Drift velocity of cadmium, V_{dr}, and the radially averaged vapour density, $\bar{n}$. (a) The dependence of V_{dr} on j for pd = 0.12 kPa cm; 1, with constant vapour density ($\Delta E_z/E_z$ = -0.06); 2, with constant T_{oven}. (b) The dependence of V_{dr} (1, 2) and $\bar{n}$(3, 4) on $(-\Delta E_z/E_z)$ for $j = 2$ A cm^{-2}, pd = 0.12 kPa cm (1, 3) and pd = 0.25 kPa cm (2, 4).

magnitude higher than on the axis.

In cataphoretic laser discharges, there is a difference between the metal atom density in the discharge and the density in the oven. This density difference depends greatly on tube design. For various tube designs, the optimum temperature for the Cd oven may lie at a number of different temperatures between 220 and 270 °C (which range corresponds to more than an order of magnitude change in vapour pressure), although the metal atom density near the tube axis must remain constant for basic physical reasons. The difference between densities in the discharge and in the oven is a result of cataphoresis, which means that, even for a fixed metal evaporation rate, the on-axis atom density will depend on discharge parameters such as current and/or buffer gas pressure.

2.2.6 Electrophoresis of the Buffer Gas

As well as the transport of metal atoms towards the cathode, a flow of neutral buffer-gas atoms in the direction of the anode appears, which causes an undesirable pressure difference, Δp, between the ends of the discharge tube [37–39]. The increase in pressure near the anode is due to the momentum of longitudinally directed electrons being imparted to the gas atoms, the process being only partly compensated for by momentum transferred to the atoms by ions. This leads to a nett force on the gas atoms in the opposite direction to E_z. In [37] it was found that for intermediate gas pressures (when the ion mean free path related to momentum transfer, λ^+, is less than the tube radius) this 'volumetric' force is a maximum on-axis and reduces towards the wall, as does the charged particle density. At such pressures, the volumetric forces acting on the gas are much stronger than the forces acting on the gas in the wall boundary layer (which can be considerable at low pressures), and they yield a pressure gradient [38]

$$\Delta p / l = 37eE_zN^+\lambda^+\Delta'(v_r^+, v_t^+) / r , \tag{2.30}$$

where v_r^+ and v_t^+ are the radial drift and thermal velocities of the ions, and Δ' is an increasing function [38].

In a gas mixture, momentum is transferred from electrons and two types of ions to the gas atoms. As metal vapour enters the discharge, buffer-gas ions are substituted by metal ions due to the difference in ionization potentials (see Figure 2.4). The heavier metal ions pass to the gas atoms at each

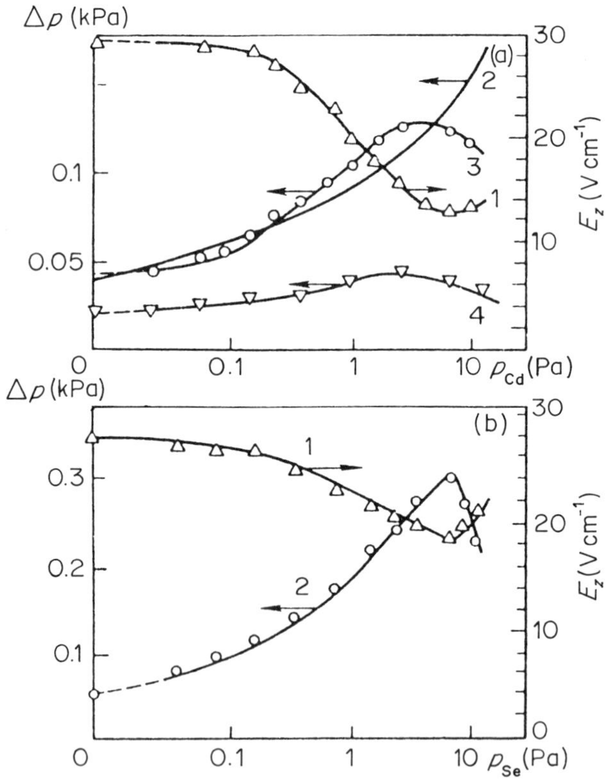

Figure 2.9 Buffer-gas electrophoresis in mixtures with (a) Cd and (b) Se vapours in a positive-column discharge. (a) A He–Cd mixture (1–3) and a Ne–Cd mixture (4), with pd = 0.12 kPa cm and j = 1.5 A cm^{-2}: 1, E_z; 2, $\Delta p(n_{Cd})$ without allowing for Cd–Cd$^+$ collisions; 3, 4 – Δp (experiment). (b) A He–Se mixture, with pd = 0.4 kPa cm and j = 3.7 A cm^{-2}: 1, E_z; 2, Δp (experiment).

collision a much smaller fraction of their momentum (due to the greater difference in their masses), i.e. the mean free paths, λ^+, of the ions increase, leading to an increase in Δp [39] (Figure 2.9). Clearly, when a heavier buffer gas is employed, the metal ions transfer their momentum in a smaller number of collisions, yielding a shorter λ^+. Furthermore, as a rule, heavier buffer gases have lower ionization potentials, which reduces the degree of ion substitution and reduces E_z in the PC. Therefore, in such mixtures, the pressure gradient, Δp, is smaller, and a weaker increase in Δp is observed as the metal atoms enter the discharge (e.g. see Figure 2.9a for a Ne-Cd mixture). In the wall boundary layer, λ^+ is sharply reduced due to radial cataphoresis, and the role of those forces is markedly reduced there too [39]. At high pressures (>2 Pa) of metal vapours, it is necessary to consider the effect of collisions of metal ions with metal atoms. This effect explains the extremum in the dependence of Δp on vapour pressure (Figure 2.9).

Hence, when a buffer gas is mixed with metal vapours, the process of movement of the buffer gas towards the anode is intensified, and the pressure gradient increases if lighter buffer gases are used. The relative value of the pressure gradient in both the pure buffer and a gas mixture is greatest at low gas pressures and high currents. By connecting the anode and cathode ends of the tube with a bypass channel, Δp can be reduced due to gas movement in the loop with a velocity of 10's of cm s^{-1} (Figure 2.10) in the direction opposite to the cataphoretic flow, thus partially decreasing the waste of metal from the oven.

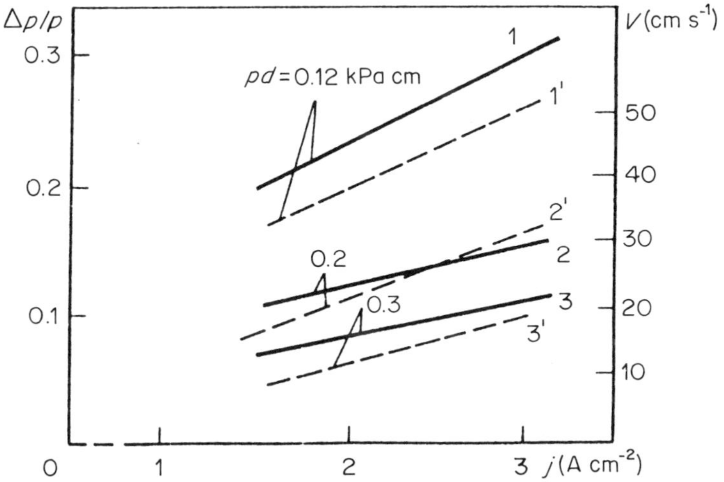

Figure 2.10 Dependence of $\Delta p/p$ in a tube without a bypass channel (1–3) and of the flow velocity of helium in a tube with a bypass channel (1′–3′) on j in a He–Cd mixture at $\Delta E_z/E_z$ = -0.1, pd = 0.12 kPa cm (1, 1′); 0.2 kPa cm (2, 2′) and 0.3 kPa cm (3, 3′). The length of each tube was 50 cm, with diameters 3 mm (for the capillary) and 6 mm (for the bypass channel).

2.3 PULSED POSITIVE-COLUMN DISCHARGES

Metal vapour ion lasers that utilize pulsed discharges with 'cold' electrodes typically have tube bores of 0.3–2 cm, require buffer-gas pressures of 0.01–80 kPa, metal vapour pressures of 0.1–100 Pa, microsecond currents of up to 1000 A amplitude and electric fields of up to 300 V cm^{-1}. The electron density is one to three orders of magnitude higher than in a stationary PC discharge and can reach 10^{15} cm^{-3} or more.

The evolution of a pulsed longitudinal discharge can be divided into three stages: breakdown, the high-current stage and the afterglow. The first two stages are important for self-terminating MVILs, while the second and third stages are important for recombination lasers.

2.3.1 Gas Breakdown [40]

The breakdown stage consists of several substages which are observed in oscillograms of voltage, current (Figure 2.11) and spontaneous emission. In the first substage ($t_1 < t < t_2$), the longitudinal electric field is established as a result of the movement of a wave of potential from the high-voltage electrode to the ground one, and back again in the reverse direction. In the second substage ($t_2 < t < t_3$), a dense glow discharge begins to form. The field is concentrated mainly near the cathode, which as yet does not have a cathode spot, and around the cathode a narrow dark strip is observed. In the third substage ($t_3 < t < t_4$), another potential wave starts from the cathode (with a speed of 5×10^9 cm s^{-1}) with a very fast (< 5 ns) increase in the number of electrons produced by field and thermionic emission at the

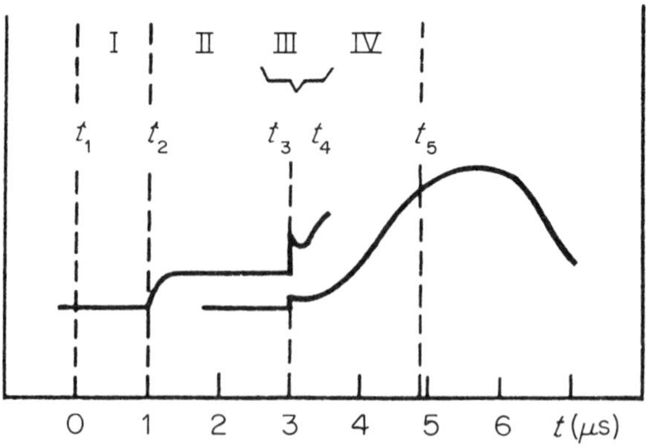

Figure 2.11 The typical waveform of a discharge current pulse. The first stages (I–III) are shown enlarged.

cathode spot already created. In the fourth substage ($t_4 < t < t_5$), a high-current pulsed discharge forms, characterized by a rapid increase in current and in spontaneous emission. The field is now uniform, and is determined by the external applied voltage.

With an increase in tube voltage, the sequence of stages does not alter, but the duration of each stage, particularly of the first and second, reduces markedly. With a further increase of voltage, the duration of the first three stages may shorten to 10's of nanoseconds, and for every gas there is a pressure for which it is a minimum (a desirable condition for a self-terminating MVL).

2.3.2 The High-Current Stage of the Discharge

The processes that characterize the high-current stage are transient. The change in n_e during a current pulse can be described by

$$\dot{n}_e = (z_i - \nu_d)n_e \,, \tag{2.31}$$

where ν_d is the depletion frequency of ions and electrons due to ambipolar diffusion. In the stationary state, $\dot{n}_e = 0$, i.e. the ionization frequency, z_i, compensates for the loss, by diffusion, of charged particles to the wall, and therefore determines the necessary electron temperature. When the field is determined by an external source, a significant overvoltage can arise (usually $z_i >> \nu_d$) and according to (2.31), n_e will rise exponentially:

$$n_e = n_e(t{=}0)\exp(z_i - \nu_d)t \,. \tag{2.32}$$

The current density, j, follows the same type of dependence, since

$$j = \sigma E_z \; = \; n_e e^2 E_z/(m_e \nu_e) \,,$$

where ν_e is the frequency of collisions of electrons with atoms and ions. For a helium buffer gas, with only elastic collisions of electrons with atoms accounted for, the plasma conductivity is

$$\sigma \approx 10^4 \, n_e \,/\, (N\sqrt{T_e}) \,, \tag{2.33}$$

where σ is in units of $\text{ohm}^{-1}\,\text{cm}^{-1}$, n_e and N are in cm^{-3} and T_e is in electron-volts.

The current and density of charged particles will keep rising near-exponentially until the field E_z is practically constant, and the ion composition of the plasma does not change markedly (when the degree of ionization of the gas mixture may still be quite low). Upon complete discharge of the storage capacitor in the excitation circuit, the exponential rise of current will be limited by the fall in potential across the tube or, when a vacuum tube is used to modulate the partial discharge of a capacitor, by the internal resistance of the tube.

In a pulsed PC discharge, as in a stationary one, the introduction of metal vapour reduces the electron temperature and increases the electron density. The time to establish a Maxwellian EEDF depends on electron–electron Coulomb collisions, and can be evaluated using (2.1)–(2.3). For example, for a typical electron density of 10^{14} cm^{-3}, $T_e = 5$ eV, and we find that $(\nu_{ee})^{-1} = 4$ ns, which is usually much less than the times of variation in j and E_z. Hence the EEDF is similar to that of a steady-state discharge.

The electron temperature during a current pulse can be found using [40]

$$\frac{d}{dt}\frac{3}{2}(T_e n_e) = \sigma E^2 - \frac{3}{2}\delta^* \nu_{in} n_e (T_e - T_g)\,. \tag{2.34}$$

Taking $T_e = 5$–10 eV and $N = 10^{18}$ cm^{-3}, we have $\nu_{in} = 10^8$–10^9 s^{-1} and $\delta^* = 1$. The time characterizing a change in T_e in response to a change in the electric field is only 10^{-9}–10^{-10} s, which is much less than the relaxation time of the discharge (the time that changes can occur in E_z in the high-current substage). Consequently, the right-hand side of (2.34) can be set approximately to zero, and, since $\sigma \sim n_e/N$, it follows that T_e is not dependent on n_e, but is determined by E/N (or E/p). Since E falls during the pulse, due either to the decrease in voltage on the storage capacitor or to the fall of voltage on the switch, there is a decrease in T_e. Because of the strong increase in plasma conductivity σ by the end of the pulse, the decay of T_e becomes faster.

2.3.3 Relaxation of the Plasma in the Afterglow

Upon termination of the current pulse, the electrons rapidly cool to $T_e \approx 1$ eV via inelastic collisions with a time constant of $1/\delta^* \nu_{in} \approx 1$ ns. Further cooling of the electrons occurs due to elastic collisions of electrons with atoms and ions. In addition, recombination and superelastic electron collisions take place with excited atoms and ions (electron de-excitation).

The density of charged particles in the afterglow will decrease with time mainly due to ambipolar diffusion and recombination:

$$\dot{n}_e = -\,\nu_d n_e - \alpha n_e^3\,. \tag{2.35}$$

If $n_e < 5 \times 10^{13}$ cm^{-3} and the gas pressure is less than 1 kPa, ambipolar diffusion will be the major process. Its frequency in a helium afterglow, for example, can be calculated using

$$\nu_d = D_a/\Lambda^2 = 1.47 \times 10^8 T_e^{3/2}(1 + T_e/T_g)/(pd)\,, \tag{2.36}$$

where Λ is the diffusion length for ambipolar diffusion, and T_e and T_g are in electron-volts, p is in Pa and d is in cm. Note that this frequency is an order of magnitude higher during the pulse (when T_e is high) than in the afterglow. For $n_e = 10^{14}$–10^{15} cm^{-3}, three-body recombination becomes important in the decay of charged particles[1]. Hence, n_e reduces slowly with time in the afterglow (Figure 2.12).

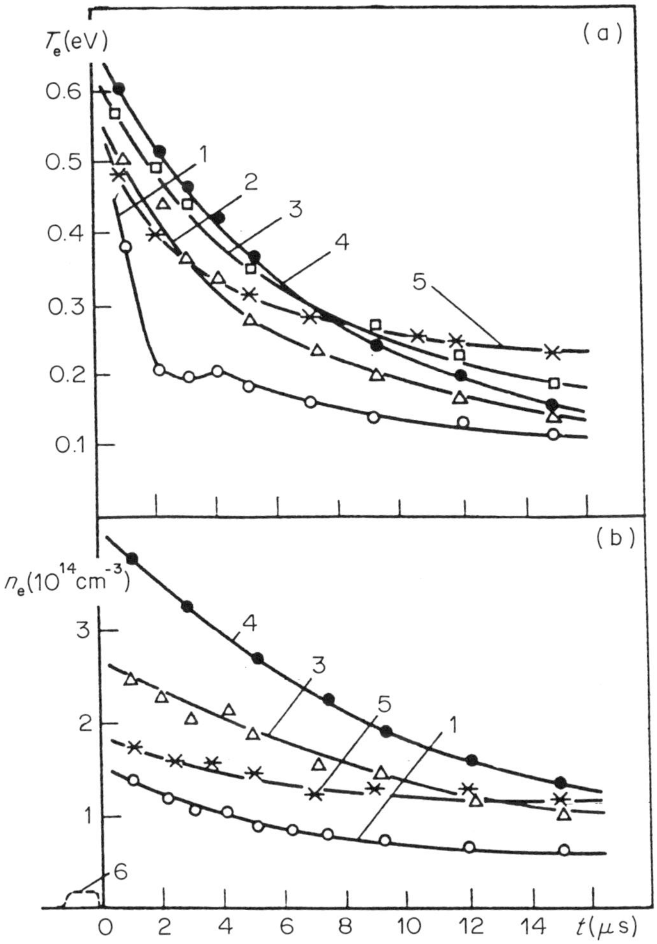

Figure 2.12 Temporal behaviour of T_e and n_e in the discharge afterglows of He-Cd and Ne; j = 100 A cm^{-2}, d = 0.5 cm, p_{He} = 0.35 kPa, p_{Cd} = 0 (1); 1.3 Pa (2); 13.3 Pa (3); p_{He} = 0, p_{Cd} = 13.3 Pa (4); p_{Ne} = 0.35 kPa (5); 6, current pulse.

The electron temperature in the afterglow of a discharge in helium can be found approximately from the equation [41]

$$\frac{\mathrm{d}}{\mathrm{d}t}\left(\frac{3}{2}T_e n_e\right) = \varepsilon_m k_{m0} N_m n_e + \varepsilon_r \alpha n_e^3 - \frac{3}{2}\delta\,\nu_{el}(T_e - T_g)n_e\,, \qquad (2.37)$$

where ν_{el} is the frequency of elastic collisions, ε_m is the energy of metastable atoms, and ε_r is the energy received by free electrons at each recombination event.

For $T_e \leq 1$ eV, electron cooling is due to elastic collisions:

$$T_e = T_e^0 \exp(-t/\tau_{el});\ \tau_{el} = 1/(\delta_a\,\nu_{ea} + \delta_i\,\nu_{ei})\,, \qquad (2.38)$$

where ν_{ea} and ν_{ei} are the frequencies of collisions of electrons with atoms and ions. For helium, these frequencies have the form

$$\nu_{ea} = q_{ea}N\sqrt{8T_e/(\pi m_e)} \approx 3.6\times 10^{-8}N\sqrt{T_e}\,; \qquad (2.39)$$

$$\nu_{ei} = \sqrt{8\pi e^4}\,L_c N^+/(3T_e\sqrt{T_e m_e}) \approx 8.3\times 10^{-6}\,N^+/T_e^{3/2}\,, \qquad (2.40)$$

where N and N^+ are the concentrations of helium atoms and ions in cm^{-3}, T_e is in electron-volts, and the Coulomb logarithm, L_C, $= 5.7$ (typical conditions in an afterglow). Estimates made using (2.38–2.40) with $N = 10^{17}$–10^{18} cm^{-3}, $N^+ = 10^{14}$–10^{15} cm^{-3} and $T_e = 0.5$–1 eV yield $\tau_{el} \approx 0.1$–1 μs. Thus, the initial decay of T_e via elastic collisions is fast enough to promote recombination. Later, T_e approaches a quasi-steady-state value, $(T_e)_{eq}$, much higher than T_g (Figures 2.12–2.15).

By setting (2.37) equal to zero, and considering nearly all the recombination flux to pass through the metastable levels, we find that

$$T_e - T_g = \frac{2}{3}(\varepsilon_m + \varepsilon_r)\alpha n_e^2/(3\delta\nu_{el})\,. \qquad (2.41)$$

Using (1.24) for the coefficient of three-body recombination, α, and taking account of elastic collisions of electrons with helium atoms only, we obtain

$$(T_e)_{eq} \approx 4.2\times 10^{-3} n_e^{1/3}/N^{1/6}\,. \qquad (2.42)$$

As an example, for optimum conditions in the afterglow of Sr and Ca recombination MVILs ($n_e \approx 3$–5 ($\times 10^{14}$) cm^{-3} and $N \approx 3\times 10^{18}$ cm^{-3}), $(T_e)_{eq} = 0.25$ eV, i.e. after a fast decay it equilibrates at a relatively high value (Figure 2.15).

Just after a current pulse through He there is sometimes a minimum in T_e, before $(T_e)_{eq}$ is approached (see Figures 2.12 and 2.14). This is more

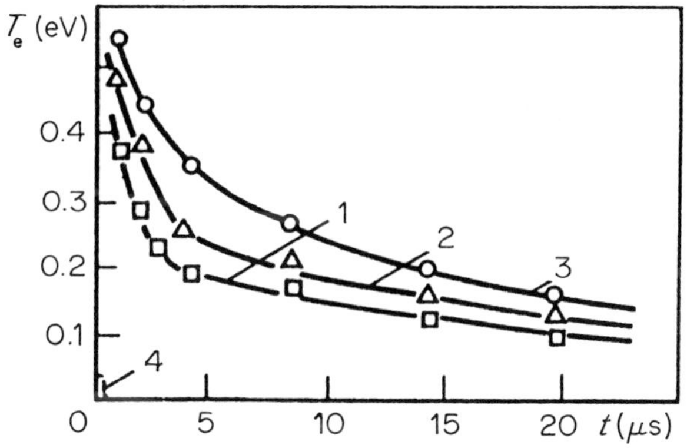

Figure 2.13 Temporal evolution of T_e in the afterglow of a He–Sr mixture (measured by the double-probe technique); p_{He} = 1.3 kPa, d = 0.8 cm, l = 50 cm, current = 50 A, p_{Sr} = 0 (1); 0.4 Pa (2); 2.7 Pa (3); 4, current pulse.

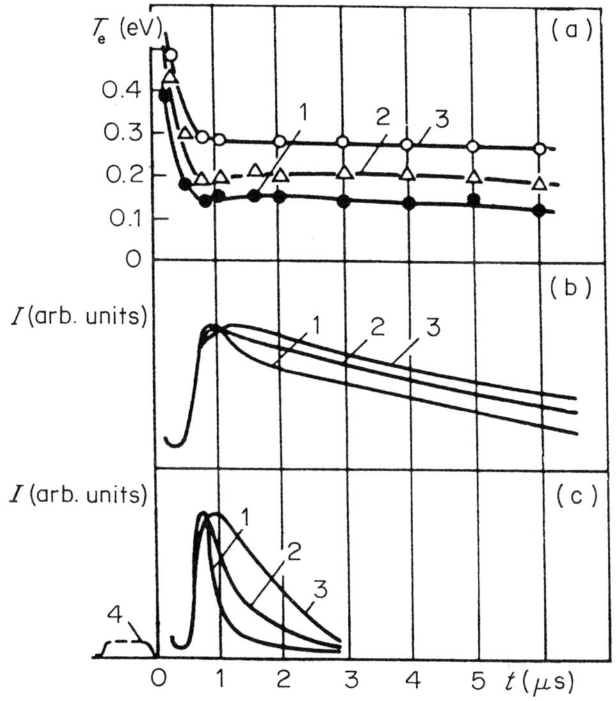

Figure 2.14 Temporal behaviour of (a) T_e, (b) the He I 447.1 nm line intensity, and (c) the Sr II 430.5 nm line intensity in the discharge afterglow of a He–Sr mixture (T_e was measured using a spectral method); p_{He} = 1.2 kPa, U_c = 6 kV, d = 1.1 cm, l = 30 cm, i = 45 A, p_{Sr} = 0.05 Pa (1); 0.67 Pa (2); 4 Pa (3); 4, current pulse.

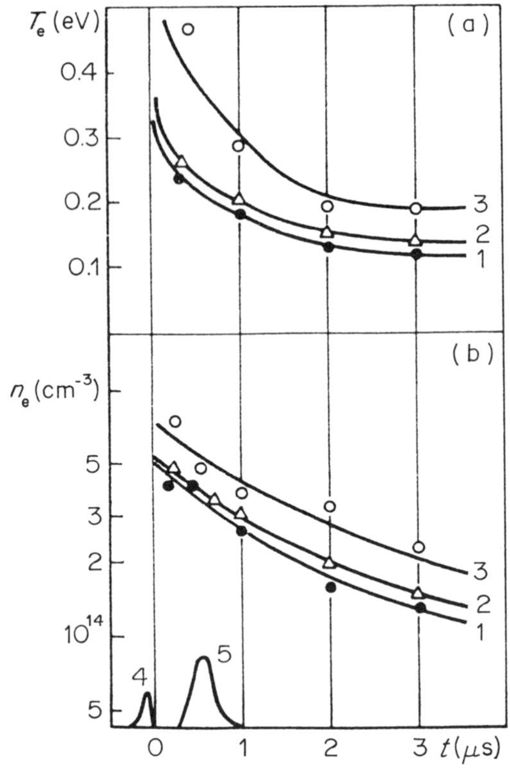

Figure 2.15 Temporal behaviour of T_e and n_e in the discharge afterglow in a mixture of helium (33 kPa) and strontium vapour (measured using a spectral method). Conditions were $U_c = 20$ kV, $C = 825$ pF, $d = 1.1$ cm, $l = 30$ cm, $p_{Sr} = 0$ (1), 13 Pa (2), 133 Pa (3), current pulse (4), laser pulse of the Sr II 430.5 nm line at $p_{Sr} = 13$ Pa (5).

pronounced for helium pressures below 2 kPa. The behaviour of T_e in He afterglows was predicted in [13], and can be explained by the de-excitation of high-lying (≈20 eV) helium metastables by electron collisions. The energy received by the electrons from the metastables is at least 5 times larger than from the de-excitation of other levels (down to the metastable states). Initially, the density of helium metastables is below the equilibrium value specified by the values of n_e and T_e that exist, and recombination heating is relatively low during the period of most rapid decrease in T_e. However, as the metastable populations build up, T_e will begin to increase towards $(T_e)_{eq}$.

When easily ionized metal vapour is admitted to a helium discharge, the trough in T_e disappears, since the helium metastable levels are less populated due to quenching by the metal atoms; metal ions substitute for helium ions, and T_e decays more slowly. As a result, the role of helium metastables in determining the electron energy diminishes, and the metastable densities follow T_e.

In addition, as metal vapour is introduced, the electron density increases and δ_i decreases due to the substitution of light helium ions by heavy metal ions. Thus, the influence of electron–ion collisions in cooling the electrons is sharply reduced. However, the effective coefficient of recombination, α, [42] increases, since for atoms with a configuration of relatively uniformly spaced levels (as have alkali and alkaline earth metals), $\alpha \sim T_e^{-9/2}$ is valid up to 0.8 eV. For helium, α falls abruptly for $T_e > 0.25$ eV. In addition, a great number of doubly charged ions of the metal are created, for which α is an order of magnitude higher than that of singly charged ions (1.24).

Thus, when metal vapours are added, $\delta\nu$ decreases, and α and n_e both increase which leads, from (2.38) and (2.41), to an increase in τ_{el} and in $(T_e)_{eq}$ (Figures 2.12–2.15)[2]. The durations of the initial falls in T_e and $(T_e)_{eq}$ also rise when helium is substituted by the heavier noble gases (Figure 2.12).

In powerful pulsed discharges, the concentration of metal atoms in the ground state can be reduced by an order of magnitude or more due to ionization [43], which diminishes the trapping of resonance radiation. In decaying plasmas, strong waves may appear [44]. In the high pulsed recurrence frequency (PRF) regime, acoustic oscillations can occur, and standing waves of gas density with frequencies that are multiples of the resonant frequency may exist [45]. As a result of this, the output powers of pulsed lasers can be a complicated function of PRF. In addition, convection vortices of gas can be present, carrying gas from the tube axis to the wall [46], hence intensifying the cooling of the gas medium, which is important for self-terminating lasers.

2.4 STEADY-STATE HOLLOW-CATHODE AND TRANSVERSE-RF DISCHARGES

Long before the invention of the laser, it was noticed that the relative intensity of the 'spark' spectrum (radiation from high-lying levels) in the discharge of a cold (i.e. non-heated) hollow cathode is much greater than in longitudinal PC discharges. In discharges through helium, the atomic singlet lines (with broad maxima of excitation at 100's of eV) and ion lines (e.g. the 468.6 nm He II line with a threshold for excitation of 75.6 eV) are much stronger in hollow-cathode discharges, and are insensitive to the addition of low ionization potential species.

Inside a hollow cathode, the negative glow (NG) zones belonging to each part of the concave surface overlap and produce a rather uniform plasma region which is surrounded by the layer of the cathode dark space (CDS).

The discharge voltage fall is mainly across the CDS [47]. In general, electrons are emitted from the cathode surface as a result of bombardment with photons, ions and neutral atoms; at medium pressures, ions play the main role. In steady-state HC discharges, the CDS voltage, U_0, is a function of both gas pressure and plasma composition. The effect of the hollow cathode, and the accompanying efficient ionization and excitation, arises as a result of a shortening in diffusion length for ions moving out of the NG into the CDS, and also due to the superposition of many NG zones inside the cathode at certain gas pressures and internal dimensions of the cathode. For example, for a cylindrical hollow cathode of diameter d, with $pd \approx 1$–2 kPa mm, the regime is characterized by a minimum in tube voltage under constant current conditions, or a maximum in current at constant voltage (Figure 2.16).

When the pressure is reduced below that at which the HC effect is present, U_0 rises due to the increase in the CDS which becomes comparable with the

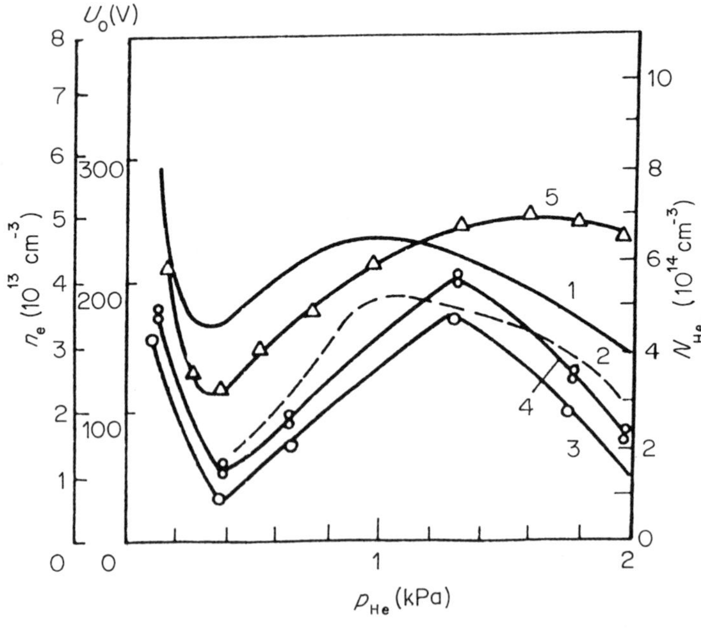

Figure 2.16 Dependence of the parameters of a steady-state HC discharge in helium on gas pressure: 1, electrode voltage (cathode potential fall); 2; electron concentration n_e, d = 6 mm, j = 8.8 mA cm^{-2}, measured by the probe method [62]; 3, electron concentration n_e, d = 6 mm, j = 5 mA cm^{-2}, measured using Stark line broadening [69]; 4, electron concentration n_e, d = 6 mm, j = 5 mA cm^{-2}, measured by a laser heterodyne technique [69]; 5, concentration of He 2 ^3S metastable atoms, d = 6 mm, j = 7 mA cm^{-2}, measured by absorption [70].

radius of the hollow cathode, and the volume of the NG, where ions are produced, diminishes. The discharge constricts and the CDS voltage increases to compensate for the lowering of the ionization rate in the NG in order to ensure the necessary cathode electron emission and the discharge current. However, with a rise in pressure, U_0 increases more slowly with the suppression of oscillations of electrons between the space charges on opposing walls, and at higher pressures still increases with the reduction in ions reaching the wall as a result of volumetric processes such as conversion, recombination and scattering. It should be noted that the optimum helium pressure for excitation of ion levels in HC discharge lasers may not correspond to the optimum pressure for the HC effect and, as a rule, is somewhat higher; the current corresponds to that of an abnormal glow discharge.

It is noticeable in HC and TRF discharges that the acceleration zone (the CDS) is spatially separated from the NG, i.e. the CDS is a source of external ionization for the NG. However, each zone is dependent on the other because the cathode electron emission is brought about by the ions produced in the NG and accelerated through the CDS. We shall show how the ion composition of the NG in a mixture of helium and metal vapour influences the CDS. The condition for sustaining a HC discharge is for the electron emission from the cathode surface to be accounted for by the bombardment of buffer and metal ions [48]:

$$\gamma_B \psi_B N^+ + \gamma_M \psi_M n^+ = 1 , \tag{2.43}$$

where γ is the secondary emission coefficient, i.e. the number of electrons emitted per bombarding particle; usually $\gamma_B > \gamma_M$ (e.g. for $U_0 = 100$–1000 V, $\gamma_{He} = 0.25$ and $\gamma_M = 0.05$–0.1 [5, 6]); ψn^+ is the number of ions entering the CDS from the NG per emitted electron. Clearly, in order to maintain the cathode emission for a given current and gas composition, the number of ions, their energies and the CDS potential must increase with the addition of metal vapour (Figure 2.17). This is due to the substitution of light buffer ions with heavy metal ions which have smaller γ. Note the slight rise and subsequent saturation of U_0 with increasing current over the entire range of vapour pressures in Figure 2.17. These processes lead to the positive dynamic impedance that is typical of anomalous discharges (Figure 2.17b).

In the so-called γ regime of TRF discharges, strong stationary fields close to the wall [4] cause electrons to be emitted from dielectric surfaces (usually quartz) next to external electrodes [3], or from the surfaces of internal metal electrodes [49]. Hence the near-axial zone of a TRF discharge is similar to the NG of a HC discharge, and the spark spectrum is intense and insensitive to the character of gas admixtures. Also, a process called electrode sputtering occurs [50] which resembles cathode sputtering [51].

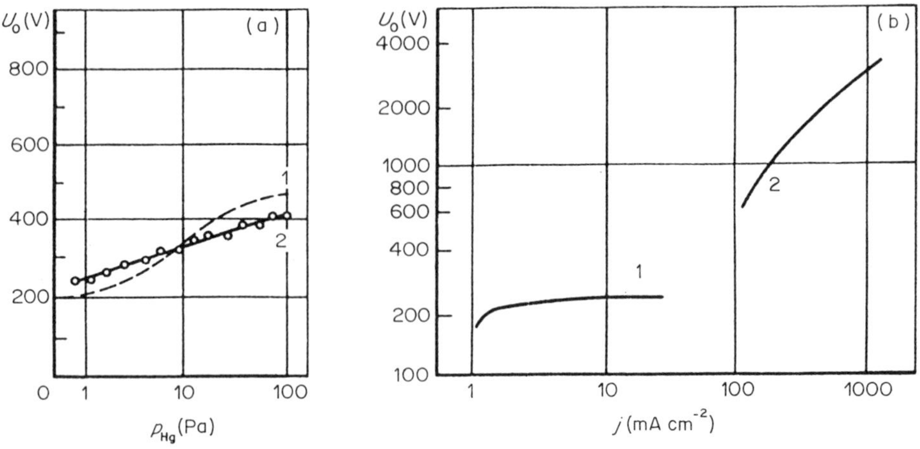

Figure 2.17 Cathode fall U_0 in a HC discharge. (a) The dependence on Hg vapour pressure; p_{He} = 0.5 kPa, j = 10 mA cm^{-2}; 1, calculation; 2, experiment. (b) Typical current–voltage characteristics of stationary (1) and pulsed (2) HC discharges (pulse duration 0.5 μs, p_{He} = 1.5 kPa).

2.4.1 Electron Energy and Concentration

In a PC discharge, the EEDF is formed during the acceleration of thermal electrons produced by ionization of gas atoms in a relatively weak electric field ($E_z \approx 10$ V cm^{-1}) in the presence of energy losses due to ionization and excitation. The EEDF is near-Maxwellian, or has a slight deficit of fast electrons. In the NG of a HC discharge, however, the EEDF is non-Maxwellian and has an excess of fast electrons. Mechanisms of generation of fast electrons and of the specific profile in various parts of the NG of their EEDF, which determines the excitation and ionization rate constants of atoms, are among the main questions under investigation in the physics of HC discharges [7, 52]. The difference between the EEDFs is due mainly to the fact that in the HC regime, the zones where electrons gain energy from the electric field, and lose it in various collision processes are spatially separated. In the CDS, which for $p > 0.1$ kPa is concentrated near the cathode surface, electrons emitted from the cathode gain energy in the cathode fall of 1000–10 000 V cm^{-1}, with negligible collisions with heavy-body species, indicated by the absence of a glow in the CDS (ionization does not occur there, presumably because its probability is less than that of excitation [5, 53]). Upon passing into the NG region, the electrons take part in intense inelastic processes of excitation and ionization, and the field there is very weak (except in the cases we shall mention).

This mechanism of formation of the EEDF leads to the existence of three groups of electrons in the plasma. The first group (on the boundary of the CDS and the NG) is formed by the initial mono-energetic electrons having energy equal to the cathode fall, eU_0. The second and third groups form in the NG. In the second group are fast electrons with an energy between ε_1 and eU_0 (where ε_1 is the first excitation potential of the mixture), which are initial electrons that have suffered one or more collisions with gas atoms. The third group is made up of thermal (or plasma) electrons with energies less than ε_1, produced as a result of ionization of gas atoms by the first two groups. The thermal electrons suffer only elastic and electron–electron collisions, and form a Maxwellian group, since there is no field in the NG region.

As in PC discharges, the typical conditions for the operation of lasers with transverse discharges (i.e. small diameters of the hollow cathode (1–5 mm), medium gas pressures, and a high current density (up to 0.1 A cm^{-2} in a stationary discharge, and 100 A cm^{-2} in a pulsed one)) differ greatly from those previously studied in HC discharges [4, 12, 47].

The spatial (transverse) profile of the intensities of spectral lines in HC and TRF discharges depends strongly on gas pressure (Figure 2.18). At low pressures, the profile is bell-like. With increasing pressure, the profile

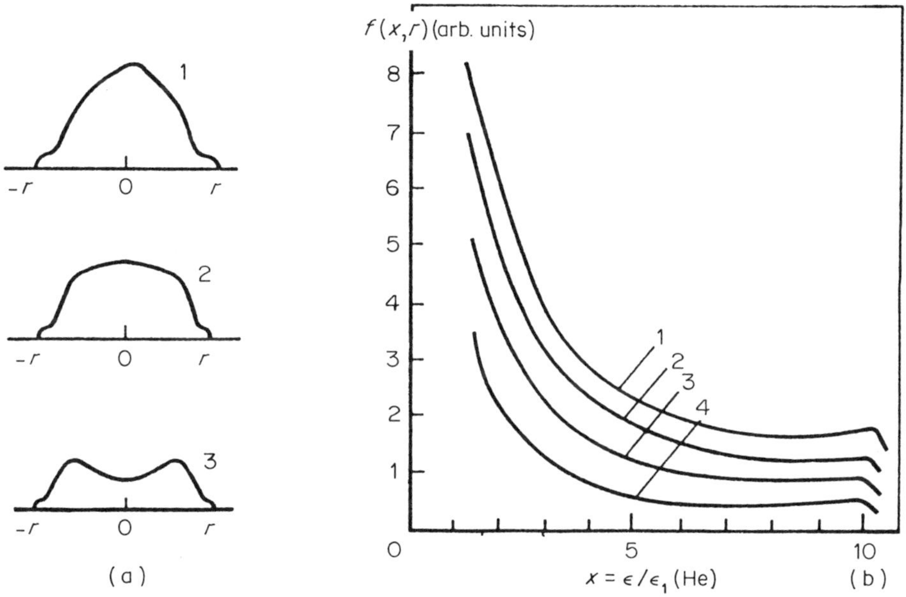

Figure 2.18 (a) Radial profiles of ionization rate as functions of helium pressure $p_1 < p_2 < p_3$. (b) The typical electron energy distributions in HC and TRF discharges at different distances, l, from the boundary of the CDS: $l = 0$ (1), λ_0 (2), $2\lambda_0$ (4) (where λ_0 is the mean free path of electrons with energy eU_0).

flattens, while at higher pressures still, the peak shifts towards the cathode wall, and there is an on-axis minimum [54–57]. This behaviour can be explained by the kinetics of fast electrons in HC and TRF discharges. From the kinetic equation [19, 54, 55], it can be shown that the local EEDF, $f(\varepsilon,r)$, of fast electrons throughout the NG region (assuming they lose energy only in inelastic collisions) is given by

$$\mathrm{div}\left\{v^2\nabla_r[n_e f(\varepsilon,r)\,/\,3v]\right\} = \frac{\partial}{\partial\varepsilon}[\nu_{\mathrm{eff}}\,\varepsilon n f(\varepsilon,r)] + \rho(\varepsilon,r)\,, \qquad (2.44)$$

where $\nu_{\mathrm{eff}} = a_{\mathrm{ex}}\nu_{\mathrm{ex}} + a_{\mathrm{i}}\nu_{\mathrm{i}}$ is the effective frequency of inelastic processes [54]; $a_{\mathrm{ex}} = \varepsilon_{\mathrm{ex}}/\varepsilon$ and $a_{\mathrm{i}} = \varepsilon_{\mathrm{i}}/\varepsilon$ are the relative energies of excitation and ionization respectively, and ν_{ex} and ν_{i} are the frequencies of excitation and ionization respectively.

If the scattering of fast electrons by gas atoms is isotropic, the effect of the initial (primary) directed flux of electrons from the CDS is equivalent to the effect of an electron source with an isotropic velocity distribution function and a density $\rho(\varepsilon,r)$ that falls from the CDS boundary towards the centre. The external electrodes of a TRF discharge can be simplified to parallel, flat electrodes of separation $2r_0$, and the density ρ will decrease to the axis (without accounting for edge effects) as [54]:

$$\rho(\varepsilon,r_0) = \sqrt{\frac{m}{2eU_0}}\,\frac{n_e'}{s}\left[\exp\frac{r-r_0}{\lambda} + \exp\left(-\frac{r+r_0}{\lambda}\right)\right]\delta(\varepsilon - eU_0)\,, \quad (2.45)$$

where n_e'' is the number of primary electrons emitted from the cathode surface per unit time, s is the area of the cathode plate, and λ is the mean free path of the primary electrons. In order to find the form of $\rho(\varepsilon,r)$ for a cylindrical cathode, we shall divide its surface into pairs of centro-symmetric elemental strips having the source density described by (2.45). To avoid infinitely great electron densities at the centre due to the focusing of electrons that have not been scattered, it is necessary to account for their initial angular distribution. Integrating over the entire area of the cathode surface, we find [55] at each point, except for the small volume around the axis with a radius $r_{\min}$ (where $r_{\min}/r_0 = v_0/(v_0 + \sqrt{2eU_0/m})$, that

$$\rho(\varepsilon,r) = \sqrt{\frac{m}{2eU_0}}\,\frac{n_e'}{s}\left[\exp\frac{r-r_0}{\lambda} + \exp\left(-\frac{r+r_0}{\lambda}\right)\right]\delta(\varepsilon - eU_0)\,, \quad (2.46)$$

where r_0 and s are the radius and area of the cathode, and v_0 is the initial speed of the electrons. Equation (2.44) can be solved by the method of separation of variables, and from consideration of (2.46) with the boundary condition $\nabla_r[nf(\varepsilon,r_0)] = 0$.

Figure 2.18 shows the EEDF at various distances from the boundary of the CDS. It is seen that the number of fast electrons reduces with increasing distance from the CDS. The fast electrons are focused on the axis of the NG due to the negative curvature of the cathode emitting surface, and they are depleted due to inelastic processes that take place between the CDS and the cathode axis. At lower pressures, the focusing is stronger, and the electron concentration increases around the axis. With an increase in gas pressure, the electron mean free path reduces, the axial region is depleted of electrons, the profile of the rate of ionization becomes flat, and finally, the peak moves towards the boundary of the CDS. For slow electrons, the pattern is the same, but is shifted to higher (by 1.5–2 times) pressures.

In order to calculate experimentally observed quantities such as excitation rates, ionization rates, etc., it is convenient to use the integral of the EEDF, $F(\varepsilon)$, which gives the number of fast electrons with energy between ε and $\varepsilon + d\varepsilon$, in the full volume of the NG, and which is related to $f(\varepsilon,r)$ in the cylindrical geometry of the NG by

$$F_0(\varepsilon) = \int_V f_0(\varepsilon,r)\, 2\pi rl\, \mathrm{d}r\,. \tag{2.47}$$

During the calculation of the EEDF over the middle range of energies $\varepsilon_1^M < \varepsilon < \varepsilon_1^B$, we must consider inelastic collisions with metal atoms and elastic collisions with buffer-gas atoms, as well as electron–electron collisions (here, ε_1^M and ε_1^B are the first excitation potentials of metal and buffer-gas atoms, respectively). In the range of low energies, $\varepsilon < \varepsilon_1^M$, only elastic and electron–electron collisions are possible, and for $n_e > 10^{12}\ \mathrm{cm}^{-3}$, the frequency of electron–electron collisions is higher than that of elastic collisions.

Integrating the two parts of the kinetic equation [19]

$$-\frac{v_e}{3r}\frac{\mathrm{d}}{\mathrm{d}r}\left(r\lambda\frac{\partial f_0}{\partial r}\right) = \left(\frac{\delta f}{\delta t}\right)_c, \tag{2.48}$$

(where v_e and λ are the electron velocity and mean free path, and $(\delta f / \delta t)_c$ is the collision integral) over the volume of the hollow cathode , we find that

$$I(\varepsilon) = \delta F_0(\varepsilon) / \delta t\ , \tag{2.49}$$

where $I(\varepsilon)$ is the intensity of the 'beam' of primary electrons entering the NG from the CDS, which according to [58], is equal to

$$I(\varepsilon) = m^{3/2} I_0 \delta(\varepsilon - eU_0) / 4\pi\sqrt{2eU_0}\,, \tag{2.50}$$

(where I_0 is the number of electrons emitted from the cathode surface per second). From (2.49) and (2.50), we find for a two-component gas mixture that

$$\sum_{\kappa=1,2} N_\kappa S_\kappa(\varepsilon) = -I(\varepsilon) \,. \tag{2.51}$$

The collisional integrals S_κ, in which are included $F_0(\varepsilon)$, are found in [59].

The exact solution of (2.51) can be found only for a new function, $\phi(x)$ (where $x = \varepsilon / \varepsilon_1^{\mathrm{B}}$ is a new variable) related to F_0 by

$$\phi(x) = 2\pi e^4 N z' F_0(x) / (m^2 I_0) \,, \tag{2.52}$$

where z' is the number of valence electrons. However, we can use an approximate form, $\phi_1(x)$, which differs from $\phi(x)$ only for fast electrons in the CDS, given by [58] as

$$\begin{aligned} \phi_1(x) &= A(x_0)\delta(x - x_0) && \text{for } x_0 - 1/\zeta_0 \leq x \leq x_0 \,; \\ \text{or } \phi_1(x) &= \ln x + \frac{z_{\mathrm{M}}' n}{z_{\mathrm{B}}' N} \ln \zeta_0 x && \text{for } 1 < x < x_0 - 1/\zeta_0 \,. \end{aligned} \tag{2.53}$$

where $\zeta_0 = \varepsilon_1^{\mathrm{B}} / \varepsilon_1^{\mathrm{M}}$ is the ratio of the first excitation potentials of the buffer and metal atoms, and $A(x_0)$ is a function of the cathode fall and the first excitation potential of the gas [59].

The function $F_0(x)$ is found from (2.52) and I_0 is determined from the total anode current:

$$i_{\mathrm{a}} = e(I_0 + W^+) \,, \tag{2.54}$$

where W^+ is the total number of ionizations in the volume of the NG:

$$W^+ = \sum_\kappa N_\kappa \int_{\varepsilon_i}^{eU_0} F_0(\varepsilon) \frac{2\varepsilon}{m^2} q_\kappa(\varepsilon) d\varepsilon \,. \tag{2.55}$$

It should be noted that formulæ (2.51)–(2.55) can be used for the calculation of the EEDF in three (or more) component gas mixtures by increasing the number of terms under the summation signs to account for each mixture component.

In the high-energy region of the EEDF (Figure 2.19), $\phi(x)$ peaks at the energy of the CDS potential, due to the electrons that are accelerated, but not scattered. The broader second peak, shifted from the effective potential by $\varepsilon_1^{\mathrm{M}}$, occurs due to electrons that have effectively undergone one collision. The superimposed peaks for electrons that have undergone more collisions form the 'tail' of the EEDF, which is shown in Figure 2.19. Calculations for the NG predict a large group of thermal (or plasma) electrons with energies corresponding to $T_{\mathrm{e}} = 0.1$–1 eV.

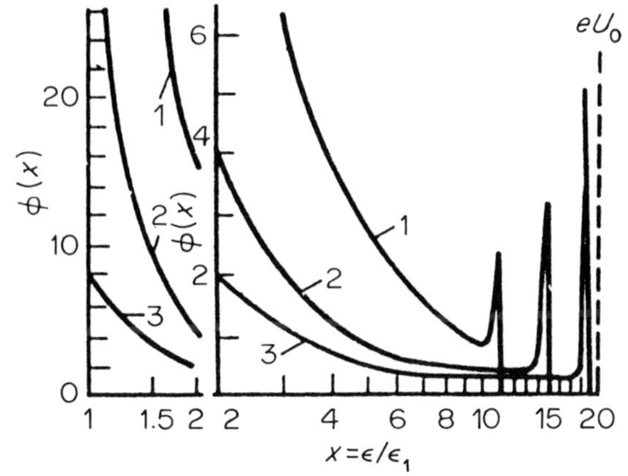

Figure 2.19 Typical calculated dependences of the EEDF $\phi(x)$ for p_{He} = 0.6 kPa and various cadmium pressures: p_{Cd} = 0.1 Pa (1), 6 Pa (2) and 40 Pa (3); ε_1 = 19.8 eV.

An increase in the pressure of the low ionization potential (metal atom) component of the mixture causes, on the one hand, an increase in the CDS potential (Figure 2.17a) and, on the other hand, a reduction in the number of electrons with energies equal to and greater than the excitation potentials of the metal atoms. It is noteworthy that, at different helium pressures, the form of the EEDF remains the same when the metal vapour density is altered. In Figure 2.19, it is seen that in general the EEDF is fairly insensitive to the pressure of the metal vapour, even at concentrations up to 5–10% of the main gas.

For calculations of the excitation and/or ionization rates in NG plasmas, it is convenient to use a value κ, which is equal to the number of ions or excited atoms produced by a single fast electron. Using the EEDF, it is found that these constants for ionization κ_i and excitation κ_e for helium in a mixture with metal vapours (having ionization potentials of between 8 and 11 eV) can be approximated by [54]:

$$\kappa = a(eU_0 / \varepsilon + b) / (1 + cn / N), \tag{2.56}$$

where ε is the energy of ionization (or excitation). For the ionization of helium, a = 0.5, b = −1 and c = 15; for excitation, a = 0.2, b = 0.6 and c = 30. Evaluations based on (2.56) show that one fast electron in the NG can ionize about six helium atoms and two metal atoms. The kinetic equation for electrons can also be solved by the Monte-Carlo procedure, although this consumes much computer time [60, 61]. Besides those cited above, calculations and measurements of the EEDF have been performed in [62–68] too.

The majority of measurements of the EEDF and electron density in the near-cathode region of the discharge have been performed in pure helium in the absence of metal atoms. In [63], the EEDF was measured in the range 0 to eU_0 in a flat cathode discharge with an electrostatic analyser. In [62, 64], it was measured by the probe technique using the second derivative of the probe current in a HC discharge very similar to that used in lasers. The findings qualitatively confirm the theoretical results.

The low-energy peak of the EEDF strongly depends on the distance from the cathode, and shifts to higher energies when approaching the CDS. The average electron energy lies in the range 0.5–1 eV and rises with the increase in current and as the helium pressure decreases [62].

As mentioned earlier, thermal electrons make up the bulk of the electron population in the NG. For those electrons, the assumptions that $\nu_{ee} > \nu_{in}$ and $\nu_{ee} > \nu_{el}$ are valid, and their energy distribution is formed in the absence of a field and is Maxwellian in form. For that reason the EEDF and the electron density in the low-energy regime (up to $\varepsilon = \varepsilon_1^{(M)}$) have been successfully measured using the probe technique [62]. A two-wavelength laser interferometer, and a Stark broadening technique have also been used [69]. Although there are some discrepancies, the dependence of electron density on current is nearly linear, and for the same input power per unit volume, the density of charged particles in the NG of a HC discharge is one to two orders of magnitude higher than in the PC discharge regime. The greatest electron density (and He ion density) occurs for helium pressures of around 1 kPa (Figure 2.16) which is somewhat higher than optimum for the HC effect.

The density of thermal electrons in a NG is determined by the rate of ionization of the gas mixture by electron impact and by the rate of loss of the electrons. For $T_e \approx 0.5$ eV and $n_e < 10^{13}$ cm^{-3}, the dominant electron loss mechanism is diffusion, which is strongly dependent on the design and construction of the electrodes. Thus, in segmented cathodes, diffusion is mainly to the sides, and in long, transverse constructions the diffusion is towards the anode [69]. If we compare the density of fast electrons ($\varepsilon > \varepsilon_1$), obtained by calculation and by experiment [59, 62, 65], with the total electron number density, i.e. with the concentration of thermal electrons, then we find that the fraction of fast electrons is about 10^{-4}.

2.4.2 Density of Helium Metastable Atoms

The results of absorption measurements of the densities of helium atoms in the metastable 2 $^{1,3}S$ states and low-lying excited states 2 $^{1,3}P$ [70, 71] differ greatly among various investigators. Nevertheless, it can be inferred that the densities of metastable atoms in HC and PC discharges are of the same order, that the density of 2 ^3S is an order of magnitude higher than that of

2 ^1S, and that its saturation occurs more slowly with current. The similar dependences on helium pressure of the 2 ^3S density (Figure 2.16), the discharge voltage U_0 and the density of thermal electrons released in ionization events imply that the ionization and excitation of the gas take place via direct electron impact from the atomic ground state due to primary electrons with energy eU_0.

2.4.3 Excitation Processes in Hollow-Cathode Discharges

The insensitivity of the EEDF to gas additives leads to a weak influence of the density of metal vapour on the processes of excitation and ionization of the buffer gas; direct processes of excitation and ionization therefore dominate over stepwise ones in the NG. Thus, for example, the dependences on metal vapour pressure of the rates of production of helium ions and metastables can be observed in the intensity behaviours of the singlet He spectral lines. The inaccuracy of the method is not great if the forms of the excitation and ionization functions in the vicinity of the threshold energy regions (23.74 eV for the 396.5 nm He I line; 24.56 eV for ionization) are taken into account, and the similarity of excitation functions is used [14]. In Fig 2.20 it can be seen that when metal vapour is introduced, the intensities of the He lines degrade one to two orders of magnitude more slowly in a HC discharge than in a PC discharge. For example, when there is a vapour pressure of 10 Pa, the ionization of helium is practically absent in a PC discharge (see Figure 2.4),

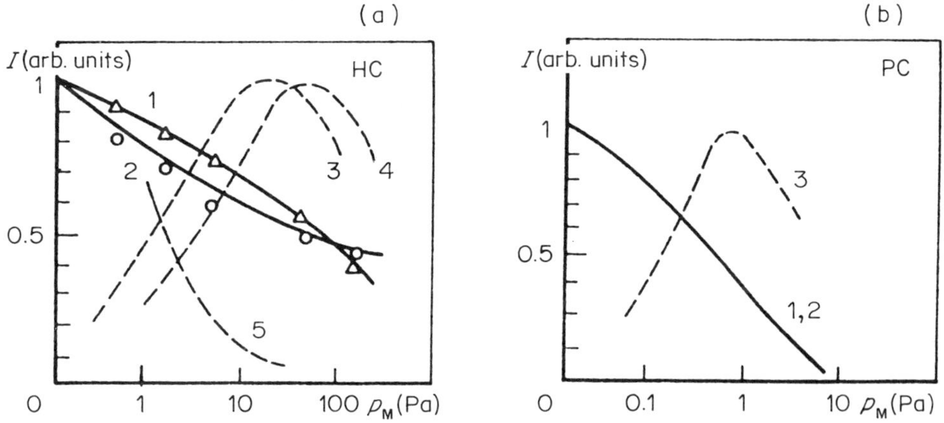

Figure 2.20 Dependences of the intensities of helium and cadmium lines on Cd vapour pressure in a HC discharge and (b) a PC discharge. 1, He I 396.5 nm singlet line; 2, He I 471.3 nm triplet line; 3, Cd II 441.6 nm; 4, Cd II 537.8 nm; 5, helium ion density, N^+.

whereas in the NG of a HC discharge helium ionization is less than halved, i.e. the number of fast electrons persisting in the NG is greater even though the relative metal vapour pressure may be 5–10%.

This implies first that there is a particular mechanism of formation of the EEDF from the higher energy electrons (due to the losses in the NG of the initial fast electrons with energies equal to the potential of the CDS). This would permit considerably higher metal vapour pressures to be reached in a HC (and TRF) discharge than in a PC discharge. Second, with an increase in metal atom concentration, the decay of fast electrons in the NG is partly compensated for by an increase in the energies of primary electrons due to a rise in U_0 (see Figures 2.17 and 2.19). However, an increase in vapour density in a PC discharge leads to a lowering of E_z (see Figure 2.3). The high-energy group of electrons in a NG mainly ionizes the buffer gas atoms, since the energy losses of electrons in the range 10's–100's eV for excitation and ionization of the low-potential metal atoms are relatively low (Figure 2.21). In contrast, the EEDF in PC discharges is formed as a result of the acceleration of very slow electrons, and hence chiefly metal atoms are ionized.

The higher (by one to two orders of magnitude) metal atom densities in

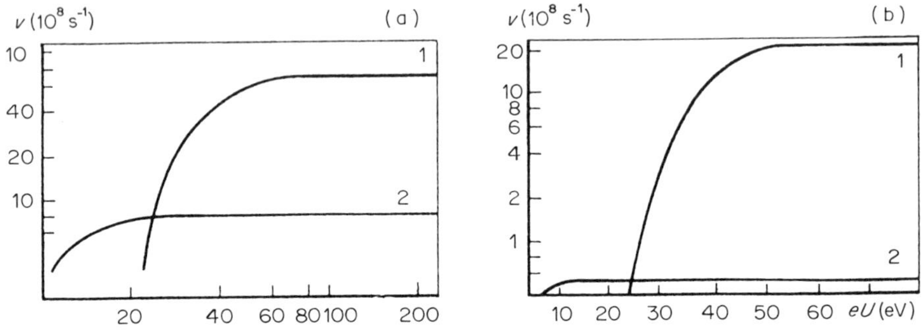

Figure 2.21 Dependences on electron energy of the frequencies of inelastic collisions of electrons with helium atoms (1) and metal atoms (2) in: (a) a HC discharge (1.3 kPa He, 13 Pa metal), and (b) a longitudinal PC discharge (0.67 kPa He, 0.4 Pa metal).

the NG regime allows the energies stored by buffer-gas ions and metastables to be utilized more efficiently in Penning and charge-transfer processes and, as a result, the pump rates to metal ion levels are larger [55]. The charge-transfer pump rate to level i (the partial cross-section is $\xi_i Q_c$) is given by

$$W_c^i = N^+ n\xi_i K_c = W_B^i n\xi_i K_c / (\nu_B^+ + nK_c)\,, \tag{2.57}$$

where ν_B^+ is the sum of the frequencies of ion depletion by diffusion, volume recombination and other processes. In a NG at 'laser' helium pressures (1–10 kPa) and optimum vapour pressures (10–40 Pa), $nK_c \gg \nu_B^+$, i.e. helium ions are neutralized primarily through charge transfer. Then (2.57) can be written as

$$W_c^i = \xi_i W_B^+ , \qquad (2.58)$$

i.e. the total pumping rate of all the levels of the metal ions populated by charge transfer is equal to the rate of helium ionization.

Thus, with an increase in n in the NG of a HC discharge, the initial increase in the rate of charge transfer is determined by the rate of increase in n, and saturation corresponds to the complete deactivation of buffer ions by metal atoms. The subsequent decrease in the charge-transfer rate exactly follows the fall of the ionization rate of the buffer gas due to the decrease in the density of fast electrons with the increase in metal vapour pressure. The full deactivation of buffer ions by metal atoms leads to an increase in the frequency of ion decay (in the denominator of (2.57)) in the mixture compared with the 'pure' gas (by between 5 and 15 times in various gas mixtures), and to an abrupt (usually 10 to 30 times) decrease in the concentration of buffer-gas ions in the plasma of the NG (Figure 2.20). In that case, $n^+ \gg N^+$, and we can assume that under 'laser' conditions $n_e \approx n^+$. There is a similar situation with the metastable atoms of the buffer gas, which decay by the Penning process, making stepwise ionization of helium in HC and TRF discharges extremely unlikely.

The calculations carried out in [55] have shown that the total rate of ionization of the buffer-gas atoms in the NG is greatest at the He pressure for which the radial dependence of that rate is nearly uniform. Hence, when various metals are added to helium, the optimum helium pressure is different for each metal, because each metal has its own characteristic cross-section for interaction with fast electrons.

Since in the plasma of a NG, most of the electrons have energies much lower than in a PC, the processes of de-excitation of levels by electron collision occur not only in the afterglow of a current pulse, but also during the current pulse, as well as in steady state discharges. In the last case for example, the electronic de-excitation of high-lying levels yields pumping of $F \longrightarrow D$ transitions of Cd II in HC and TRF discharges.

Since the ionization of buffer-gas atoms, which determines the rate of charge transfer, is produced from the energy of primary electrons accelerated in the CDS, the excitation rate is a function of the potential of the CDS (Figure 2.22). The saturation of the rate at $U_0 > 400$ V is due to the decrease in the number of primary electrons despite the increase in their energy (see relation (2.50)). For any one particular gas, the highest values of U_0 can be attained using a graphite [48] or dielectric-coated cathode, or a cathode with a restricted inner surface [72].

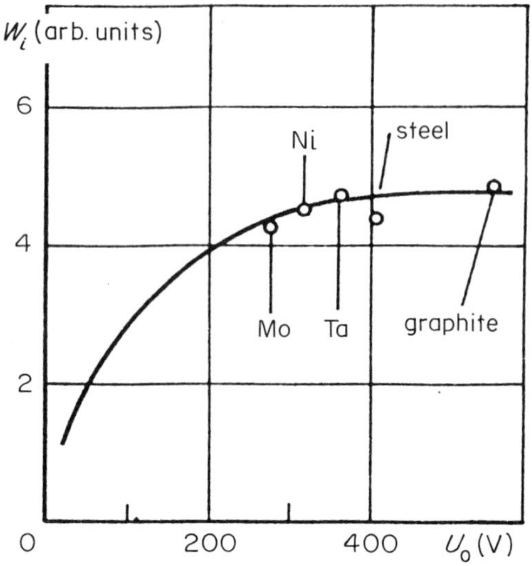

Figure 2.22 Dependence on U_0 of the rate of laser-level pumping by charge transfer, based on calculation and experiment with cathodes of various materials.

Various designs have been suggested in order to maximize U_0, e.g. a grid anode tube, or internal anodes (the so-called hollow-anode cathode tube), mounted in the form of a 'squirrel' wheel, at a distance less than the width of the CDS [64, 73, 74]. Under these conditions, the HC effect disappears also. The tube voltage sharply rises with discharge current, gas pressure and the number of anode rods. The current–voltage characteristic strongly differs from the ordinary HC discharge (Figure 2.23), being somewhere between a HC discharge and a discharge with flat electrodes (compared with a HC discharge there is less current at higher dynamic resistance) [75].

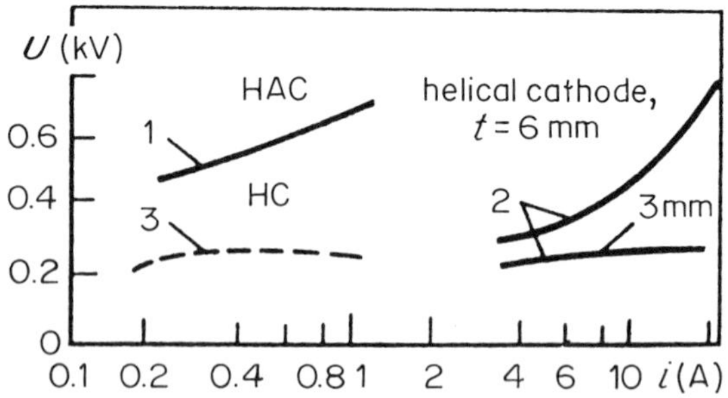

Figure 2.23 Voltage–current characteristics of discharges with a hollow-anode cathode (HAC) (1) [74], a helical cathode (2) [76] and a normal hollow cathode (HC) (3) (here, t is the pitch of the helix).

We should also note here the helical cathode [76], where a layer of the CDS directly covers each winding of the helix. The NG zones of neighbouring windings comprise a more intensely glowing region within the helix, compared with outside it. If the total current is held constant, a larger pitch of the helix leads to an increase of the current density for a reduction in cathode area. Hence a greater voltage U_0 is needed (Figure 2.23), which also increases cathode sputtering. A magnetic field can be easily provided inside the helix by directing an electric current through it; this also leads to an increased U_0, as stated in [77].

Compared with the discharge of a HC, discharges containing a grid anode or internal anodes have increased losses of ions bombarding the cathode. If the potential of the anode is positive with respect to the NG, ions are repelled by the anodes, and if it is negative, ions are attracted to them. In both cases, the decrease in ion flux to the cathode leads to a decrease in ψ and hence to a rise in U_0 (Figure 2.23). In other words, in all discharges with increased U_0, the discharge is more 'anomalous', which, however, does not lead to any increase in the effectiveness of the excitation of MVILs.

2.4.4 Cathode Sputtering

This effect is utilized in MVILs as a means of metal vapour production and is due to the bombardment of the cathode surface by plasma ions, accelerated in the CDS. The sputtered atoms diffuse to the NG zone of the HC or TRF discharge where they undergo excitation, or in cataphoretic lasers they fill the volume of the evaporator and then flow along the capillary discharge tube.

In [51] data are compiled on the sputtering coefficients of metals, which increase with the increase in ion energy and gas atom mass (there is about an order of magnitude change in going from He to Ar). The sputtering coefficients also depend on the surface quality, the presence of oxide films, etc. Sputtering increases in the following order: Ti, W, Nb, Zr, Ta, V, Mo, Al, Ge, Fe, Co, Cr, Pt, Ni, Cu, Au, Ag. The less-sputterable metals are used as cathode materials, and the easily sputtered metals make good sources of vapour. When sputtered metal atoms appear in the discharge, the enhanced effective mass of the bombarding ions leads to an increase of the voltage, as shown earlier, and hence increases the energies of the ions and the sputtering rate.

In a HC discharge with cathode sputtering, where there is a linear dependence of electron density on current and the density of sputtered metal atoms is proportional to the ion density, we can write an expression for the buffer-gas ion density in the NG by analogy with (2.57):

$$N^{+} \sim jk_{0i} / (\nu_{B}^{+} + \gamma_{s} N^{+} K_{c}) \,, \tag{2.59}$$

where γ_s is the number of sputtered atoms liberated per bombarding ion.

In a HC discharge with sputtering, the metal atom density, as a rule, does not rise above 10^{14} cm^{-3}. At low metal atom densities, there is a nearly linear fall in the ionization rate constant with metal vapour pressure (see Figure 2.20). Hence, with an increase in current density the buffer-gas ion density saturates. In the case of a discharge in neon with a copper cathode, the buffer-gas ion density saturates at around 30 mA cm^{-2}, for example [78]. The spatial distribution of metal atoms is normally non-uniform, and it has a maximum near the cathode surface and a minimum around the axis.

Cathode sputtering is seen to be a useful technique for seeding a discharge with the vapour of a refractory element, and for some metals is the only feasible means for doing so. However, since the sputtering rate depends on many factors, the vapour density is difficult to control, and the gas ion density tends to saturate.

2.5 PULSED HOLLOW-CATHODE DISCHARGES

The pulsed current in a PC discharge is limited chiefly by the external power-supply circuit. On the contrary, in a HC discharge the current is limited by the transition of the high-voltage glow discharge to a low-voltage arc plasma, when the discharge constricts strongly and the discharge volume reduces by some orders of magnitude. To date, pulsed HC discharges have not been extensively studied, but they are more stable than stationary HC discharges for time periods of several microseconds. Current densities of around 10 A cm^{-2} can be obtained in pulsed HC discharges [79–82]. As in a pulsed PC discharge, a steady-state regime cannot be established in a pulsed HC discharge (Figure 2.24). Hence there is a continuous rise in the concentration of plasma electrons, n_e, during the pulse (with a simultaneous increase in the conductivity, σ). Also, the initially high cathode fall, U_0, (up to 2–3 kV), decreases monotonically, and as a result the energies of the thermal electrons decrease. A feature of the pulsed HC discharge is a strong field in the NG of 10^2 V cm^{-1} during the first few microseconds [82]. As in a steady-state HC discharge, the EEDF is significantly non-Maxwellian and develops from the high-energy electrons that are accelerated across the CDS (Figure 2.25).

The large electron density (~10^{14} cm^{-3}) together with a typical energy in the NG of around 1 eV leads to high rates of electron de-excitation, not only in the afterglow, but also during the current pulse, which strongly affects the distribution of excited level populations (Figure 2.26). The number of fast electrons with energies higher than the first buffer-gas excitation potential in a pulsed HC discharge is great at the beginning of the discharge, due to a high cathode fall and NG field (i.e. due to the high energies of the primary electrons). However, after 1–2 μs the number of fast electrons saturates, indicated by saturation of the emission intensities of transitions from high-lying levels [83]. It is therefore better to produce short current pulses (0.1–1 μs) which contain more fast electrons, but fewer thermal electrons (compared with the plasma of a steady-state discharge) which would otherwise act to reduce an inversion. The high cathode fall (and high energy of primary electrons) ensures effective ionization and excitation in cathodes of significant diameter (up to 3–5 cm). The voltage–current characteristics of pulsed HC discharges are positive (see Figure 2.17), which is typical of the constricted (anomalous) discharge.

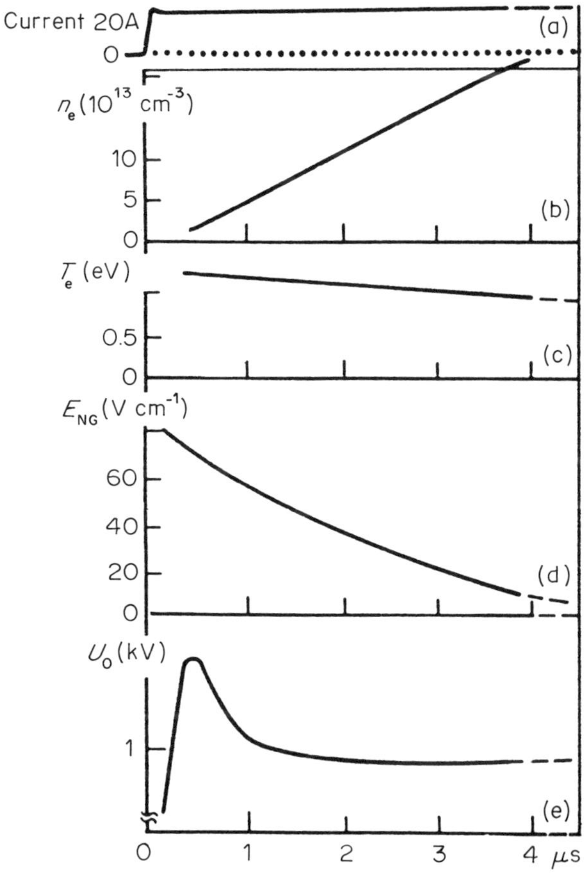

Figure 2.24 Evolution of parameters in a HC discharge during the excitation pulse: (a) current; (b,c) density and temperature of thermal electrons; (d) electric field of the NG; (e) tube voltage.

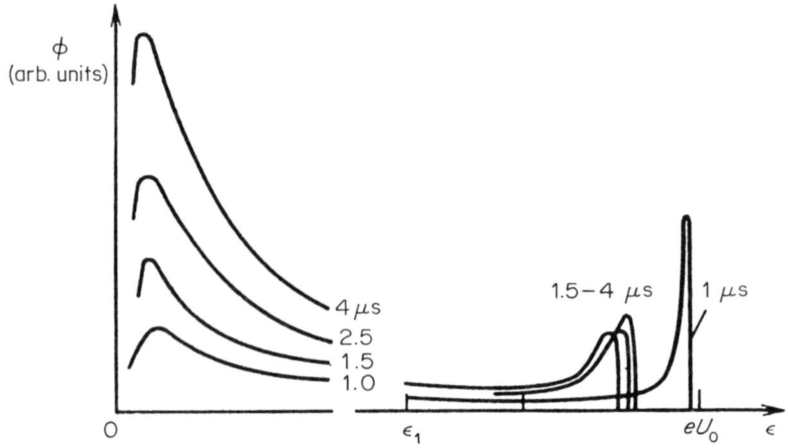

Figure 2.25 Form of the EEDF in a pulsed HC discharge at different delays from the beginning of the current pulse.

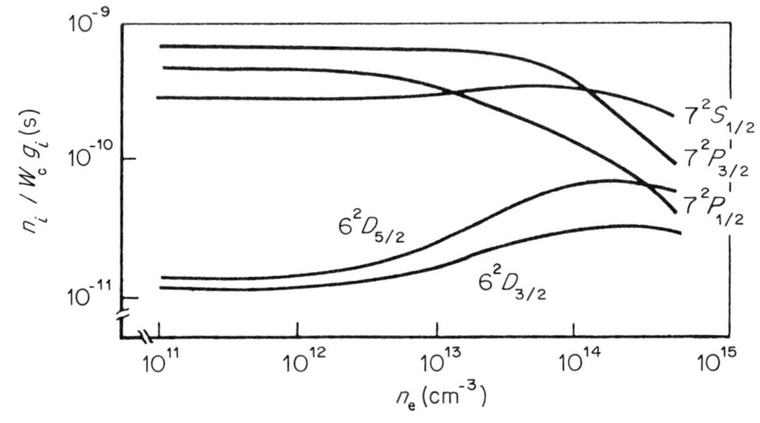

Figure 2.26 The ratio of the populations of Hg ions in various levels n_i to the pumping rate by charge transfer W_c, versus electron density.

In the repetitively pulsed regime, the parameters of the plasma slowly tend towards quasi-steady-state values after a few tens of pulses. The voltage drops with each pulse, and the plasma conductivity increases as a result of the increase in remanent electron density between pulses (Figure 2.27) [84].

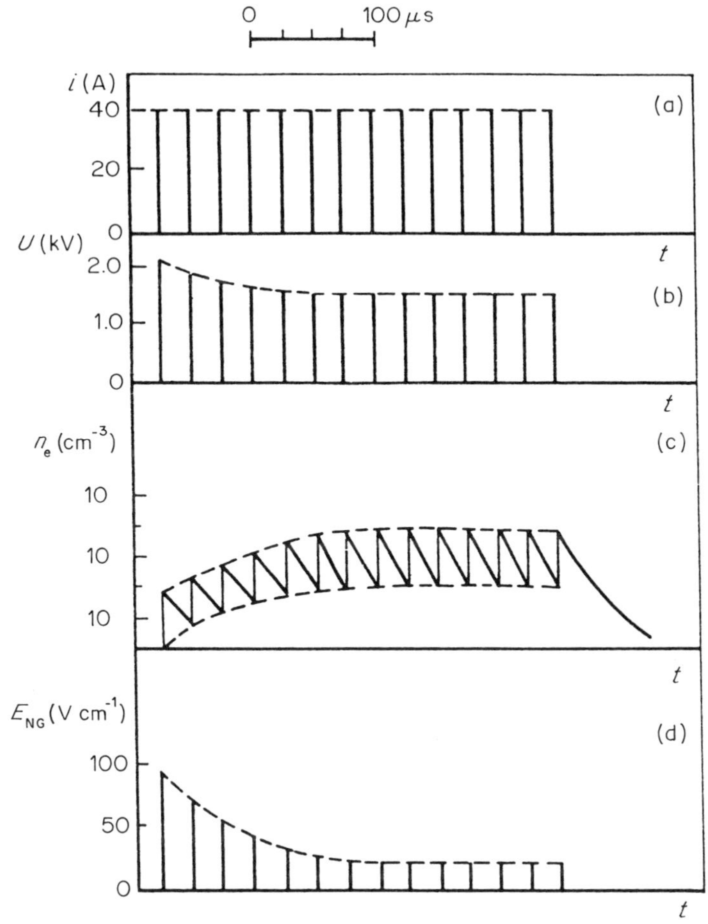

Figure 2.27 Oscillograms of (a) the current pulses, (b) the tube voltage, (c) the thermal electron density and (d) the electric field in the NG, with excitation in burst mode.

2.6 EFFICIENCY OF EXCITATION OF METAL VAPOUR ION LEVELS IN POSITIVE COLUMNS AND NEGATIVE GLOWS

In the steady-state regime, the efficiency of conversion of discharge energy to the energy of metal ions in the upper laser level due to second-kind collisional excitation (charge transfer and Penning ionization) has the form [85]

$$\eta^* = \eta_1 \eta_2 \eta_3 = \eta_d \eta_3 , \tag{2.60}$$

where η_1 is the fraction of discharge energy transferred to fast electrons; η_2 is the fraction of energy of fast electrons transferred to buffer-gas ions or metastable atoms; η_3 is the fraction of energy transferred from the buffer gas to the metal atoms by second-kind collisions. The product $\eta_1\eta_2 = \eta_d$, determined basically by the discharge parameters, is called the discharge efficiency.

The coefficient η_1 for a PC discharge can be expressed as the ratio of the power (per unit volume) spent in creating fast electrons with energy $\varepsilon > \varepsilon_i^B$ (or the energy lost by fast electrons per second in inelastic collisions with the gas atoms) to the discharge input power (per unit volume)

$$\eta_1 = \left(\sum_l \sum_m \varepsilon_m^l n^l n_e k_{0m}^l\right) \Big/ (E_z j) , \tag{2.61}$$

where n^l is the density of heavy particles and k_{0m}^l is the excitation rate constant of the energy level ε_m excited by electrons with energy $\varepsilon > \varepsilon_m$; j is the discharge current density averaged across the PC cross-section. The summation is made over all levels m for both components l of the gas mixture ($l = 1,2$).

Provided excitation of the laser transition in the PC occurs via charge transfer, then the fraction of energy of fast electrons which is transferred to the buffer-gas ions (denoted by coefficient η_2), is given by the ratio of power lost by fast electrons in ionization of the buffer gas to the power lost in the production of fast electrons:

$$\eta_2 = \varepsilon_i^B W_i^B / \sum_l \sum_m \varepsilon_m^l n_m{}^l n_e k_{0m}^+ , \tag{2.62}$$

where W_i^B is the ionization rate of the buffer gas. If the laser transition is excited via Penning ionization, W_{B*} should be used in (2.62) in place of W_i^B.

The coefficient η_3 in a PC can be expressed as the ratio of charge transfer frequency $\nu_c\xi_i$ to a certain level, i, to the frequency of deactivation of the buffer ions:

$$\eta_3^{(i)} = \nu_c\xi_i / (\nu_c + \nu_B^+) , \tag{2.63a}$$

where ν_B^+ is the frequency for ambipolar diffusion of the buffer-gas ions. A similar expression can be written for Penning ionization:

$$\eta_3^{(i)} = \nu_p\xi_i / (\nu_p + \nu_d + \nu_i) , \tag{2.63b}$$

where ν_P, ν_d and ν_i are the deactivation frequencies of the excited atoms of the buffer gas by Penning collisions, by diffusion to the tube wall and by ionizing electron collisions respectively. Estimates of (2.63a,b), taking account of the frequencies ν (see section 2.1) give $0.5\xi_i$ and $0.4\xi_i$, respectively, for η_3.

In a HC discharge, ions are accelerated in the CDS from the NG into the cathode surface. Ion bombardment is accompanied by electron emission, i.e. $j = \gamma i$ (here, j, i are densities of the fast electron and ion currents, respectively). The emission coefficient of the cathode due to ion bombardment $\gamma \ll 1$, so that the ion flux is much higher than the flux of emitted electrons. The same is true in TRF discharges, where emission from the dielectric surface is very low.

Consider electrons and ions to be accelerated in the CDS without any collisions:

$$\eta_1 = j / (j + i) = \gamma / (1 + \gamma) . \qquad (2.64)$$

If each fast electron accelerated in the CDS produces k_i ions in the NG region, and a fraction, ψ^*, of the electrons is lost by diffusion into the CDS, then we can write

$$i = j / (k_i^B + k_i^M)\psi^* \qquad (2.65)$$

($\psi^* < 1$ and depends only on the electrode geometry). Hence, we find another expression for η_1 [85]:

$$\eta_1 = 1 / [1 + (k_i^B + k_i^M)\psi^*] . \qquad (2.66)$$

For various ions, $\gamma = 0.05$–0.4, and hence $\eta_1 \approx 0.05$–0.3, i.e. a small part of the energy deposited in the discharge is used for electron acceleration, but most is used in acceleration of ions (and then lost by bombardment of the electrodes or tube wall).

The coefficient η_2 can be expressed as

$$\eta_2 = \varepsilon_i k_i / (eU_0) \text{ (for ionization);} \qquad (2.67a)$$

$$\eta_2 = \varepsilon_a k_a / (eU_0) \text{ (for excitation).} \qquad (2.67b)$$

When estimating (2.66) and (2.67), the constants for ionization k_i and excitation k_a of helium can be found by approximation (see (2.56)). Note that for $eU_0 \gg \varepsilon_i$, η_2 for ionization is approximately 1/2, i.e. half of the average energy lost by electrons is due to ionization events [13]. When helium is excited from a metastable state, η_2 is less than 0.2.

The effectiveness of charge transfer (and Penning ionization) in TRF discharges depends, as in PC discharges, on its rate compared with the total rate of decay of ions (metastables) in the discharge volume. Using (2.58), we find

$$\eta_3 = \xi_i . \qquad (2.68)$$

The partial cross-section of the reaction ξ_i should also take into account other mechanisms of pumping the upper level, such as population by cascade from higher levels. Thus (2.60) for HC and TRF discharges will have the form

$$\eta^* = \eta_d \eta_3 = a\xi_i \gamma / (1 + \gamma) . \tag{2.69}$$

In Table 2.2, η_1, η_2 and η_d are given for longitudinal PC discharges, and for HC and TRF discharges under typical 'laser' conditions.

Table 2.2 The effectiveness of various types of discharge for ionization and excitation of metastable atoms (2 $^{1,3}S$) of helium.

Process	Discharge type	η_1	η_2	η_d
Ionization	PC	0.05	0.30	0.015
	HC (stationary)	0.13	0.45	0.058
	HC (pulsed)	0.25	0.48	0.12
	TRF	0.29	0.45	0.13
Excitation	PC	0.075	0.4	0.03
	HC (stationary)	0.15	0.24	0.036
	HC (pulsed)	0.30	0.2	0.06
	TRF	0.29	0.24	0.072

The data of Table 2.2 show that HC and TRF discharges have higher efficiencies of transfer of discharge energy to the energy of fast electrons capable of exciting and ionizing helium atoms (cf. η_1 = 0.13–0.30 with 0.07 in a PC discharge). This is the result of the formation of an EEDF having, in comparison with a Maxwellian EEDF, an excess of fast electrons, due to its evolution from a high-energy electron source. The optimum metal vapour density in HC and TRF discharges is about two orders of magnitude higher than in a PC discharge, which leads to an enhanced effectiveness of charge transfer and Penning ionization in the former cases. Large densities of fast electrons and lasant metal atoms are features that permit MVILs with HC and TRF discharges to be highly effective for second-kind collisional pumping. This is the main reason for the great number of laser media and laser lines (especially those pumped by charge transfer) of HC and TRF discharges, and their better characteristics.

In HC and TRF discharges with microsecond excitation, ionization is more effective, with their higher emission coefficients, γ, due to U_0 being 5 times larger in magnitude, and due to the additional heating of electrons in the electric field of the NG. Since η_1 depends on the emissivity of the cathode (see (2.64)), it can be raised by utilizing a cathode (in a HC discharge) or dielectric material for a wall (in a TRF discharge) with high emittance (hence, in [86] for a cathode of $\gamma = 2.5$, $\eta_1 = 0.7$ was reported).

The coefficient η_1 can be increased by using more complicated electron-beam guns, in which space-charge regions of e-beam formation and gas/e-beam interaction are separated [86–89]. Typical e-beam energies of hundreds of keV are developed in a vacuum acceleration zone, followed by injection of the e-beam into the gas through a separating foil. However, e-beam energies of only tenths of a keV are optimum for excitation of MVILs. Furthermore, part of the energy is wasted in the foil, and hence the average e-beam energy is limited by the foil's damage threshold.

E-beams with optimum energy for MVIL excitation are produced in gas-filled guns of the glow-discharge type [86–88], and in open discharges in the regime of 'escaping' electrons [89]. The latter regime is obtained in helium, for example, when $E/p > 38$ V m^{-1} Pa^{-1} (Dreiser's condition) and permits η_1 to be as high as 0.5–0.8.

REFERENCES

[1] W. E. Bell, A. L. Bloom and J. P. Goldsborough, *IEEE J. Quantum Electron.*, **1**(9), 400 (1965).

[2] E. A. Verni, L. D. Mash and B. M. Rabkin, *Zh. Prikl. Spektrosk.*, **16**(1), 156–7 (1972) (in Russian).

[3] E. L. Latush, V. S. Mikhalevskii, M. F. Sem, G. N. Tolmachev and V. Ya Khasilev, *JETP Lett.*, **24**(1), 69–71 (1976).

[4] V. A. Godyak, *Fiz. Plazmy*, **2**(1), 141–51 (1976) (in Russian).

[5] J. R. Acton and J. D. Swift, *Cold Cathode Discharge Tubes*, London: Heywood, (1963).

[6] V. A. Granovskii, *Electric Current in Gases. Breakdown*, Moscow: Nauka (1971) (in Russian).

[7] Yu P. Raizer, *Gas Discharge Physics*, Berlin: Springer-Verlag (1991).

[8] J. P. Goldsborough, *Appl. Phys. Lett.*, **15**(6), 159–161 (1969).

[9] T. P. Sosnowski, *J. Appl. Phys.*, **40**(13), 5138–44 (1969).

[10] J. R. Fendley Jr, I. Gorog, K. G. Hernqvist and C. Sun, *RCA Rev.*, **30**(3), 422–8 (1969).

[11] M. F. Sem, V. F. Keidan and V. C. Mikhalevskii, *Elektronnaya Tekhnika. Ser. 1 Elektronika SVCh*, No. 1, 69–76 (1971) (in Russian).

[12] *Spectroscopy of Gas-Discharge Plasmas: Collected Papers*, ed. S. E. Frisha, Leningrad: Nauka (1970) (in Russian).

[13] L. I. Gudzenko and S. I. Yakovlenko, *Plasma Lasers*, Moscow: Atomizdat (1978) (in Russian).

[14] L. A. Vainshteyn, I. I. Sobelman, E. D. Yukov, *Excitation of Atoms and Spectral Line Broadening*, Moscow: Nauka (1979) (in Russian); *Cross-sections for Electronic Excitation of Atoms and Ions*, Moscow: Nauka (1973) (in Russian).

[15] I. G. Ivanov, V. S. Mikhalevskii, M. F. Sem and S. P. Zinchenko, Patent No. 999919 Russ. Fed., Kl. HO1s 3/22, 8.04.81.

[16] V. E. Privalov, *Gas-Discharge Lasers in Metrology*, Leningrad: Sudostroenie (1989) (in Russian).

[17] V. E. Privalov and E. A. Smirnov, *Opt. Spectrosc.,* **46**(1), 16–19 (1979); **48**(5), 522–5 (1980).
[18] O. V. Kozlov, *Electrical Probes in Plasmas,* Moscow: Atomizdat (1969) (in Russian).
[19] V. E. Golant,, *Microwave Methods for Investigating Plasmas* Moscow: Nauka (1968) (in Russian).
[20] I. G. Ivanov and M. F. Sém, *Sov. Phys. - Tech. Phys.,* **17**(7), 1234–5 (1973).
[21] I. G. Ivanov, E. L. Latush, V. F. Papakin and M. F. Sém, *Sov. Phys. J.,* **15**(8), 1149–53 (1972).
[22] J. K. Mizeraczyk, *IEEE J. Quantum Electron.,* **11**(5), 218–20 (1975).
[23] M. H. Dunn, *J. Phys. B.,* **5**(3), 665–72 (1972).
[24] P. G. Browne and MH. Dunn, *J. Phys. B.,* **6**(6), 1103–17 (1973).
[25] W. T. Silfvast, *Phys. Rev. Lett.,* **27**(22), 1489–92 (1971).
[26] M. K. Dyatlov, E. P. Ostapchenko, V. A. Stepanov, *Opt. Spectrosc.,* **29,** 539–40 (1970).
[27] V. S. Aleinikov, *Opt. Spectrosc.,* **28**, 15–17 (1970).
[28] A. R. Turner-Smith, J. M. Green and C. E. Webb, *J. Phys. B.,* **6**(1), 114–30 (1973).
[29] V. I. Bylkin, *Opt. Spectrosc.,* **29**, 552–5 (1970).
[30] V. S. Aleinikov and V. V. Ushakov, *Opt. Spectrosc.,* **29**, 111–12 (1970); R. F. Pottie, *J. Chem. Phys.,* **44**(3), 916–22 (1966).
[31] S. P. Varshavskii, A. A. Mityureva and N. P. Penkin, *Opt. Spectrosc.,* **29**, 341–3 (1970).
[32] L. D. Schearer and F. A. Padovani, *J. Chem. Phys.,* **52**(3), 1618–19 (1971); L. A. Riseberg and L. D. Schearer, *IEEE J. Quantum Electron.,* **7**(1), 40–1 (1971).
[33] L. A. Riseberg and L. D. Schearer, *Phys. Lett. A,* **35**(4), 269–70 (1971).
[34] G. N. Tolmachev, E. L. Latush and V. S. Mikhalevskii, *Sov. Phys. – Tech. Phys.,* **19**(8), 1120–1 (1975).
[35] L. D. Mash, B. M. Rabkin and B. V. Rybakov, *JETP Lett.,* **13**(5), 169–71 (1971).
[36] E. L. Latush, V. S. Mikhalevskii and G. N. Tolmachev, *Sov. Phys. – Tech. Phys.,* **22**(3), 320–2 (1977).
[37] C. C. Leiby Jr. and H. J. Oskam, *Phys. Fluids,* **10**(9), 1992–6 (1967).
[38] A. N. Chester, *Phys. Rev.,* **169**(1), 172–84 (1968).
[39] V. V. Zhukov, I. G. Ivanov and V. F. Keidan, *Sov. Phys. – Tech. Phys.,* **18**(7), 958–9 (1974).
[40] I. N. Kinyazev, *Tr. Fiz. Inst., Akad. Nauk SSSR,* **56,** 119–90 (1971) (in Russian).
[41] V. V. Zhukov, V. S. Kucherov, E. L. Latush and M. F. Sem, *Sov. J. Quantum Electron.,* **7**(6), 708–14 (1977).
[42] L. M. Biberman, V. S. Vorobev and I. T. Yakubov, *Nonequilibrium Kinetics of Low Temperature Plasmas,* Moscow: Nauka (1982) (in Russian).
[43] V. S. Egorov, V. N. Skrebov and A. M. Shukhtin, *Opt. Spectrosc.,* **20**, 211–13 (1966).
[44] A. M. Shukhtin, V. N. Skrebov and M. N. Polyanskii, *Opt. Spectrosc.,* **27**, 308–11 (1969).
[45] V. F. Kravchenko, É K. Karabut, A. A. Gudkov and V. E. Bogoslavskii, *Sov. J. Quantum Electron.,* **12**(2), 143–6 (1982).
[46] V. G. Ilyushko, É K. Karabut, V. F. Kravchenko and Mikhalevskii, *Sov. J. Quantum Electron.,* **15**(10), 1442–4 (1985).
[47] B. I. Moskalev, *Hollow Cathode Discharges,* Moscow: Energiya (1969) (in Russian).
[48] V. V. Vainer, I. G. Ivanov and M. F. Sem, *Proc. XV Int. Conf. on Phenomena in*

Ionized Gases, Part 2, pp. 869–70, Minsk: Institute of Physics, BSSR Academy of Sciences (1981).
[49] G. P. Strokan and G. N. Tolmachev, *Optoelectronics, Instrumentation and Data Processing,* No. 1, 56–8 (1984).
[50] V. S. Mikhalevskii, M. F. Sem, G. N. Tolmachev and V. Ya Khasilev, *J. Appl. Spectrosc.,* **32**(4), 321–3 (1980).
[51] N. V. Pleshivtsev, *Cathode Sputtering,* Moscow: Atomizdat, (1968) (in Russian).
[52] Yu P. Raizer, *High Temp.,* **24**(5), 744–54 (1986).
[53] E. W. McDaniel, *Collision Phenomena in Ionized Gases,* New York: John Wiley (1964).
[54] V. Ya Khasilev, V. S. Mikhalevskii and G. N. Tolmachev, *Sov. J. Plasma Phys.,* **6**(2), 236–9 (1980).
[55] G. A. Kalinchenko, V. V. Vainer and I. G. Ivanov, *Summaries of Papers. Population Inversion and Oscillation on Atomic and Molecular Transitions,* Part 1, ed. O. V. Ravodina and I. I. Murav'ev, p. 247, Tomsk: Institute of Atmospheric Optics, (Sept. 1986) (in Russian); *Sov. J. Plasma Phys.,* **16**(4), 263–7 (1990).
[56] P. Mezei, P. Apai, M. Jánossy and K. Rózsa, *Opt. Commun.,* **78**(3,4), 259–62 (1990).
[57] V. V. Vainer, I. G. Ivanov, G. A. Kalinchenko and M. F. Sem, *Metal Vapor Lasers and Their Applications: CIS Selected Papers,* ed. G. G. Petrash *Proc. SPIE,* **2110**, 128–49 (1993).
[58] Yu M. Kagan, R. I. Lyagushchenko and S. N. Khvorostovskii, *Opt. Spectrosc.,* **33**, 232–5 (1972).
[59] V. V. Vainer, I. G. Ivanov and M. F. Sem, *Zh. Tekh. Fiz.,* **49**(8), 1604–8 (1979).
[60] J. P. Boeuf and E. Marode, *J. Phys. D.,* **15**, 2169–87 (1982).
[61] S. Hashiguchi and M. Hashikuni, *Jpn. J. Appl. Phys.,* **27**(10), 2007–8 (1988).
[62] J. Mizeraczyk and W. Urbanik, *J. Phys. D.,* **16**(11), 2119–33 (1983).
[63] P. Gill and C. E. Webb, *J. Phys. D.,* **10**(3), 299–311 (1977).
[64] A. N. Soldatov, *Opt. Spectrosc.,* **31**(2), 97–101 (1971).
[65] F. J. de Hoog, J. R. McNeil, G. J. Collins and K. B. Persson, *J. Appl. Phys.,* **48**(9), 3701–4 (1977).
[66] G. J. Fetzer and J. J. Rocca, *IEEE J. Quantum Electron.,* **28**(9), 1941–56 (1992).
[67] B. Shi, G. J. Fetzer, Z. Yu, J. D. Meyer and G. J. Collins, *IEEE J. Quantum Electron.,* **25**(5), 948–54 (1989).
[68] R. R. Arslanbekov, A. A. Kudrjavzev and I. A. Movchan, *Zh. Tekh. Fiz.* **62**(4), 63–70; **62**(6) 116–125; **62**(10) 65–73 (1992) (in Russian).
[69] I. K. Belal and M. H. Dunn, *J. Phys. D.,* **11**(3), 313–23 (1978).
[70] A. I. McIntosh, M. H. Dunn and I. K. Belal, *J. Phys. D: Appl. Phys.,* **11**(3), 301–11 (1978).
[71] D. C. Gerstenberger, R. Solanki and G. J. Collins, *IEEE J. Quantum Electron.,* **16**(8), 820–34 (1980).
[72] T. Iijima, *Jpn. J. Appl. Phys., Part 1,* **21**(12), 1732–5 (1982); *Phys. Lett. A.,* **85**(8,9), 436–8 (1981).
[73] J. Mizeraczyk and M. Neiger, *Appl. Phys. B.,* **33**(1), 17–21 (1984).
[74] K. Rosza, *Magy. Fiz. Foly.,* **34**(1), 1–56 (1986).
[75] V. G. Il'yushko, V. F. Kravchenko, V. S. Mikhalevskii and V. A. Polunin, *J. Appl. Spektrosc.,* **28**(1), 106–7 (1978).
[76] M. Grozeva and N. V. Sabotinov, *Opt. Commun.,* **41**(1), 57–8 (1982); **51**(6) 417–9 (1984).
[77] Z. Zhang, N. D. Perry and R. C. Tobin, *J. Appl. Phys.,* **71**(11), 5338–43 (1992).

[78] B. E. Warner, K. B. Persson and G. J. Collins, *J. Appl. Phys.*, **50**(9), 5694–703 (1979).
[79] M. F. Sem and V. S. Mikhalevskii, *Zh. Prikl. Spektrosk.*, **6**(5), 668–9 (1967) (in Russian); *Elektron Tekh., Ser. 1 Elektronika SVCh,* No. 9, 98–102 (1969) (in Russian).
[80] E. K. Karabut, V. F. Kaydan, V. F. Kravchenko *et al., Elektron. Tekh., Ser. 1 Elektronika SVCh,* No. 9, 103–8 (1969) (in Russian).
[81] I. G. Ivanov and M. F. Sem, *Elektron. Tekh., Ser. 4 Elektrovakuumiye i Gazorazryadniye Pribori,* No. 2, 12–16 (1974) (in Russian).
[82] V. V. Vainer, S. P. Zinchenko, I. G. Ivanov and M. F. Sém, *Sov. J. Quantum Electron.*, **10**(5), 581–6 (1980).
[83] S. P. Zinchenko, I. G. Ivanov, E. L. Latush and M. F. Sem, *Opt. Spectrosc.*, **58**(2), 180–3 (1985).
[84] S. P. Zinchenko, I. G. Ivanov and M. F. Sém, *Sov. J. Quantum Electron.*, **10**(8), 1055–7 (1980).
[85] V. V. Vainer, I. G. Ivanov, M. F. Sém and B. Ya Khasilev, *Sov. J. Quantum Electron.*, **16**(1), 79–84 (1986).
[86] Z. Yu, J. J. Rocca and G. J. Collins, *J. Appl. Phys.*, **54**(1), 131–6 (1983) .
[87] J. J. Rocca, J. Meyer and G. J. Collins, *Phys. Lett. A.*, **87**(5), 237–9 (1982); *Appl. Phys. Lett.*, **40**(4), 300–2 (1982).
[88] J. J. Rocca and G. J. Collins, *Optoelectronics, Instrumentation and Data Processing,* No. 1, 1–16 (1984).
[89] P. A. Bokhan and A. R. Sorokin, *Sov. Phys. – Tech. Phys.*, **30**(1), 50–4 (1985).

Notes

1. Dissociative recombination in the early afterglow becomes important for $p_{He} \geq$ 10 kPa.
2 The estimated inaccuracies in the measurements of T_e and n_e in the afterglow are not more than 50%. The relative behaviours of these parameters are more reliable. In the first 1–5 μs of the afterglow, the values of T_e measured by probe and spectrographic techniques have a significant spread—therefore, more sophisticated techniques for diagnostics of the early plasma afterglow are necessary.

3

Pulsed Metal Vapour Ion Lasers with Positive-Column Discharge Excitation

3.1 LASER ENGINEERING

The longitudinal positive-column (PC) discharge is the most prevalent active medium used for gas lasers, because of the simplicity of scaling this medium in length. In a pulsed PC discharge, the instantaneous values of current can be high (10's–100's A), as can the electron density, the densities of metastable atoms and ions and the rates of level excitation. Since the average current can be low, there is no problem with overheating or damage to laser tubes. In addition, the pulsed regime of operation makes inversion possible due to transient processes during the rising and falling edges of current pulses, as well as in the discharge afterglow.

In the majority of cases, the attainment of laser oscillation was achieved in the pulsed regime before it was obtained in the stationary regime, since pulsed operation gives higher instantaneous rates of pumping. The pulsed regime also permits the observation of spontaneous and stimulated emission in time, which makes determination of the excitation mechanisms easier.

The discharge tubes used for pulsed MVILs are relatively simple in design (Figure 3.1). The lasant metal is placed in the central region, or in side-arms, and the tube is normally filled with a noble gas. The metal vapour is generated and admitted to the discharge region by the use of either external heating or so-called self-heating. The output windows are attached to the ends of the tube, which are cool, and metal vapour is impeded from reaching them by the presence of the noble gas. For that reason, the noble-gas pressure should not be less than 50–100 Pa. The central part of the laser tube can be made of glass, silica or alumina, depending on the lasant metal used—the choice of tube material depends on the temperature required for metal evaporation, and the chemical inertness between it and the metal. If metal pieces are placed along the tube floor (Figure 3.1a), and the pulse recurrence frequency (PRF) is greater than about 1 kHz, then the tube temperature, and therefore metal vapour pressure, will depend not only on external heating

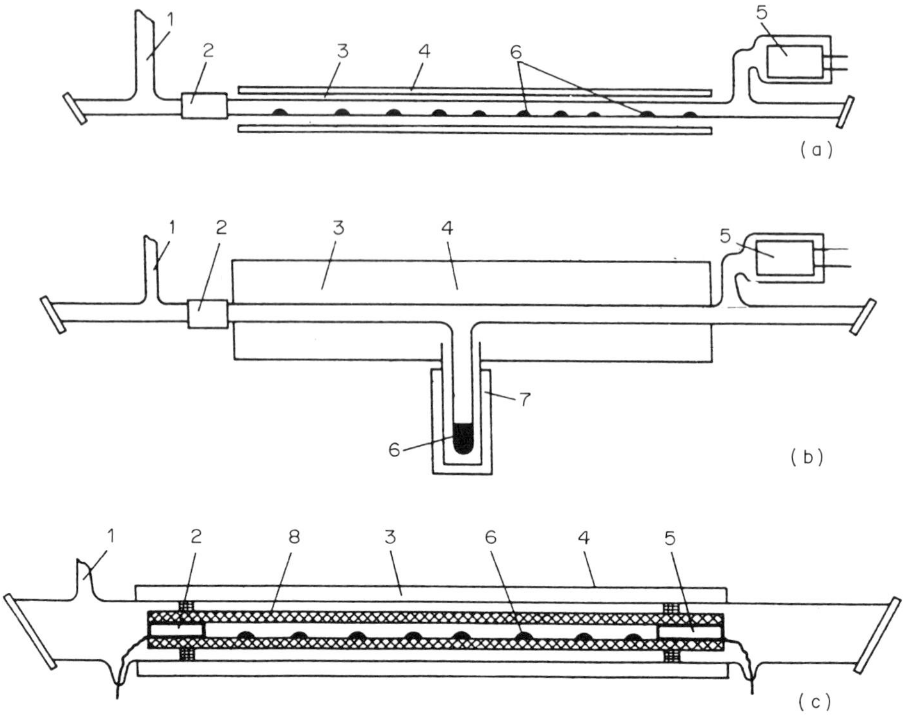

Figure 3.1 (a–c) Three typical constructions of discharge tube used for pulsed MVILs with longitudinal excitation: 1, Pump outlet; 2, anode; 3, central heated section; 4, main oven; 5, cathode; 6, metal lasant; 7, side-arm heater; 8, ceramic insert tube.

but also on the heat released in the discharge. The latter makes optimization of the vapour pressure for laser oscillation more difficult. When the metal is placed in side-arms (Figure 3.1b), the vapour pressure in the discharge is a function only of the side-arm heater temperature, provided the discharge tube wall remains at a higher temperature. A uniform vapour pressure along the tube cannot be achieved using cataphoresis if the number of side-arms is too small. In either case, a uniform vapour distribution is inhibited due to diffusion of the vapour to the colder ends of the tube. If no buffer gas is used, then the entire tube must be heated to the operating temperature, which the window-to-tube seals must be able to withstand.

Hot oxide cathodes are not usually required. The cathodes can be cold, of small size, and should be made of a refractory metal such as Ni, Ta or Mo. Explosive field emission of electrons ensures the high currents needed. The cathode and anode are usually cylinders, mounted close to the tube wall, and can be located in the heated zone. The electrical connexions to the electrodes are usually in the cold zones. For normal operation of MVILs, the metal vapour pressure is in the range 0.1–10 Pa, and the central section of the tube must be heated to the temperature that corresponds to that pressure. Glass tubes cannot be heated above around 500 °C, and for silica tubes the maximum temperature is 1100 °C. Ceramic tubes (Al_2O_3, BeO) are used up to 2000 °C, and are chemically resistant. A ceramic insert in a silica tube can be used with chemically aggressive vapour, when the temperature is not very high. This construction is simpler than joining a ceramic tube to a silica one, since no high-temperature ceramic–silica seals are necessary.

At high PRFs, the heat released in the discharge is often sufficient for vaporization of the metal. The oven may be removed (self-heating regime) and the construction of the laser is much simplified. However, in order to maintain the optimum vapour pressure for a given input energy, the PRF should be held constant.

The pulsed power supply for longitudinal-discharge MVILs [1] is normally based on the full discharge of a capacitor through a thyratron switch (a 'line-type' pulser, Figures 3.2a–c), or on the partial discharge through a modulating vacuum tube (a 'hard-tube' pulser), the second resulting in near-rectangular current pulses (Figure 3.2d). There are several versions of line-type pulsers. The direct discharge of a storage capacitor through the tube (Figure 3.2a), together with a peaking capacitor (Figure 3.2b), and the

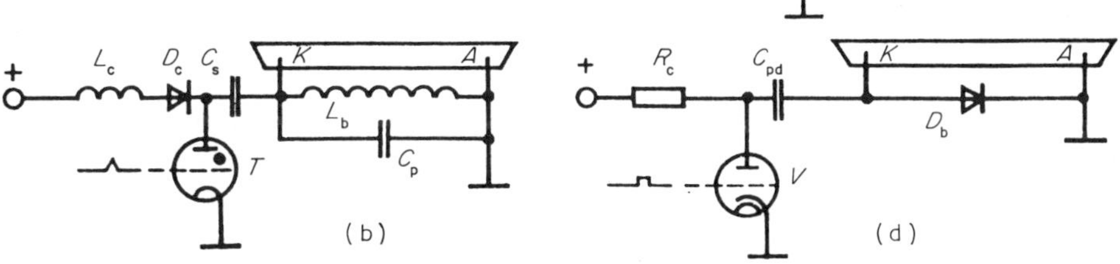

Figure 3.2 Excitation circuits for pulsed lasers: (a) with direct capacitor discharge; (b) with a peaking capacitor; (c) of the Blumlein type; (d) with partial discharge of a capacitor through a valve. *A*, Anode; *K*, cathode; *T*, thyratron; *V*, valve; L_c, charging inductor (0.1–0.5 H); D_c, charging diode; L_b, bypass inductor (~1 mH); L_h, hold-off inductor (1–0.01 μH); C_s, storage capacitor (0.5–5 nF); C_p, peaking capacitor (~0.3 C_s); C_{pd}, partial-discharge capacitor (~0.1 μF); R_c, charging resistor; D_b, bypass diode.

Blumlein-type circuit (Figure 3.2c) are the most common [2]. The Blumlein circuit increases the voltage on the tube above that initially present across the storage capacitor, and increases the rate of rise of the current pulse, which is of importance for self-terminating lasers. The storage capacitor is normally charged between pulses via a charging inductor and a diode to a voltage which is double that supplied by the high-voltage rectifier. As a result, the tube voltage in the Blumlein circuit can be up to 4 times the rectifier voltage. In hard-tube pulsers, for producing rectangular pulses, high-power vacuum tubes (such as type GMI-26 or GMI-29 (Russian manufacture)) may be used to switch pulsed currents of up to 300 A peak, which is much less than that which is possible with line-type pulsers that employ thyratron switches (usually up to 5 kA). Rectangular pulses can also be produced using a thyratron in a line-type pulser when a transmission line is used as both a pulse-forming network and an energy storage element. An advantage of the hard-tube pulsers is the easily controlled shape and duration of the current pulse, which is of special importance in searches for new laser transitions.

In this Chapter, we shall deal with pulsed ion lasers, whose upper levels are pumped by electron impact and by energy transfer in heavy-body collisions. Recombination lasers are considered separately in Chapter 4.

3.2 LASERS EXCITED BY ELECTRON IMPACT

With electron-impact pumping the most populated level is the resonance level, and hence the best results are obtained with self-terminating transitions, from the resonance level of an ion to a metastable level which lies between it and the ion ground state. To date, laser oscillation has been obtained on such transitions in ions of Ca, Sr, Ba, Hg, Eu, Yb and Pb (Table 3.1).

For all the ions of Table 3.1, other than Pb, the excitation of the upper laser levels is produced by electron impact from the ion ground state. This leads to a delay in the onset of laser oscillation of 50 to 100 ns relative to the onset of the current pulse, during which time the ground-state ion population greatly increases [3, 4]. This stepwise excitation scheme is evinced in [5, 6] by the observation of shifts in the laser frequency profiles, of the order of a Doppler width, which may be attributed to the drift in the discharge field of long-lived ion states, from which excitation takes place. An exception is the Pb II line, 1159 nm, where upper level excitation is by direct electron impact from the atomic ground state [7].

The most intense laser lines belong to ions of Ca, Sr and Eu. Many investigations have focused on increasing the efficiencies of these lasers to those typical of atomic self-terminating lasers. The quantum efficiencies of

Table 3.1 Self-terminating transitions in metal vapour ion lasers.

Element	Transition	λ (nm)
Ca II	$4p^2P^o_{3/2} \longrightarrow 3d^2D_{5/2}$	854.2
Ca II	$4p^2P^o_{1/2} \longrightarrow 3d^2D_{3/2}$	866.2
Sr II	$5p^2P^o_{3/2} \longrightarrow 4d^2D_{5/2}$	1032.7
Sr II	$5p^2P^o_{1/2} \longrightarrow 4d^2D_{3/2}$	1091.4
Ba II	$6p^2P^o_{3/2} \longrightarrow 5d^2D_{5/2}$	614.2
Ba II	$6p^2P^o_{1/2} \longrightarrow 5d^2D_{3/2}$	649.7
Hg II	$6p^2P^o_{3/2} \longrightarrow 6s^2\,{}^2D^*_{5/2}$	398.4
Eu II	$6p^9P_5 \longrightarrow 5d^9D^o_6$	664.5
Eu II	$6p^7P_3 \longrightarrow 5d^7D^o_4$	989.8
Eu II	$6p^9P_4 \longrightarrow 5d^7D^o_5$	1361.1
Eu II	$6p^9P_3 \longrightarrow 5d^7D^o_4$	1477
Eu II	$6p^7P_4 \longrightarrow 5d^7D^o_5$	1001.9
Eu II	$6p^7P_4 \longrightarrow 5d^7D^o_4$	1016.8
Yb II	$5p^2P^o_{3/2} \longrightarrow 5d^2D_{3/2}$	1345.3
Yb II	$5p^2P^o_{3/2} \longrightarrow 5d^2D_{5/2}$	1649.8
Yb II	$5p^2P^o_{1/2} \longrightarrow 5d^2D_{3/2}$	2437.7
Pb II	$6s6p^2 \longrightarrow 6s^27p$	1159

ion lasers relative to the atomic ground states are relatively small (e.g. 13–14% for Sr II). However, the lifetimes of metal ions in the plasma, which are determined by ambipolar diffusion and recombination, can be long (10–100 μs). Hence, as the PRF is increased, the efficiency can be expected to rise, since the pre-pulse ion density will be higher.

However, the lower laser level population must decay in the inter-pulse period. The lifetimes of metastables in neutral atoms are mainly determined by diffusion and by electron collision. For metastable ions, the latter remains effective, while ordinary diffusion is replaced by ambipolar diffusion, which is faster because its speed is approximately proportional to $(1 + T_e/T_g)$. Other processes occur also: dielectronic and three-body recombination; dissociative recombination of molecular ion complexes; and resonant energy transfer to neutral metal atoms [8]. Hence, the lifetimes of metastable ions are much less than those of metastable atoms. This leads to higher attainable PRFs for ion lasers.

Electron collision is the main process responsible for quenching the lower metastable levels in Sr II [4, 6]. Thus, in the discharge afterglow, the metastable density falls sharply (τ_1 = 1–10 μs) due to the initial rapid fall in T_e, and then the decay of the metastables slows (τ_2 = 10–80 μs) to the rate of ion recombination (Figure 3.3). For He pressures of less than 1.5 kPa, ambipolar diffusion of the metastable ions to the wall is the main quenching process [4]. Experiments with double pulses [3, 4, 8] show that the peak power in the second laser pulse increases 2–3 times with a reduction in time delay to approx. 50 μs, and only for shorter time delays does it begin to fall (Figure 3.3, curve 1). In the case of the Sr I laser transition (Figure 3.3, curve 2), the peak power of the second laser pulse simply falls as the delay time is increased. The differences in behaviour between Sr I and Sr II transitions can be explained by the accumulation of ions in the ground and metastable states, which is, in fact, the major difference between self-terminating lasers utilizing atomic and ion transitions.

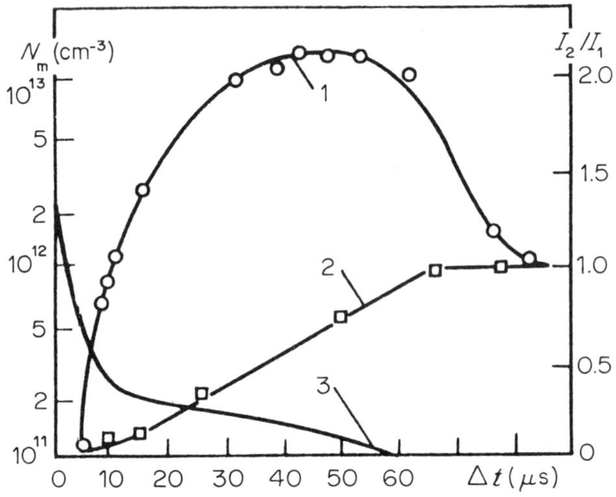

Figure 3.3 Dependence of the laser power (relative) of the second pulse on the time delay for the Sr II 1033 nm (1) and Sr I 6456.7 nm (2) transitions, and the temporal evolution of the 4 $^2D_{5/2}$ Sr II metastable population (3). T_{oven} = 530°C; p_{He} = 0.66 (1,2) and 1.3 (3) kPa; C_s = 1.1 nF; U = 10 kV; d = 10 mm; l = 40 cm [4].

In [9], laser oscillation was reported in Ca II on the lines 854.2 nm and 866.2 nm. The peak power was 5.4 kW, the average power was 0.74 W and the pulse energy was 0.11 mJ with an efficiency of 0.03–0.05%. The tube had a bore of 38 mm and was 15.2 cm long. The optimum Ca vapour pressure was found to be 54 Pa when the buffer gas (He) pressure was 1.33 kPa or 8 kPa (Ne buffer gas).

Self-terminating laser transitions are also obtained in atomic Ca and Sr. In [10], lasing was achieved simultaneously on the following lines: Sr I 6.45, 3.011 and 3.088 μm; Sr II 1.03 and 1.09 μm. The tube had an internal diameter of 10 mm and a length of 38 cm, and He pressures of up to 53 kPa were used (the optimum He pressure was 10.7 kPa). The total laser power was 1.2 W (5% was on Sr II lines). Under these conditions, oscillation is also observed on the 416.2 and 430.5 nm Sr II lines (see Chapter 4). Thus, the Sr laser can operate 'simultaneously' on a set of lines from the violet to the mid-infrared.

The Eu II self-terminating laser [11, 12] is of particular interest, since for helium pressures above 15 kPa, quasi-steady-state laser oscillation was obtained for the first time. At low He pressures (≈100 Pa), lasing begins first on the 1.36 μm line (Figure 3.4), and then on the 664.5 nm line. With an increase in He pressure, the laser power rapidly decreases. Such behaviour is typical of self-terminating lasers, and is caused by the decrease in the ratio of electric field to gas pressure, E/p. As He pressure is increased from 4 to 25 kPa, weak emission on the 1.36 μm transition becomes stronger, and

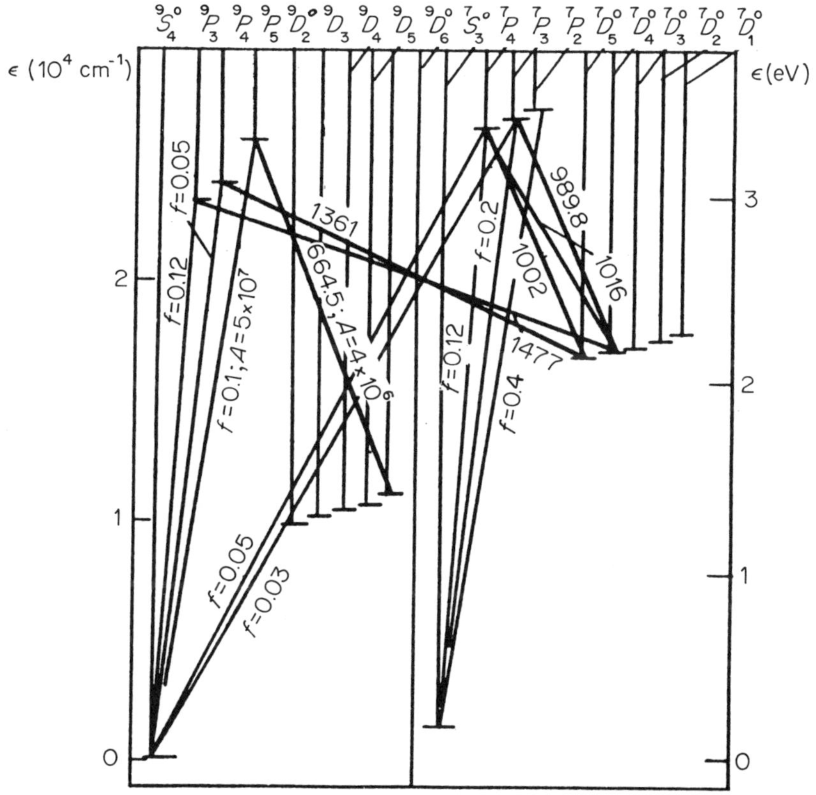

Figure 3.4 Diagram of lower Eu II levels [12]. Here and in later figures the laser transitions are drawn as thick lines; wavelengths are in nanometres, and the oscillator strength f is given for some transitions.

oscillation appears at 1.0019 μm ($^7P_4 \longrightarrow {}^7D^o_5$) and on other transitions between ^7P and ^9P (Figure 3.5). By this stage, the form of the laser pulse follows the discharge current pulse. By lengthening the current pulse, the laser pulse duration can be increased to 150 μs, but further temporal scaling is prevented by the appearance of discharge contraction at later times.

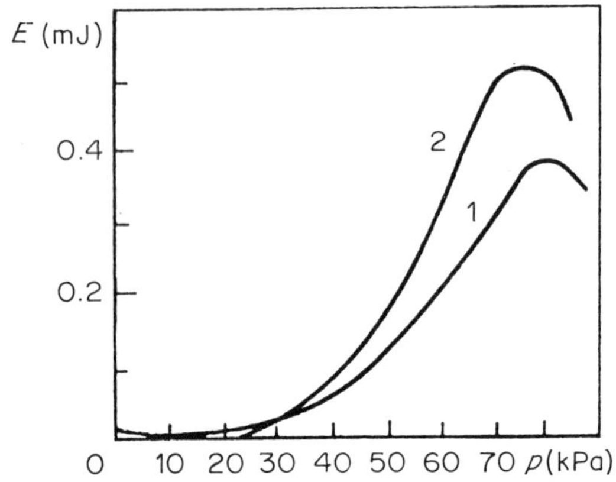

Figure 3.5 Laser pulse energy of Eu II laser lines versus helium pressure for a tube with d = 0.7 cm and l = 50 cm. 1 – λ = 1361 nm; 2 – λ = 1002 nm and λ = 1016.6 nm [11].

It has been established [12, 13] that inelastic collisions of excited Eu ions with He atoms is the basic mechanism for maintaining a population inversion in the quasi-stationary regime of the Eu II laser. Due to effective mixing of the upper levels $^7P_{2\text{–}4}$, $^9P_5$ in collisions with He, these levels can be considered to be a single level. The lower laser levels (including $^7D^o_5$) decay by collisions with helium atoms, despite the large energy defect (0.71 eV). The lifetimes of these levels can be as short as 25 ns. In three-body collisions involving two helium atoms and a Eu^+ ion in the lower laser level, a $HeEu^+$ complex may form, which recombines dissociatively with electrons [14].

To date, on the 1.002 μm Eu II line at atmospheric pressure, the following parameters have been reported [15]: at a PRF of 6.5 kHz, an average output power of 11.8 W, a specific output pulse energy, $\mathcal{E}_{sp}$, of 11.6 μJ cm^{-3}, and an efficiency of 0.24%; at a PRF of 9.5 kHz, 12.7 W, 8.5 μJ cm^{-3} and 0.21%, respectively. These results were obtained using a BeO plasma tube of 50 cm active length and 27 cm inner diameter (290 cm^3 active volume). With an increase in helium pressure, the specific output energy continues to rise [16], at least up to 3.5 atm. of helium, attaining an average output power P_{av} = 4 W, a specific average output power P_{sp} = 165 W.cm^{-3}, and a specific output pulse energy $\mathcal{E}_{sp}$ = 21.1 μJ.cm^{-3} at a PRF of 2.2 kHz with an efficiency of

0.13% from a laser tube of 15 cm length and 2.7 cm inner diameter. The reduction in efficiency is explained in [16, 17] by absorption in the near-electrode parts of the tube.

Besides laser transitions from resonant to metastable levels, many other laser lines have been obtained on the transitions of metal ions due to favourable lifetimes and pumping rates of the upper and lower laser levels (see Appendix 4). In general, these transitions lie high in ion spectra and require high-energy electrons to excite them. They are populated stepwisely via the ion ground state, as in noble-gas ion lasers [18]. For that reason, laser oscillation is obtained at low vapour and buffer-gas pressures, with high voltages across the discharge tube (which gives a high T_e in the positive column) and a high current density.

Although there is a great number of high-lying laser lines of the latter type which are pumped by electron impact, covering a wide spectral range, none of these lines has found practical application. This can be explained partly by the fact that the highest powers, of 0.1–0.2 W (which, however, could be improved by further power scaling), are obtained from very aggressive elements such as Cl and Br. High efficiencies and high powers cannot be expected from these lasants (compared with noble-gas ion lasers), and, as far as we know, studies of such lasers have not continued.

In the case of lasers other than the self-terminating type, there are the relatively low-lying ion laser transitions 441.6 and 325 nm of Cd II, and 589.4 and 747.9 nm of Zn II. The main pumping mechanism is Penning ionization (see section 3.3). As was noted in Chapter 1, the upper levels of these transitions have large cross sections for electron-impact excitation from the atomic ground state ($\approx 10^{-16}$ cm^2). For that reason, they can be effectively populated via electron impact in a discharge through pure cadmium vapour. Hence, in a PC discharge without buffer gas, laser oscillation at 441.6 nm has been attained with a higher gain (120% m^{-1}) than is obtained in mixtures with helium [19]. In a discharge tube of 3 mm internal diameter and 60 cm length, the optimum discharge current was only 1.4 A at a vapour pressure of 2.7 Pa [19]. However, lasers with buffer gases are considerably more reliable than those that utilize the pure metal vapour.

3.3 LASERS EXCITED BY HEAVY-BODY COLLISIONAL ENERGY TRANSFER

3.3.1 Laser Oscillation in Hg, Cd and Zn

In pulsed discharges, laser oscillation on metal ions pumped by collisions involving buffer-gas metastables and ions has been obtained in about 20 metals. We shall discuss those devices that use Zn, Cd and Hg lasants. These elements were the first to be studied in pulsed PC discharges because their spectra were well known, the required working temperatures were low, and their chemical reactivities were low, making possible the use of glass discharge tubes.

Many pulsed MVIL transitions have been obtained, pumped by electron impact, charge transfer, Penning collisions and recombination. Those investigations have permitted the basic principles of pumping and deactivating ion levels to be determined and the inversion mechanisms to be identified, and have directed the search for transitions in other elements. This is the reason why we now pay close attention to pulsed oscillation in Zn, Cd and Hg.

Mercury vapour was the lasant used in the first gas ion laser. Laser oscillation was observed on the yellow-green 567.7 nm, red 615 nm and infrared 734.6 and 1058.3 nm transitions of Hg II by Bell in 1963 [20, 21], using a pulsed PC discharge in a mixture of Hg vapour and helium. Pulsed oscillation was first reported in Zn II and Cd II in [22], and since then there have been many papers on new laser lines in those elements in both stationary and pulsed discharges (see Appendix 4).

Zinc, cadmium and mercury are transition elements of group II of the periodic table. The configuration of the outer electron shell is $(n-1)\mathrm{d}^{10}n\mathrm{s}^2$, where $n = 4, 5, 6$ for Zn, Cd and Hg, respectively. Ion spectra can be formed in two ways (Figures 3.6–3.8): by removal of one of the outer s electrons with subsequent excitation of the remaining one—this is leads to a 'single-electron' spectrum; or by removal of an inner electron from the d shell with subsequent excitation of one of the s electrons—this leads to Beutler's spectrum. The ground state of Beutler's spectrum—the levels $(n-1)\mathrm{d}^9n\mathrm{s}^2$ ${}^2D^*_{3/2,5/2}$—in cadmium and zinc lie higher than the $np\ {}^2P^{\mathrm{o}}_{1/2,3/2}$ resonance terms of the single-electron spectrum. Hence, the optical transitions ${}^2D^*_{3/2,5/2} \longrightarrow n\ {}^2P^{\mathrm{o}}_{1/2,3/2}$ are permitted. The probabilities of these double-electron transitions are relatively low ($\approx 10^6\ \mathrm{s}^{-1}$), i.e. they are about two orders of magnitude less than the probabilities for decay from the levels $n\ {}^2P^{\mathrm{o}}_{1/2,3/2}$ to the $n\ {}^2S_{1/2}$ state. This fact favours the production of an inversion and laser oscillation on the respective lines of Cd II and Zn II. In mercury, the level $5\mathrm{d}^9\ 6\mathrm{s}^2\ {}^2D^*_{3/2,5/2}$ lies below the term $6\ {}^2P^{\mathrm{o}}_{1/2,3/2}$, and is therefore metastable. Self-terminating laser oscillation is thus obtained on the Hg II transition $6\ {}^2P^{\mathrm{o}}_{3/2} \longrightarrow 6s^2\ {}^2D^*_{5/2}$ at 398.4 nm (see Table 3.1).

On all the laser transitions of the single-electron spectra of Hg, Cd and Zn, oscillation is observed in the discharge afterglow of a pulsed PC discharge. At buffer-gas pressures above 0.15 kPa, laser emission exists only with a helium buffer gas, and on those transitions with upper laser levels lying at energies below the ionization potential of helium. Laser oscillation on $D^* \longrightarrow P$ transitions in Cd and Zn at He pressures over 0.1 kPa is also observed in the afterglow.

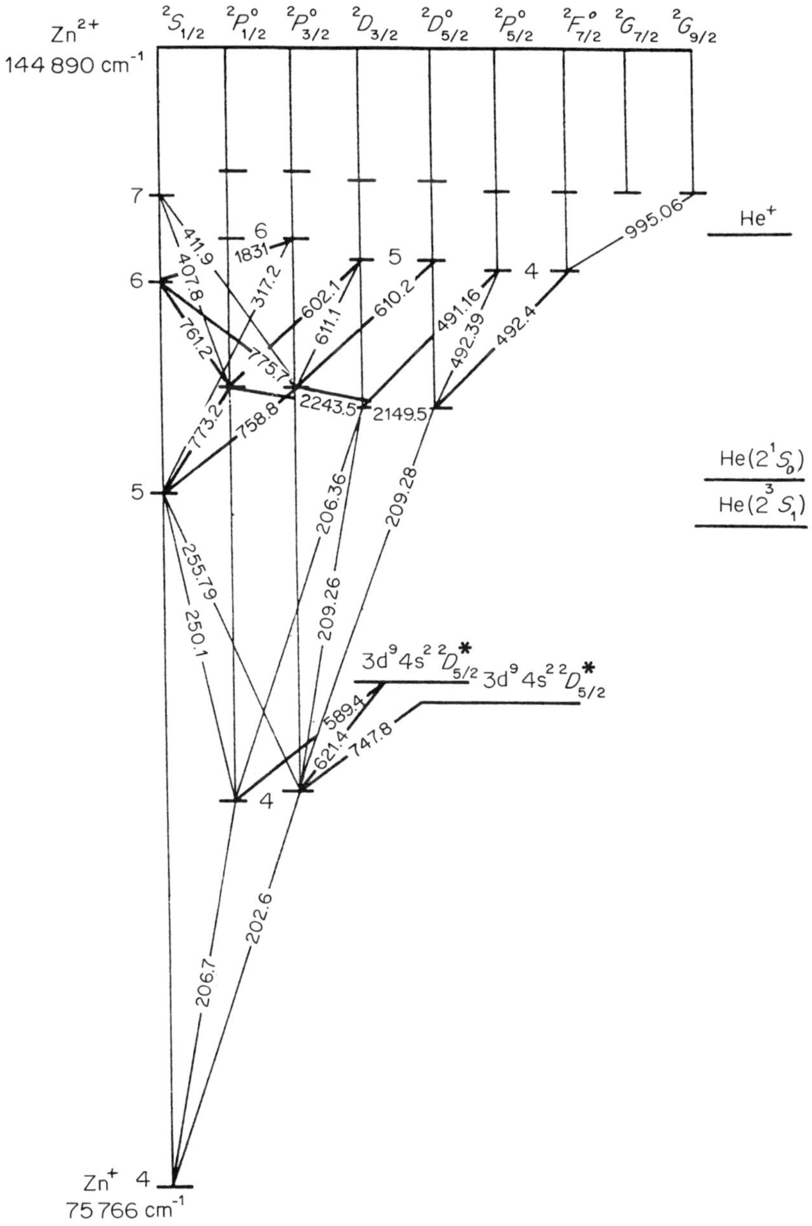

Figure 3.6 Energy-level diagram of Zn II.

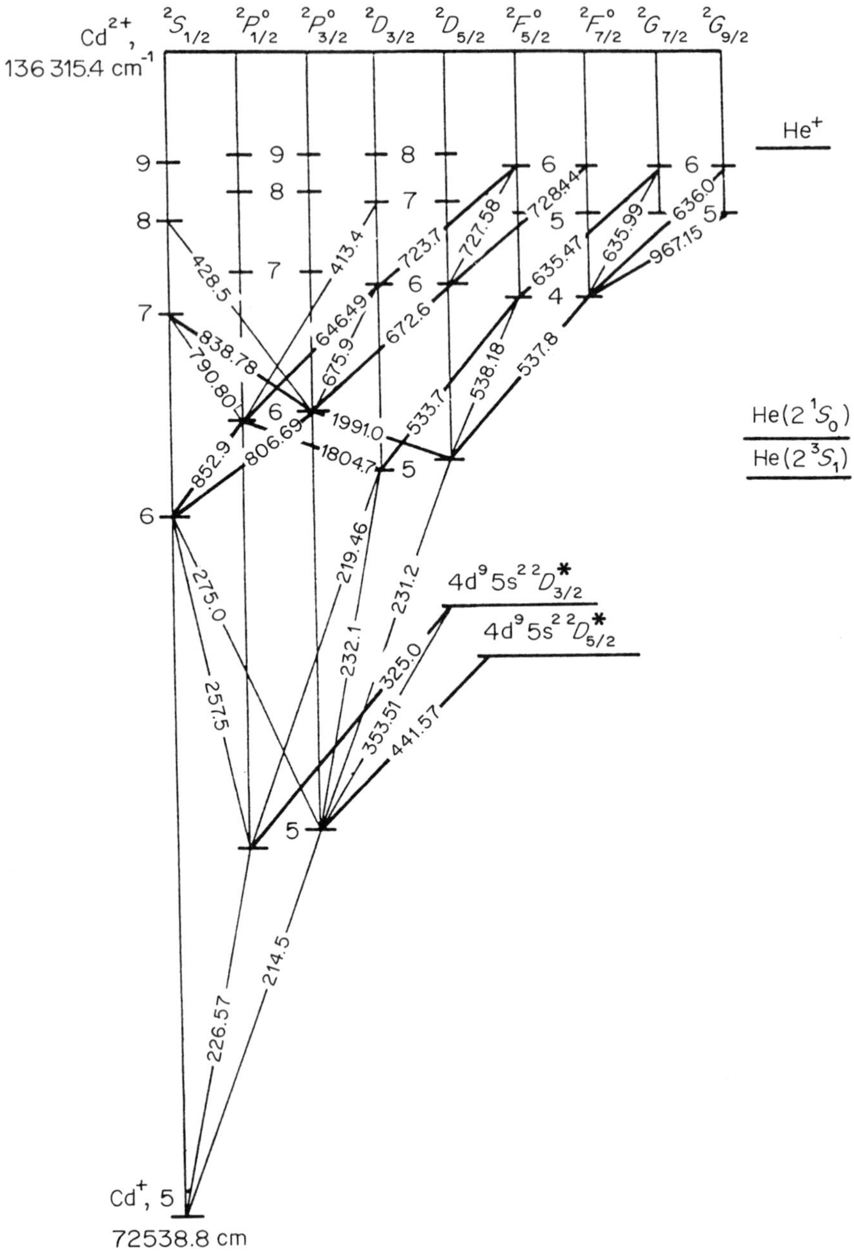

Figure 3.7 Energy-level diagram of Cd II.

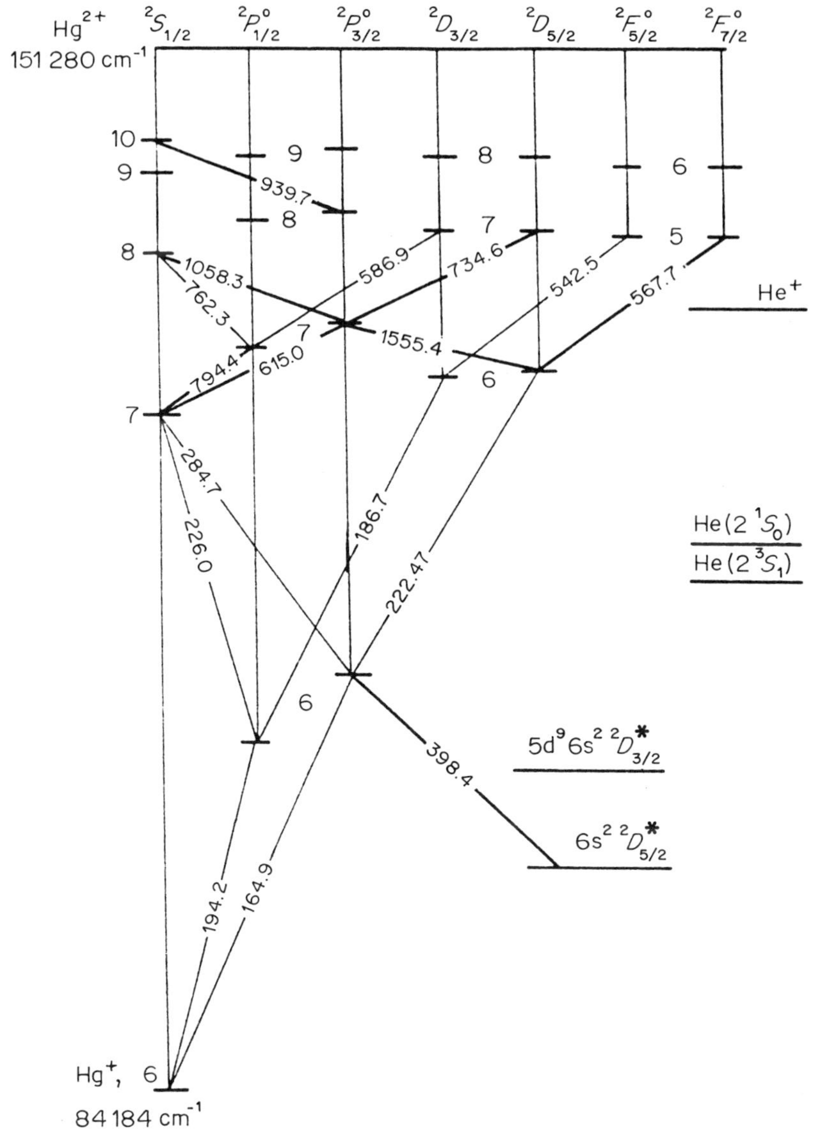

Figure 3.8 Energy-level diagram of Hg II.

The typical conditions for oscillation on single-electron transitions in mixtures with He in discharge tubes of 5–10 mm bore are: 0.1–4 kPa He pressure (the optimum is ≈0.3 kPa); 0.1–1 Pa metal vapour pressure; a current pulse length of 0.3–2 μs; a peak current of 2–10 A; and a PRF of several kHz. For $D^* \longrightarrow P$ transitions in Cd and Zn, the optimum current is less (0.5–1.5 A), and the pressure of helium is higher (10–20 kPa).

The most intense lines of the single-electron spectra of cadmium and zinc are 537.8, 533.7 nm (Cd II) and 492.4, 491.2 nm (Zn II) for which the gain coefficient is 20–30% m^{-1}. For 615 nm (Hg II) and 441.6 nm (Cd II), the gain is up to 100% m^{-1}, and the pulse power is 1–10 W. Note that for the Hg II line, 615 nm, the Cd II lines, 441.6 and 325 nm, and the Zn II lines, 589.4 nm and 747.8 nm, isotopic splitting greater than the Doppler width occurs. If only one even isotope is used, the gains on those lines can be increased by 3, 3.5 and 2 times, for Hg, Cd and Zn, respectively [23–25].

The typical temporal dependences of spontaneous emission on various lines are shown in Figure 3.9. The lines of neutral atoms (Figure 3.9b,c) have peaks during the current pulse, then short reductions in intensity and a long intense afterglow (20–100 μs). The intensity and duration of the helium afterglow reduce with increasing metal vapour pressure. Line emission during the current pulse is due to electron impact, while in the afterglow it is due to the recombination of singly charged ions, produced earlier during the current pulse. Following termination of the current pulse, T_e quickly falls and electron-impact excitation ceases. When T_e is less than 0.5 eV, electron–ion recombination begins. Thus, just after the current pulse, emission is sharply reduced, but appears again after the electrons have cooled sufficiently.

The molecular spectrum of He_2^* (Figure 3.9d) is also of interest. Its increase in the afterglow for 30–50 μs is explained by the accumulation of He_2^+ by heavy-body processes and subsequent electron–ion recombination. At typical He pressures for lasing, the intensity of molecular bands is not great, which implies a low density of He_2^+.

All lines of metal-ion single-electron spectra, which originate from levels below the potential of He^+, exhibit long afterglows. At low pulsed currents (to 2 A) the maxima of their intensities are at the end of the current pulse, after which there is an exponential decrease (Figure 3.9e), comparable in duration to the He afterglow. When the current is increased to 10 A, the shapes of the lines alter. The lines originating from levels with a large orbital-momentum quantum number, L, such as 537.8, 533.7 nm ($4\,^2F \longrightarrow 5\,^2D$ Cd II), 492.4, 491.2 nm ($4\,^2F \longrightarrow 4\,^2D$ Zn II) and 222.4 nm ($6\,^2D_{5/2} \longrightarrow 6\,^2P_{3/2}$ Hg II), have intensities that continue to rise for some microseconds following the end of the current pulse (Figure 3.9f). The lines from levels with a low quantum number L, 428.5 nm ($8\,^2S_{1/2} \longrightarrow 6\,^2P_{3/2}$ Cd II), 413.4 nm ($6\,^2D_{5/2} \longrightarrow 6\,^2P_{3/2}$ Cd II), 317.2 nm ($6\,^2P_{3/2} \longrightarrow 5\,^2S_{1/2}$ Zn II), 610.2, 602.1 nm ($6\,^2D_{5/2,3/2} \longrightarrow 5\,^2P_{3/2,1/2}$ Zn II), and 615.0, 794.4 nm ($7\,^2P \longrightarrow 7\,^2S$ Hg II), peak in intensity at the end of the current pulse, but then exhibit a rapid fall in intensity when, at the same time, the intensities of lines with large initial L are still increasing (Figure 3.9g). Some lines, such as 672.6 and 646.49 nm ($6\,^2D \longrightarrow 6\,^2P$ Cd II),

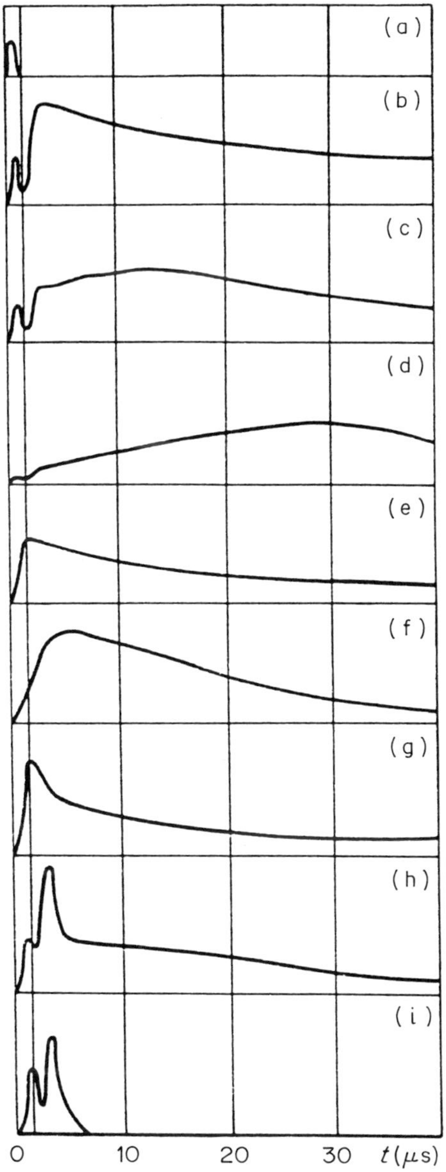

Figure 3.9 Typical oscillograms of spontaneous emission; d = 0.8 cm; p_{He} = 0.3 kPa; p_M = 0.1–0.3 Pa. (a) Current pulse. (b) He I line intensities. (c) Line intensities of neutral metal atoms. (d) He_2^* spectrum. (e)–(h) transitions of metal ion lines from levels with energies less than He^+: (e) all lines, current 1 A; (f) 492.4 nm, 491.2 nm of Zn II; 537.8 nm, 533.7 nm of Cd II; 222.4 nm of Hg II, current 5 A; (g) 317.2 nm, 610.2 nm, 602.1 nm Zn II; 413.4 nm, 428.5 nm Cd II; 615 nm, 794.4 nm Hg II, current 5 A; (h) Current 30 A. (i) Ion lines from levels that lie above the energy of He^+, current 30 A.

display no changes at the beginning of the afterglow, and the shape remains nearly exponential.

With further increases in current (up to a peak of 30 A), the shape of all afterglow lines originating from levels of energy below He^+ have the pattern shown in Figure 3.9(h), i.e. an early short peak superimposed on a long afterglow. With the increase in current, the intensity of the short peak rises sharply and its duration reduces.

For lines originating above the energy of He^+, the long afterglow is either absent or is very weak, while the short afterglow only appears at high currents, which coincides in time and shape with the short peak appearing in the afterglows of lines that originate at energies below He^+ (Figure 3.9i). Such a line is the 567.7 nm ($5\ ^2F_{7/2} \longrightarrow 6\ ^2D_{5/2}$) transition of Hg II. The short peak is due to the recombination of doubly charged metal ions with electrons (for more detail, see Chapter 4).

At low pulse currents (up to 1 A), the afterglow intensity of all lines is approximately proportional to the current. With further increase in current, the intensities of transitions with large initial L rise faster than those with lower initial L. Saturation sets in for a certain level of current, and is earlier and stronger, the closer the energy of the upper level is to the ionization potential of helium (Figure 3.10).

We shall now analyse the behaviour of the duration, τ, of the afterglow of ion lines (τ is the time period over which the emission intensity falls to 1/e of the

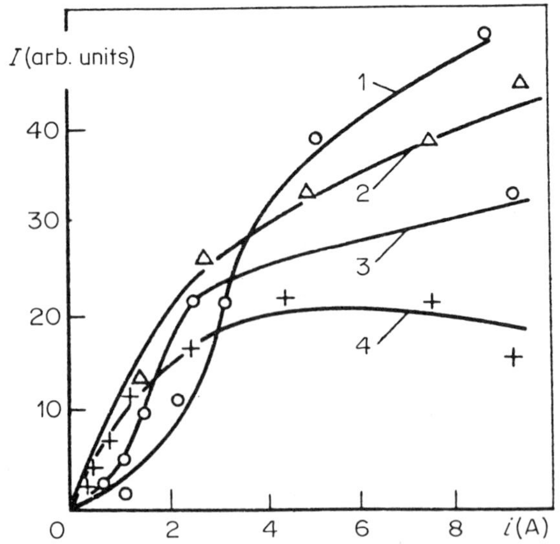

Figure 3.10 Intensity of the ion line afterglow 5 μs after the current pulse as a function of current. 1, λ = 537.8 nm Cd II, 492.4 nm Zn II; 2, λ = 610.2 nm Zn II; 3, λ = 413.4 nm Cd II; 4, λ = 636.0 nm Cd II.

initial intensity, reached after the fast initial rise, i.e. 7–10 μs after termination of the current pulse). The duration of the ion afterglow decreases with an increase in metal vapour pressure, until it disappears for vapour pressures above 10 Pa. At the same time, the duration of the recombination afterglow of helium lines decreases also. The duration of the afterglow of neutral metal atoms does not depend significantly on vapour pressure. If at low pulsed currents the values of τ for the various ion lines are nearly the same, then at some higher (medium) current they will not be (Table 3.2). For lines of the type shown in Figure 3.9(f), τ becomes smaller than for lines of the type in Figure 3.9(g). For high-lying transitions, τ is larger than that for low-lying transitions. The differences are great, especially at low vapour pressures (up to 0.2 Pa).

Table 3.2 Durations, τ, in microseconds, of the Zn II and Cd II afterglow lines.

Ion	λ (nm)	Density of metal atoms (cm^{-3})		
		2×10^{13}	10^{14}	2×10^{14}
Zn II	491.1	20	14	7
	610.2	32	15	7.5
	775.2	35	17	7.5
	317.2	40	18	8
Cd II	537.8	25	20	14
	672.6	28	24	15
	428.5	32	29	16
	413.1	30	29	16
	636.0	43	33	17

Laser oscillation on single-electron transitions occurs in the discharge afterglow, and exhibits a shift to the far afterglow with an increase in current (Figures 3.11 and 3.12). The laser pulse is delayed from the current pulse by tens of microseconds. On lines that exhibit an initially fast fall in intensity (e.g. 615 nm, Hg II), at certain currents there can be a splitting of the laser pulse into two (Figure 3.12).

The characteristics of the spontaneous emission and the laser emission on $D^* \longrightarrow P$ lines (441.6, 325 nm of Cd II; 589.4, 747.8 nm of Zn II) are similar to each other, but different from those of single-electron lines (Figure 3.13). At low currents ($\approx$0.5 A) the afterglow intensity falls exponentially. With an increase in current, the afterglow splits into two: the initial part with high intensity and rapid fall, whose duration, τ_1, reduces in inverse proportion to the current; and the second, whose duration, τ_2, does not depend on current above 2–3 A. Laser oscillation occurs during the initial fall (5–7 μs) and the optimum current is ~1 A.

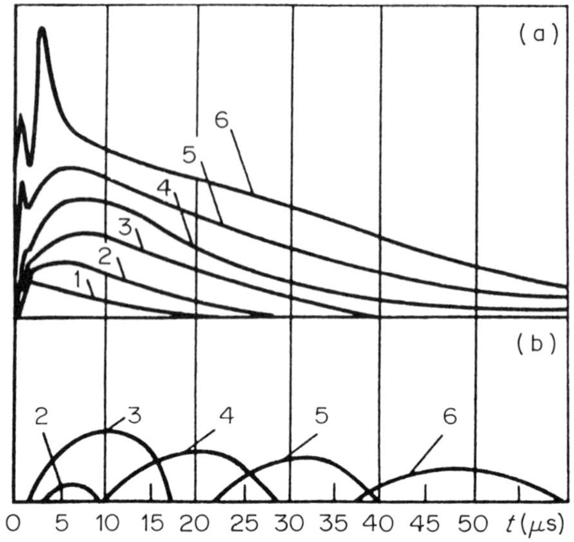

Figure 3.11 (a) Temporal behaviour of spontaneous emission and (b) laser oscillation at λ = 537.8 nm, 533.7 nm Cd II and 492.4, 491.2 nm Zn II, with p_{He} = 0.4 kPa, p_M = 0.27 Pa, as functions of current: 0.5 A (1), 1 A (2), 3 A (3), 5 A (4), 8 A (5), 10 A (6).

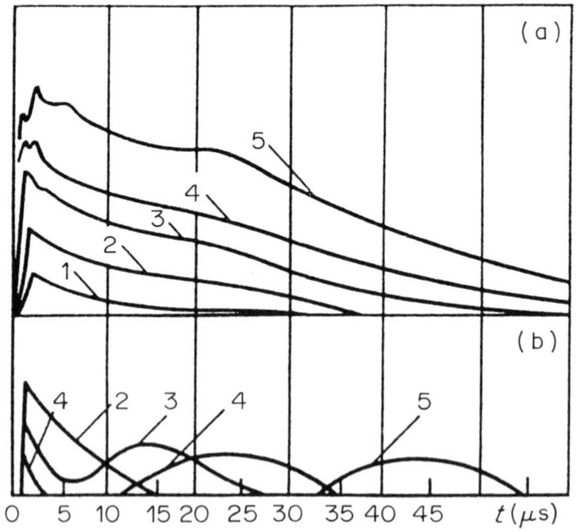

Figure 3.12 (a) Temporal behaviour of spontaneous emission and (b) laser oscillation at λ = 615 nm Hg II, with p_{He} = 0.4 kPa, p_{Hg} = 0.13 Pa, as functions of current: 0.5 A (1), 1 A (2), 3 A (3), 8 A (4), 12 A (5).

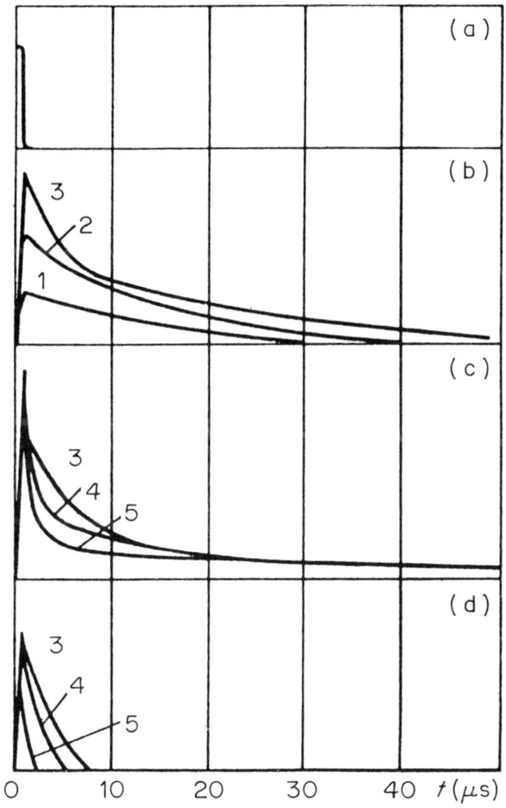

Figure 3.13 (a) Current pulse. (b), (c) Temporal behaviour of spontaneous emission and (d) temporal behaviours of laser oscillation at λ = 441.6 nm Cd II and 589.4 nm Zn II, with p_{He} = 3 kPa, p_M = 0.27 Pa, as functions of currents: 0.3 A (1), 0.7 A (2), 1 A (3), 3 A (4), 10 A (5).

Thus the temporal behaviours of spontaneous and stimulated emission of Zn II, Cd II and Hg II lines are diverse and demonstrate complex forms. We will now show how analysing them sheds light on the pumping mechanisms of metal ion levels.

3.3.2 Pumping Mechanisms of Ion Levels in the Single-Electron Spectra of Hg, Cd and Zn

Several mechanisms have been considered for pumping the levels of these ion lasers, even in the earliest reports. First, it was suggested that Hg laser transitions are excited by electron impact [20, 21]. Then Dyson [26], on the basis of observations of the spontaneous emission of Hg II lines, inferred that

$7\,^2P_{1/2,3/2}$ Hg II levels are populated by helium ions via charge transfer,

$$Hg + He^+ \longrightarrow Hg^{+*} + He\,. \tag{3.1}$$

Dyson measured the cross-section of that process to be 1.3×10^{-14} cm^2, using a now classic method (see page 103). Some time later it was proposed that the same ion levels are populated via a stepwise Penning process in collisions of Hg metastables with He metastables [27]:

$$Hg(m) + He(m) \longrightarrow Hg^{+*} + He + e\,. \tag{3.2}$$

The same processes have been considered for ion lines of the single-electron spectra of zinc and cadmium [28].

There are still differing opinions regarding the pumping mechanisms of these lasers. For example, in [29] process (3.2) is considered to be the principal pump route for the Hg ions, although in most of the literature the charge-transfer process is preferred.

Now we shall analyse the temporal characteristics of laser oscillation and the emission of Zn II, Cd II and Hg II lines. Laser oscillation in the ion spectra of Zn, Cd and Hg occurs during the afterglow, and hence the processes of direct or stepwise electron impact cannot be directly responsible for the excitation of the lines of the single-electron spectra. The effect of electron excitation during the current pulse is not great, since at currents that are optimum for laser oscillation, a spontaneous emission peak is not observed during the current pulse (see Figure 3.11). It appears at much higher currents, when laser oscillation occurs in the far afterglow if it occurs at all. The long duration of the afterglow of the ion lines can be reasonably attributed only to long-lived particles. From the energetic point of view (see Figures 3.6–3.8), it can be assumed that those particles can only be He ions.

The following experimental facts show that the long afterglow is due to charge transfer. First, the intensities of lines with upper levels that lie below the ionization potential of He are anomalously high. Second, the long afterglow of these lines is consistent with the long lifetime of He^+ ions in the plasma. The lines that lie above He^+ have either a very short afterglow or none at all. Third, the duration of the afterglow of metal ion lines and the recombination afterglow of He lines reduces with an increase in vapour pressure, which is due to a shortening of the He^+ lifetime for charge transfer. Finally, when He is substituted by Ne or Ar, the long afterglow of the ion lines of Cd, Zn and Hg disappears. Only at high currents is a short recombination afterglow present, as there is for lines that originate from levels above the ionization potential of helium (Figure 3.9i).

In process (3.2), metastable atoms of both helium and the metal take part, and their summed energy is sufficient to excite high-lying levels of Hg, Zn and Cd. However, the excitation route for these levels is not effective for a number of reasons. First, the reaction is stepwise, i.e. it requires a high concentration of both reacting metastables. Second, in order to populate the upper laser levels, the process requires the change of state of three electrons. Third, the process is not selective like charge transfer, which has an optimum energy defect of 0.1–0.6 eV [30]. Moreover, the cross-section of process (3.2) increases with increasing energy defect [31], i.e. the lower laser levels will be populated more than the upper laser levels.

In the afterglows of pulsed discharges, processes (3.1) and (3.2) can be distinguished by the addition of species that selectively quench the metastable atoms of He. The cross-section for quenching He(2 $^3S_1$) by Ar atoms is about 10^{-15} cm^2. When around 10 Pa of Ar is introduced to a helium/metal-vapour discharge, the density of He metastables is strongly reduced, but there is almost no change in the shape or intensity of the afterglow of the metal-ion single-electron spectra [28, 32].

In [33, 34], the vapours of Hg, Cd and Zn were excited in an electron gun. The addition of He to the gun caused the following effects: (a) metal ion lines that lie below the He^+ ground state increased in intensity by 5–50 times; (b) the excitation functions of those lines became similar to the ionization function of He by electrons; (c) the energy threshold for excitation became equal to the ionization energy of He; (d) with pulsed excitation, the metal-ion lines had a long ($\approx$30 μs) afterglow, and the afterglow decay constant reduced with an increase in vapour pressure. All these observations show, unambiguously, that charge transfer plays the main role in the excitation of levels in the single-electron spectra of Hg, Cd and Zn.

In [35], a double-probe technique was suggested for the monitoring of number densities in the afterglow. It was found that the intensities of the 615 and 794 nm lines and the distribution of the He^+ density have similar forms, but differ from the distribution of the density of He(2 ^3S) metastables. Experiments with flowing (velocity $\approx 10^4$ cm s^{-1}) afterglows of He atoms were conducted in [36, 37]. At the flow inlet, a discharge was initiated. Downstream ($\approx$20 cm) of the flow, i.e. about 2 ms after the discharge, metal vapours were admitted. At that place, the bright glow of metal-ion lines could be observed, which followed the flow of gases. The metal-ion lines observed belonged to the single-electron spectrum of the metal, with upper levels below the energy of He^+. Since T_e was low at the inlet of the metal vapours, metastable atoms were not produced, and process (3.2) was not active.

Hence, the dominant role of charge transfer in populating the levels of single-electron spectra of Hg, Cd and Zn has been confirmed. The following levels, lying in energy near ($\Delta\varepsilon \approx$0.1–0.6 eV) the He ionization potential, are populated directly by charge transfer: 6 ^2P, 5 ^2D, 4 ^2F Zn II; 9 ^2P, 8 ^2D, 6 ^2F, 6 ^2G Cd II; and 7 ^2P Hg II. The levels below them are populated by cascade transitions (see Figures 3.6–3.8). This conclusion is also confirmed by computer modelling of the He–Hg ion laser, taking into account stepwise population processes [38].

Since the probabilities of radiative transitions do not depend on plasma parameters, the populations of all ion levels disposed below He^+ should vary in a similar manner with a change in those parameters. However, this is only observed at low currents, insufficient for lasing in the majority of laser transitions. With an increase in current the relative forms and durations of the afterglow lines change (see Figure 3.9).

These changes in afterglow lines are accounted for by inelastic collisions with slow electrons in the decaying plasma [32, 36]. Inelastic electron collisions are enhanced by the increase in current due to the increase in n_e. In the afterglow the dominant process is electron de-excitation, because T_e falls sharply after termination of the current, and, on the one hand, the probabilities of excitation processes are low, especially for widely spaced energy levels. On the other hand, the probability of electron de-excitation increases with the decrease in T_e and in energy interval, $\Delta\varepsilon$; the probability is proportional to $1/(\Delta\varepsilon\sqrt{T_e})$ (see section 1.2). As a rule, the probabilities of radiative transitions sharply decrease with reducing $\Delta\varepsilon$.

A characteristic feature of the spectra of Zn, Cd and Hg ions is that the terms are divided into groups of closely spaced levels. The energy within these groups decreases from levels with small quantum number L to levels with large L, e.g. from level $7\,^2P$ to $6\,^2D$ and $4\,^2F$, from level $8\,^2P$ to $7\,^2D$ and $5\,^2F$ in the case of Cd; from level $6\,^2P$ to $5\,^2D$ and $4\,^2F$ in the case of Zn; from level $7\,^2P$ to $6\,^2D$ in the case of Hg. The spacings between terms inside each group are not large, and for $n_e = 10^{11}$–10^{12} cm^{-3}, transitions due to collisions of the second kind with electrons (rate constant $k \approx 10^{-5}$ cm^3 s^{-1} [36]) are important. The second-kind collisions reduce the populations of the levels with low L, and increase the populations with large L. This leads to a fall in intensity of lines originating from the former levels, and a rise in intensities from the latter levels during electron cooling.

In addition, electron de-excitation transfers population from high-lying groups to low-lying ones. Hence the intensities of lines from high-lying levels saturate at low currents (see Figure 3.10). For the same reason, at high n_e, electron de-excitation quenches the inversion on certain transitions; the lasing either disappears, or shifts to the far afterglow where n_e is lower and the conditions for inversion reappear (see Figures 3.11 and 3.12).

Quenching collisions with He atoms at the He pressures typical of MVILs do not affect the metal-ion-level populations. First, the cross-section is only just significant for very small level spacings (a few hundredths of an eV) [39]. Second, the number density of He atoms is constant with time, hence collisions with He atoms cannot lead to changes in the temporal forms of afterglow lines.

3.3.3 Afterglow Kinetics in Helium and Metal Vapour Mixtures

Using a simple mathematical model, we shall describe the excitation processes for metal ion levels more accurately. We consider charge transfer and ambipolar diffusion to be the ion loss mechanisms, which is justified for $n_e \leq 10^{14}$ cm^{-3} and $p_{He} \leq 2$ kPa. The ion densities of the metal, n^+, and of the helium, N^+, in the discharge afterglow are

$$dn^+ / dt = -\nu_M^+ n^+ + K_c n N^+ ; \quad (3.3)$$

$$dN^+ / dt = -\nu_B^+ N^+ + K_c n N^+ ; \quad 3.4)$$

$$n_e = n^+ + N^+ ; \quad (3.5)$$

where $K_c = <q_c v>$ is the charge transfer rate constant, and ν_M^+ and ν_B^+ are the frequencies of ambipolar diffusion of metal and helium ions.

If the initial densities of ions and electrons in the afterglow are n_0^+, N_0^+ and n_{e0}, then the solution of the system (3.3)–(3.5) is:

$$n^+ = \left(n_0^+ - \frac{K_c n N^+}{\nu_M^+ - \nu_B^+ - K_c n}\right)\exp(-\nu_M^+ t) + \\ + \frac{K_c n N_0^+}{\nu_M^+ - \nu_B^+ - K_c n}\exp[-(\nu_B^+ + K_c n)t] ; \quad (3.6)$$

$$N^+ = N_0^+ \exp[-(\nu_B^+ + K_c n)t], \quad (3.7)$$

and the density N^+ decays with a time constant

$$\tau = 1 / (\nu_B^+ + K_c n) . \quad (3.8)$$

If the metal atom density is high, i.e. $\nu_M^+ - \nu_B^+ \ll K_c n$, then using (3.5), we find for the metal atom density

$$n^+ = n_{e_0} \exp(-\nu_M^+) - N_0^+ \exp[-(\nu_B^+ + K_c n)t] , \quad (3.9)$$

which shows that n^+ initially increases due to charge transfer, before falling. The emission that is observed from He atoms arises from recombination of He^+ ions and electrons, i.e. the intensity of He lines is proportional to

$$I_{He} \sim \alpha n_e^2 N^+ , \quad (3.10)$$

where α is the recombination rate constant.

If we assume that $\nu_B^+ \approx \nu_M^+ = \nu^+$, then we have $n_e = n_{e_0} e^{-\nu^+ t}$, and for the intensity of He lines we find

$$I_{He} \sim a n_{e_0}^2 N_0^+ \exp[-(3\nu^+ + K_c n)t] . \quad (3.11)$$

In the case where metal vapour is absent, we have

$$I_{\mathrm{He}} \sim \exp(-3\nu^{+}t)\,. \tag{3.12}$$

For the intensity of radiation emitted by neutral metal atoms, we find

$$I_{\mathrm{M}} \sim \alpha n_{\mathrm{e}0}^{3} \exp(-3\nu^{+}t) - \alpha n_{\mathrm{e}0}^{2} \exp[-(3\nu^{+} + K_{\mathrm{c}}n)t]\,. \tag{3.13}$$

The temporal courses of the densities of ions, electrons and the afterglow lines calculated from (3.9)–(3.13) are shown in Figure 3.14. These time dependences are in agreement with experiments (see Figure 3.9).

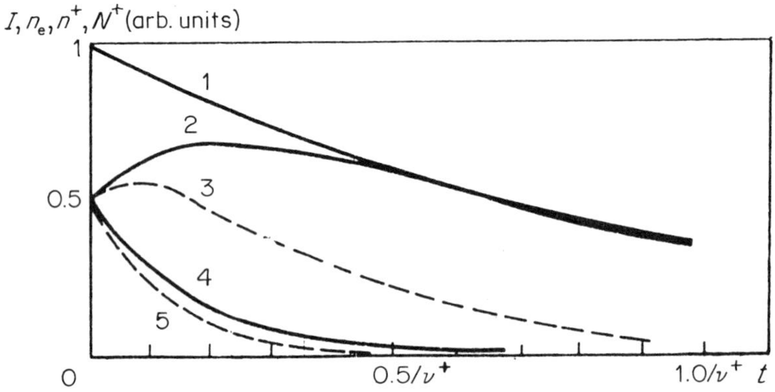

Figure 3.14 Characteristics of the afterglow, calculated for $N_0^{+} = n_0^{+} = 0.5n_{\mathrm{e}}$ and $K_{\mathrm{c}}n = 5\nu^{+}$. 1, n_{e}; 2, n^{+}; 3, emission from neutral metal atoms; 4, N^{+}; 5, helium line emission.

The equations for the populations of the metal ion levels, which are populated by charge transfer, have the form

$$\frac{\mathrm{d}n_i}{\mathrm{d}t} = W_{\mathrm{c}}^{(i)} - \sum_{k<i} n_i A_{ik} + \sum_{n>i} n_n A_{ni} - \sum_{k\neq i} k_{ik} n_{\mathrm{e}} n_i + \sum_{k\neq i} k_{ki} n_{\mathrm{e}} n_k = 0, \tag{3.14}$$

where n_i is the population of level i; A is the Einstein coefficient; $W_{\mathrm{c}}^{(i)}$ is the partial rate for pumping level i via charge transfer; $k_{ik} = \langle q_{ik} v \rangle$ is the rate constant for electron excitation or de-excitation. The radiative life-times of metal ion levels are normally much less than the lifetime of He^{+}. For that reason, n_i follows any changes in pumping by charge transfer, i.e. there is a quasi-stationary state, and the right-hand side of (3.14) can be set to zero.

If in (3.14) transitions induced by electron collision can be neglected, then

$$n_i = \left(W_c^{(i)} + \sum_{n>i} n_n A_{ni} \right) \Big/ \sum_{k<i} A_{ik} \,. \tag{3.15}$$

Hence, it can be seen that the intensities of lines excited by charge transfer have the dependence

$$I^{(i)} \sim K_c^{(i)} n N^+ \,. \tag{3.16}$$

With the exception of N^+, none of the parameters in (3.16) is time dependent. (It is assumed that the degree of ionization of metal atoms is low.) Thus the temporal dependences of all ion lines should follow the exponential decrease in He^+ density (3.7), which has been established experimentally for low discharge currents, i.e. for low electron density (see Figure 3.9b,e). By measuring the dependence of the time constant of the decay of the ion afterglow, τ, on the metal atom density, we can determine the rate and the cross-section of the charge-transfer process. It can be seen from (3.8) that K_c is equal to the tangent of the slope of $1/\tau = f(n)$. This method was applied by Dyson [26] to measuring the cross-section for charge transfer from helium ions to mercury atoms. Since that time, this technique has been used for measuring the cross sections of charge transfer and other inelastic collision processes involving heavy bodies in mixtures of various metal vapours with noble gases [28].[1]

The cross sections for charge transfer can be measured from the helium line intensities also. From (3.11) for the time of decay of the helium line intensities, we have $1/\tau_B = 3\nu^+ + K_c n$, i.e. the slope of the dependence is the same as for the ion lines.

Measurements of the cross sections for charge transfer from the durations of the lines of Zn II, Cd II, Hg II and of the lines of He yield similar values, which is further evidence that the ion levels of the metal are populated via charge transfer from helium ions.

3.3.4 The Role of Electron De-excitation in Single-Electron Ion Spectra

With an increase in discharge current, n_e increases also, and in (3.15) inelastic collisions with electrons must be considered along with radiative transitions. Using a technique for finding the collisional probabilities [32, 40], we can calculate the populations of metal ion levels due to charge transfer as functions of n_e and T_e, taking account of electron de-excitation [32, 41]. Using these calculations, the ratio of various level populations, n_i, to the sum of the pumping rate, W_c, have been evaluated as functions of electron density, and are shown in Figures 3.15 and 3.16. It has been assumed that in Zn, the three terms closest to the He^+ term—6 *P*, 5 *D* and 4 *F*—are populated via charge

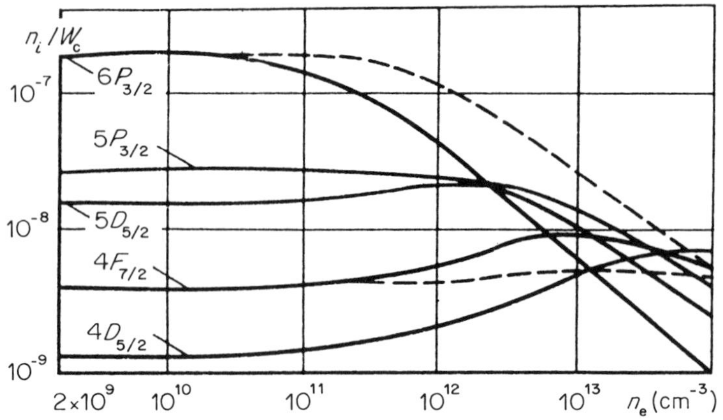

Figure 3.15 Dependences of the populations of Zn II levels n_i (normalized to the pump rate W_c) on electron density n_e: solid lines, T_e = 0.124 eV; broken lines, T_e = 1.24 eV.

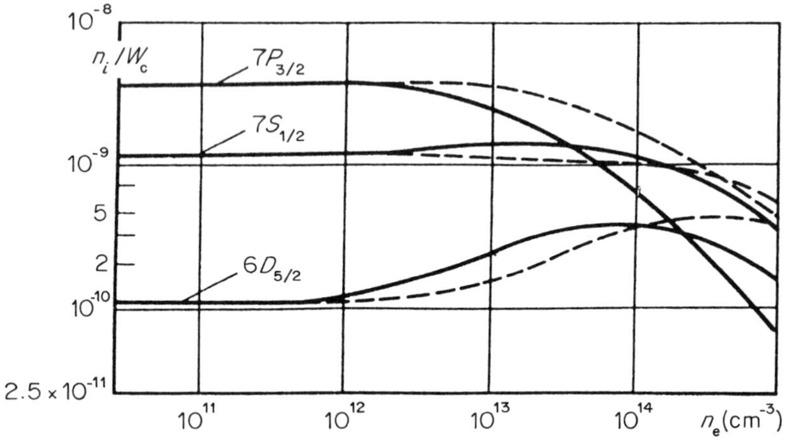

Figure 3.16 Dependences of the populations of Hg II levels n_i (normalized to the pump rate W_c) on electron density n_e: solid lines, T_e = 0.124 eV; broken lines, T_e = 1.24 eV.

transfer in the ratio 3:5:1 [37], and in Hg, the pumping is distributed between the levels 7 $^2P_{3/2}$, 7 $^2P_{1/2}$, 6 $^2D_{5/2}$ and 6 $^2D_{3/2}$ in the ratio 78:12:5.8:4.2 [30, 42].

For $n_e \le 10^{11}$ cm^{-3}, inelastic collisions with electrons play only a small part, and n_i/W_c is independent of n_e (Figures 3.15 and 3.16). In this situation, an inversion is possible on some transitions due to favourable differences in radiative lifetimes, e.g. 492.4 nm (4 $F_{7/2} \longrightarrow$ 4 $D_{5/2}$ Zn II), 2243.5 nm (5 $P_{3/2} \longrightarrow$ 4 $D_{5/2}$ Zn II) and 615 nm (7 $^2P_{3/2} \longrightarrow$ 7 $^2S_{1/2}$ Hg II).

With an increase in current, W_c increases in proportion to N^+ or n_e, and hence the ion level populations will rise linearly with increasing electron density to $n_e \approx 10^{11}$ cm^{-3}. With a further increase in n_e, the populations of 6 $^2P_{3/2}$ Zn II and 7 $^2P_{1/2,3/2}$ Hg II saturate due to electron quenching of the levels, which is seen as a sharp fall in n_i/W_c with an increase in n_e.

In contrast, the populations of the levels 5 ^2D and 4 ^2F Zn II and 6 $^2D_{3/2,5/2}$ Hg II increase faster than linearly with n_e due to electron collisional excitation from the levels 6 ^2P Zn II and 7 ^2P Hg II. The populations then saturate; in the case of Zn II, 5 $^2D_{5/2}$ saturates before 4 $^2F_{7/2}$. In Hg II, electron de-excitation appears at a value of n_e an order of magnitude higher than in Zn II, because the Hg II levels lie lower in energy, and their radiative probabilities are higher.

Around $n_e = 10^{12}$ cm^{-3}, the population of Zn II (5 $^2P_{3/2}$) saturates, which facilitates creation of an inversion on the transition 610.2 nm (5 $^2D_{5/2} \longrightarrow$ 5 $^2P_{3/2}$) starting from $n_e = 2 \times 10^{12}$ cm^{-3}. The Zn II (5 $^2P_{3/2}$) ions are made to transfer their energy by electron impact, mainly to the 4 $^2D_{5/2}$ level. For that reason, starting from $n_e = 10^{11}$ cm^{-3}, the population of the upper level increases much faster than can be attributed to charge transfer alone, and the inversion on the 492.4 nm transition, 4 $^2F_{7/2} \longrightarrow$ 4 $^2D_{5/2}$, terminates for $n_e > 6 \times 10^{13}$ cm^{-3}. An analogous effect is observed on the 615 nm transition (7 $^2P_{3/2} \longrightarrow$ 7 $^2S_{1/2}$) of Hg II. This leads to a shift of the laser pulse in the afterglow away from the current pulse, where n_e is lower.

We shall denote by n_e^* and t^* the electron density and the corresponding time at which the inversion ceases. If the electron density decays exponentially in the afterglow, with a characteristic time $t^* = 1\mathrm{n}(n_{e0} / n_e^*) / \nu^*$, then the higher the initial electron density, the further into the afterglow the laser pulse moves. If the He pressure is raised with n_{e0} held constant, ν^+ falls, and the onset of the laser pulse is further delayed.

For Cd levels, electron de-excitation plays a greater role in shaping the afterglow characteristics, since charge transfer populates higher lying Cd ion levels with low radiative probabilities. We consider the 636 and 537.8 nm transitions. Both lines peak in the afterglow, but with an increase in current, the intensity of the 636 nm line saturates before that of the 537.8 nm line. This results in laser oscillation beginning at 636 nm at lower currents, and with increasing current the laser pulse moves further into the afterglow, but reduces in amplitude. The same happens to the 537.8 nm laser pulse, but at higher currents.

In [43] a different explanation was proposed for the shift of the laser pulses further into the afterglow in the case of the 537.8 and 533.7 nm lines of Cd II. It was suggested that pumping by Penning ionization due to He metastables causes overpopulation of the lower laser level. This is unlikely, however, because delaying of the laser pulse is observed on many other lines whose lower levels cannot be populated via the Penning process from energy considerations.

Electron de-excitation can be used to explain the different durations of the afterglow of various ion lines (see Table 3.2) that are pumped by charge transfer [32]. For this reason, when using the technique of [26] for measuring the cross sections for charge transfer, it is necessary to work with low electron densities, because the differences in τ for the various lines are then at a minimum.

3.3.5 The Influence of Electron Temperature on Laser Ion Spectra

Changes in T_e have significant effects on the populations of levels pumped by charge transfer. They explain alterations in the form of the afterglow when electrons cool after the current pulse.

As can be seen in Figures 3.15 and 3.16, when T_e falls the populations of Zn II (6 $^2P_{3/2}$, 5 $^2D_{5/2}$) should decrease while that of Zn II (4 $^2F_{7/2}$) should increase. Hence, when T_e falls after termination of the current pulse, the intensity of the 492.4 nm line increases and peaks (Figure 3.9f), and the intensities of the lines at 317.2 and 610.2 nm fall away (Figure 3.9g). Similarly, the intensities of the 615 and 794.4 nm lines of Hg II decrease, while that of the 222.4 nm transition increases.

The above remarks can be confirmed in experiments where electrons are reheated in the afterglow by pulses of low current (≤ 1 A) and short duration (≈ 1 μs). According to (1.24), when such pulses are imposed, the intensities of the recombination afterglows of He I, Cd I, Zn I and Hg I lines reduce sharply (Figure 3.17b). At the same time, the second pulse does not change the form of the afterglow intensity of metal ions that are populated by charge transfer from He^+ ions when the current in the main pulse is low; the typical exponential decay of the metal ion afterglow remains the same (Figure 3.17c).

When the current in the main pulse is increased, electron de-excitation and deformations of the afterglow intensity of the metal ion lines appear. In this case, the imposition of a second pulse does affect the forms of the ion afterglow lines pumped by charge transfer. Lines such as 537.8 and 533.7 nm of Cd II, 491.1 and 492.4 nm of Zn II and 222.4 nm of Hg II, which originate from levels with large L and rise in intensity in the early afterglow, decrease in intensity during the second current pulse (Figure 3.17e). However, lines that exhibit marked decreases in intensity in the early afterglow (317.2 and 610.2 nm of Zn II, 413.4 nm of Cd II, and 615 nm of Hg II) intensify their emission during the second pulse (Figure 3.17d).

Note that increasing the duration of the second pulse to 5–7 μs for $p_{He} \leq 0.1$–0.3 kPa results in the disappearance of the recombination afterglow of neutral atomic lines, as well as ion lines produced by charge transfer (Figures 3.17b–e). This is attributed to ambipolar diffusion of charged

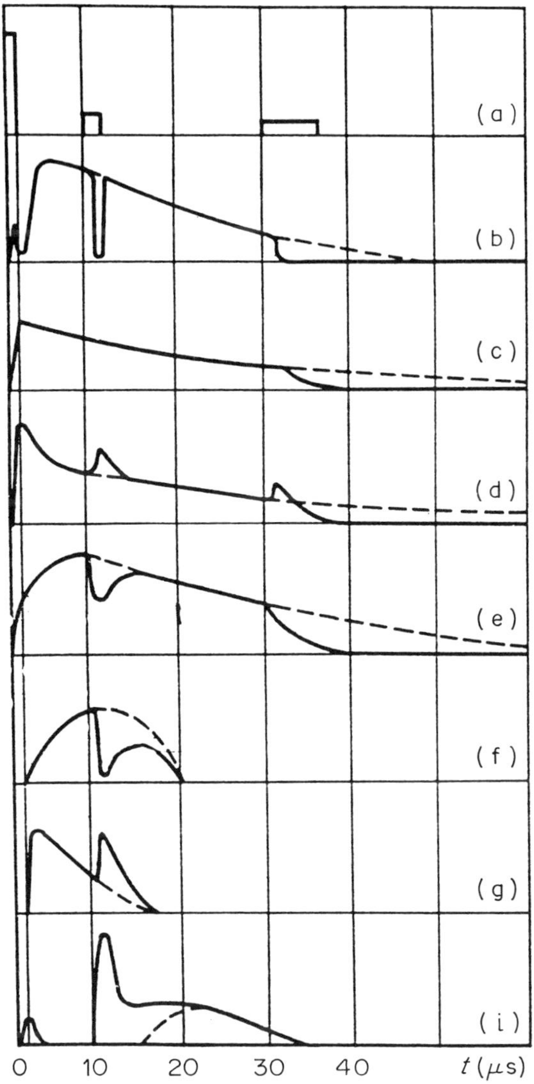

Figure 3.17 The influence of additional current pulses on the afterglow and laser oscillation; p_{He} = 0.3 kPa, p_M = 0.1–0.3 Pa, d = 0.8 cm. (a) Current pulses. (b) He I lines. (c) Metal ion lines, main current pulse 1 A. (d),(e) Metal ion lines, main current pulse 3–5 A. (f) Laser oscillation at 492.4 nm Zn II and 537.8 nm Cd II, current 4 A. (g) Laser oscillation at 615 nm Hg II, current 2 A. (h) Laser oscillation at 615nm Hg II, current 7 A. The waveform prior to the additional pulses is shown as a broken line.

particles to the wall, which is enhanced, in accordance with (2.36), and is also consistent with the population of the ion levels by charge transfer.

From calculations (Figures 3.15 and 3.16) it follows that an increase in T_e should reduce the inversion and laser power on the 492.4 nm Zn II line also; in contrast, the 615 nm Hg II line should increase in power and lase over a greater range of n_e, which has been shown experimentally (Figures 3.17f–h). This is responsible for the splitting of the Hg II laser pulse at 615 nm (Figures 3.12 and 3.17h). Thus, experiments with electron reheating demonstrate the importance of electron de-excitation in the production of inversion on ion transitions of Cd, Zn and Hg.

3.3.6 Kinetics of Laser Lines Originating from the $d^9s^2\,{}^2D^*_{3/2,5/2}$ Beutler Levels of Cd and Zn

The lines 441.6 and 325 nm of Cd II and 589.4 and 747.8 nm of Zn II originate from the Beutler levels, $d^9s^2\ {}^2D^*_{3/2,5/2}$. Since these lines have long afterglows, they should be excited by long-lived particles (Figure 3.13). From an energy point of view, the long-lived particles could be either helium ions or metastables. The pumping process has been attributed to Penning collisions [44], and charge transfer from atomic [45] and molecular [46] ions. The majority of experimental evidence supports the first process. Indeed, when small amounts of Ar (around 10 Pa) are admitted, and He metastables are quenched as a result, the afterglows of the relevant lines practically disappear. The addition of Ne permits us to determine which of the two metastable states contributes most to the pumping. It is known that some Ne levels are selectively populated by either 2 ${}^1S_0$ or 2 ${}^3S_1$ He metastables. By comparison of the shapes of the afterglow intensities of the various Ne lines (e.g. 544.8 nm as an indicator of He(2 ${}^1S_0$) and 966.5 nm for He(2 ${}^3S_1$)) with those of the metal ion lines, the dominant He metastable state can be ascertained.

It was found that the 544.8 nm afterglow is weak and short (1–2 μs), and is independent of the current, except for very low currents ($\approx$0.1 A), which is insufficient for the generation of laser oscillation on the metal ion lines. However, the 966.5 nm line has a long afterglow, similar in form to the afterglows of the 441.6 and 325 nm (Cd II) and 589.4 and 747.8 nm (Zn II) lines.

By the addition of various amounts of Ne to the discharge, we can alter the relative densities of the metastable He atoms. The rate constant for the quenching of He(2 ${}^1S_0$) by Ne is an order of magnitude higher than for the quenching of He(2 ${}^3S_1$) (see Appendix 3). For $p_{Ne} = 4$ Pa, the density of He(2 ${}^1S_0$) is reduced many times, but that of He(2 ${}^3S_1$) hardly changes. The duration and intensity of the 544.8 nm Ne I afterglow reduce, but the afterglows of the 966.5 nm Ne I line and the relevant metal ion lines do not alter, which implies excitation by only the He(2 $^3S_1$) metastables. Note that a Ne pressure of 1 Pa causes the He_2^* emission to vanish, since Ne effectively depletes He_2^+ (which by recombination leads to He_2^*), while the intensities of the metal ion lines do not change appreciably. Thus, charge transfer due to He_2^+ can be ruled out as the source of pumping of the relevant metal ion lines.

Although Zn II and Cd II ions can be excited to Beutler's levels via charge transfer with subsequent cascade transitions down to D^* (proved experimentally by e-beam pumping [45] and in a flowing gas afterglow [37]), the contribution of this mechanism does not exceed 10% due to its relatively low cross-section (10^{-17}–10^{-16} cm^2 [45]) and large energy defect ($\approx$1 eV) [37].

The importance of Penning collisions in the population of the D^* levels was demonstrated in a transverse discharge [47], formed between a heated oxide cathode and a Mo anode, which were 100 mm long and separated by 4 mm. There is a group of mono-energetic electrons with the energy of the anode voltage. Unlike in the experiments of [33, 34], the He pressure can be high (0.5 kPa). The intensity and the excitation function of the Cd II 441.6 nm line were measured. With the introduction of He to the Cd vapour, the line intensity rose by two orders of magnitude, and the shape of the excitation function altered to the form of the excitation function of He metastable atoms. Using the intensity of the 441.6 nm line, the cross-section for Penning excitation of Cd II($^2D^*_{5/2}$) was found to be $3\pm2 \times 10^{-16}$ cm^2.

We shall now consider the nature of the marked variations in the afterglow intensities for pulsed currents >0.5 A. It is necessary to take into account electron de-excitation not only of the laser levels, but also of the metastable He 2 $^3S_1$ atoms. The number density of the latter in the afterglow is

$$\dot{N}_m = -k_m n_e N_m (\nu_m + K_P n) N_m + \alpha n_e^2 N^+ \,, \tag{3.17}$$

where $k_m n_e$, ν_m and $K_P n$ are the frequencies of electron de-excitation, diffusion and Penning collisions respectively for the metastables; α is the recombination coefficient for helium ions. Since the resonance radiation of He atoms in the discharge is almost completely absorbed, practically the entire recombination flux in He passes through the metastable levels.

Substituting $n_e(t)$ and (3.7) into (3.17), we obtain

$$\dot{N}_m = \alpha n_{e0}^2 N_0^+ \exp[-(3\nu^+ + K_c n)t] - k_m n_{e0} N_m \exp(-\nu^+ t) - (\nu_m + K_P n) N_m \,. \tag{3.18}$$

This equation can be simplified at high metal vapour pressures, since $\nu_m < \nu^+ \ll K_c n \approx K_P n$. Then, N_m decays much faster in the afterglow than n_e, and for $t \ll 1/\nu^+$, we can assume that $\exp(-\nu^+ t) \approx 1$. The resulting equation has the form

$$\dot{N}_m = \alpha n_{e0}^2 N_0^+ \exp(-K_c n t) - (k_m n_{e0} + K_P n N_m) . \tag{3.19}$$

If k_m is time-independent, then the solution of (3.19) is

$$N_m = (N_{m0} - C)\exp[-(k_m n_e + K_P n)t] + C \exp(-K_c n t) , \tag{3.20}$$

where

$$C = \alpha n_{e0}^2 N_0^+ / [k_m n_{e0} + (K_P - K_c)n] . \tag{3.21}$$

If $n_e \longrightarrow 0$, then equation (3.21) takes an exponential form:

$$N_m = N_{m0} \exp(-K_P n t) . \tag{3.22}$$

For larger n_e, the solution is a sum of two exponentials, the first with a time constant $\tau_1 = [k_m n_{e0} + (K_P - K_c)n]^{-1}$, and the second with $\tau_2 = (K_c n)^{-1}$. The first applies to the initial decay of N_m, whose duration reduces with an increase in n_e, and the second corresponds to the long decay due to the depletion of helium ions by charge transfer (Figure 3.18). For the population of the D^* levels, we can write

$$N_2(D^*) = \xi_2 W_P / (A_{21} + k_{21} n_e) . \tag{3.23}$$

In the numerator of (3.23), the pump rate for a given level is expressed as the product of the total rate for the Penning process, W_P, and the partial coefficient, ξ_2. In the denominator are the probabilities of radiative and electron collisional transitions from the upper laser level to the lower laser level. When n_e is low, the term $k_{21} n_e$ can be neglected, i.e. the population of

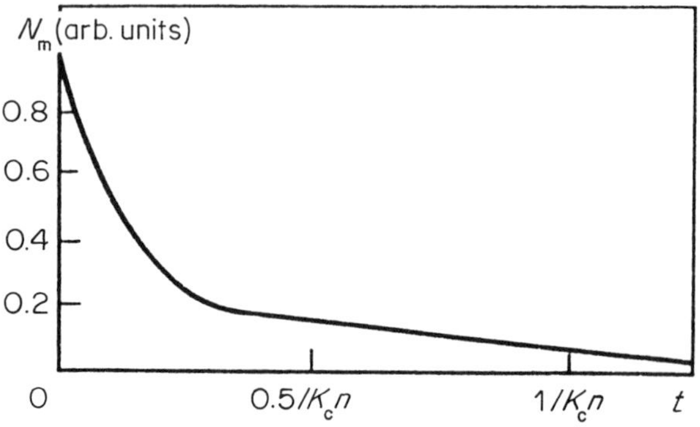

Figure 3.18 Helium metastable density in the afterglow, with $C = 0.25 N_{m0}$; $k_m n_e = 5 K_P n$.

the D^* level should follow the temporal course of the He metastable density. At high n_e, electron de-excitation of the D^* level results in some shortening of the initial decay, and in an additional saturation and extension of the long afterglow. All the processes described by (3.20)–(3.23) agree qualitatively with the behaviours of the intensities of lines that originate from the D^* levels of Cd II and Zn II. During the current pulse, a high density of He metastables is produced via electron impact. After the pulse, the metastable density falls at low current due to diffusion and Penning collisions with metal atoms, and at high currents mainly due to electron de-excitation (fast decay). The density N_m falls to an equilibrium value which is determined by the rate of production of metastables via helium ion recombination and the decay of metastables due to electron de-excitation (long decay). The form of the afterglow of lines originating from the metal ion D^* levels is similar to that of the He metastable atoms because the D^* levels are populated by Penning collisions.

Inversion on the $D \longrightarrow P$ transitions of Zn II and Cd II is the result of favourable ratios of upper and lower level radiative probabilities, and of the selectivity of Penning collisions for the level configurations (see section 1.3). The lower laser level decays radiatively to the metal ion ground state (see Figures 3.6 and 3.7). Hence a key factor in the termination of the population inversion with increasing current, besides electron de-excitation, is the trapping of radiation from the lower level. Calculations show (Figure 3.19) that n_1/W_P for the lower level increases due to self-absorption, beginning at around $n_e = 10^{11}$ cm^{-3}, while n_2/W_P for the upper level, however, decreases due to electron de-excitation for $n_e \geq 5 \times 10^{12}$ cm^{-3}. Both factors cause the inversion to be destroyed at relatively low values of $n_e \approx 10^{13}$ cm^{-3}.

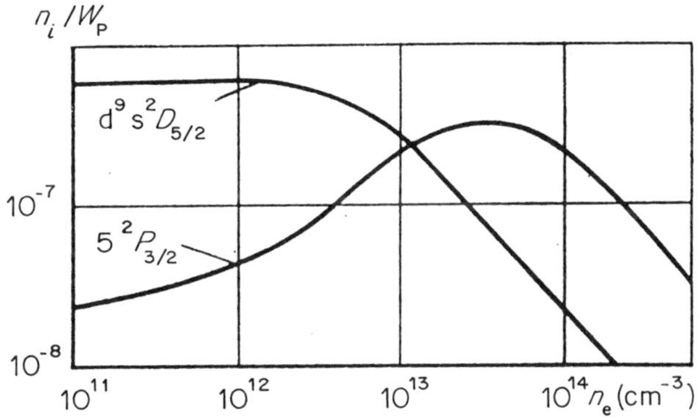

Figure 3.19 Dependences on electron density n_e of the populations of the upper and lower levels n_i of the 441.6 nm Cd II transition (normalized to the pump rate W_c); T_e = 0.124 eV.

3.3.7 Other MVILs Pumped by Charge Transfer and Penning Ionization

Pulsed metal-vapour ion lasers, other than Zn II, Cd II and Hg II, that have good characteristics include those utilizing transitions of I, Se, As, Tl, Cu, Ag, Au, Ca and Sr ions. The ranges of wavelengths available from pulsed MVILs are listed in Table 3.3. Charge transfer was proposed as an excitation mechanism for the first time in the case of the I ion laser in 1964 [48]. Lasers based on Cu, Ag and Au are attractive because of their short UV lines [49], produced mainly in the hollow-cathode discharge (see

Table 3.3 Ion laser lines excited by heavy-body collisions.

Element	Z	Number of lines	λ, $\Delta\lambda$ (nm)
		Charge transfer with helium ions	
Be	4	1	467.5
Cu	29	38	451–1970
Zn	20	12	491–5084.8
Se	34	49	447–1259
As	33	8	538–710
Ag	47	29	224–2080
Cd	48	22	537–3288.2
Te	52	16	484–873
I	53	30	448–888
Au	79	24	226–887
Hg	80	3	615.0; 794.4; 1555.5
Tl	81	8	473.7–117
		Charge transfer with neon ions	
Be	4	1	1209.6
Mg	12	2	922; 924
Al	13	5	358–747
Cu	29	14	248–2000
Ga	31	2	633.4; 719.8
Ag	47	4	318–503
Sn	50	2	1062; 1074
Te	52	19	545–938
Tl	81	5	582–935
Pb	82	3	537; 1162.8; 1322
		Charge transfer with krypton ions	
Ca	20	2	1183; 1195
Sr	38	4	1087–1245
		Penning ionization with helium metastable atoms	
Zn	30	3	589.4; 621.4; 747.8
Cd	48	2	325; 441.6

Chapter 6). Charge transfer from He^+, Ne^+ and Kr^+ ions can be employed for pumping [50–52].

Using different noble gases, it is possible to obtain different sets of laser lines from only one metal.

Electron de-excitation, which has been considered above for Zn, Cd and Hg, is also of great importance to other elements pumped by charge transfer. Penning ionization has been proposed as the pumping mechanism not only for the $D^* \longrightarrow P$ transitions of Cd and Zn, but also for some transitions in Mg II [53], Cu II, Sn II and Pb II [54, 55]. However, the arguments made in those papers for Penning ionization are not very strong (particularly the requirement for resonance between He metastables and excited metal ion levels, which, as is now known, is not needed for Penning collisions), and the investigations have not been as extensive as those into Cd and Zn. Thus it is quite possible that pumping is produced not by Penning collisions, but by electron impact or by cascade transitions from levels excited by charge transfer. Incidentally, the latter is now accepted as the principal pumping mechanism for transitions in Cu II [50]. Recent studies of pulsed laser oscillation on Pb, Sb and Mg ion transitions in the presence of He showed that the major excitation process for the ion levels is the recombination of doubly charged metal ions, produced by electron impact during the current pulse and in the afterglow by double charge transfer from He ions (see Chapter 4).

Thus, the effectiveness of the excitation of the D^* upper laser levels by Penning ionization in Cd II and Zn II is rather specific to those elements, and is due to the fact that those levels are the ground states of Beutler's term systems in those two metals.

REFERENCES

[1] A. A. Isaev and G. Yu. Lemmerman, *Tr. Fiz. Inst., Akad. Nauk SSSR,* **181,** 164–79 (1987) (in Russian).

[2] P. A. Bokhan, V. N. Nikolaev and V. I. Solomonov, *Sov. J. Quantum Electron.,* **5**(1), 96–8 (1975).

[3] L. A. Gross and M. C. Gokay, *J. Appl. Phys.,* **50**(2), 624–7 (1979).

[4] V. E. Prokop'ev and V. I. Solomonov, *Sov. J. Quantum Electron.,* **15**(6), 832–7 (1985).

[5] E. K. Karabut, V. F. Kravchenko and V. S. Mikhalevskii, *Opt. Spectrosc.,* **48**(4), 386–8 (1980).

[6] E. K. Karabut, V. F. Kravchenko and V. S. Michalevskii, *J. Appl. Spectrosc.,* **37**(4), 1123–7 (1982).

[7] L. H. Szeto, W. T. Silfvast and O. R. Wood II. *IEEE, J. Quantum Electron.,* **15**(12), 1332–4 (1979).

[8] M. C. Gokay, M. Soltanolkotabi and L. A. Cross, *IEEE J. Quantum Electron.,* **14**(12), 1004–7 (1978).

[9] T. W. Karras, *Proc. Int. Conf. on Lasers '81,* McLean, VA, USA: STS. Press, pp. 871–6 (1982).
[10] A. V. Platonov, A. N. Soldatov and A. G. Filonov, *Sov. J. Quantum Electron.,* **8**(1), 120–1 (1978).
[11] P. A. Bokhan, V. M. Klimkin and V. E. Prokop'ev, *JETP Lett.,* **18**(2), 44–5 (1973).
[12] P. A. Bokhan, V. M. Klimkin and V. E. Prokop'ev, *Sov. J. Quantum Electron.,* **4**(6), 752–4 (1974).
[13] V. M. Klimkin, V. E. Prokop'ev and S. S. Monasterev, *Efficient Gas Discharge Lasers,* Tomsk: Russian Academy of Sciences, pp. 6–12 (1978) (in Russian).
[14] V. M. Klimkin and V. E. Prokopev, *Opt. Spectrosc.,* **49**(6), 589–92 (1980).
[15] P. A. Bokhan and D. E. Zakrevsky, *Opt. Quantum Electron.,* **23**(4), S513–22 (1991).
[16] P. A. Bokhan and D. E. Zakrevsky, *Metal Vapor Lasers and Their Applications: CIS Selected Papers,* ed. G. G. Petrash, *Proc. SPIE,* **2110**, 220–34 (1993).
[17] P. A. Bokhan and D. E. Zakrevsky, *Pis'ma Zh. Tekh. Fiz.,* **17**(23), 89–91 (1991) (in Russian).
[18] *Laser handbook*, ed. A. M. Prokhorova Moscow: Sovyetskaya Radio, (1978) (in Russian).
[19] V. V. Zhukov, I. G. Ivanov and M. F. Sem, *Zh. Prikl. Spektrosk.,* **26**(3), 544–7 (1977) (in Russian) (translation in *J. Appl. Spectrosc.*).
[20] W. E. Bell, *Appl. Phys. Lett.,* **4**(2), 34–5 (1964).
[21] A. L. Bloom, W. E. Bell and F. O. Lopez, *Phys. Rev.,* **135**(3A), 578–9 (1964).
[22] G. R. Fowles and W. T. Silfvast, *IEEE J. Quantum Electron.,* **1**(3), 131 (1965).
[23] V. F. Papakin and M. F. Sém, *Sov. Phys. J.,* **13**(2). 230–1 (1972).
[24] V. G. Il'yushko, V. F. Papakin and M. F. Sém, *Sov. Phys. J.,* **16**(3). 416–17 (1974).
[25] R. L. Byer, W. E. Bell, E. Hodges and A. L. Bloom, *J. Opt. Soc. Am.,* **55**(12), 1598–602 (1965).
[26] D. J. Dyson, *Nature,* **207**, 361–3 (1965).
[27] N. Suzuki, *Jpn. J. Appl. Phys.,* **4**(6), 452–7 (1965).
[28] G. J. Collins, *J. Appl. Phys.,* **44**(10), 4633–52 (1973).
[29] H. Ninomiya, H. Osumi and S. Horiguchi, *J. Appl. Phys.,* **51**(12), 6091–4 (1980); **52**(4), 2699-704 (1981); **52**(6), 3889-91 (1981); **52**(5), 3229–32 (1981).
[30] V. I. Bylkin, *Opt. Spectrosc.,* **29,** 552–5 (1970).
[31] Yu. A. Tolmachev, *Population of excited ion levels via Penning ionization,* in *Plasma Chemistry,* **9**, pp. 80–100, Moscow: Energoatomizdat (1982) (in Russian).
[32] E. L. Latush, V. C. Mikhalevskii and M. F. Sem, *Opt. Spectrosc.,* **34**(2), 120–3 (1973).
[33] V. S. Aleinikov, *Opt. Spectrosc.* **28,** 15–17 (1970).
[34] I. P. Bogdanova, V. D. Marusin and V. E. Yakhontova, *Opt. Spectrosc.,* **37**(4), 365–8 (1974).
[35] T. Shay, H. Kano, S. Hattori and G. J. Collins, *J. Appl. Phys.,* **48**(11), 4449–53 (1977).
[36] J. M. Green and C. E. Webb, *J. Phys. B.,* **8**(9), 1484–500 (1975).
[37] A. R. Turner-Smith, J. M. Green and C. E. Webb, *J. Phys. B.,* **6**(1), 114–30 (1973).
[38] D. A. Korogodin, E. L. Latush, M. F. Sém and G. D. Chebotarev, *Sov. J. Quantum Electron.,* **20**(6), 634–8 (1990).
[39] L. M. Biberman, V. S. Vorobev and I. T. Yakubov, *Nonequilibrium Kinetics of Low Temperature Plasmas,* Moscow: Nauka, (1982) (in Russian).

[40] L. A. Vainshteyn, I. I. Sobelman, E. D. Yukov, *Excitation of Atoms and Spectral Line Broadening,* Moscow: Nauka, (1979) (in Russian); *Cross Sections for Electronic Excitation of Atoms and Ions,* Moscow: Nauka, (1973) (in Russian).
[41] S. P. Zinchenko, I. G. Ivanov, E. L. Latush and M. F. Sem, *Opt. Spectrosc.,* **58**(2), 180–3 (1985).
[42] E. L. Latush, M. F. Sem and G. D. Chebotarev, *Sov. Phys. J.,* **27**(5), 429–35 (1984).
[43] L. Csillag, M. Jánossy and T. Salamon, *Phys. Lett.,* **A31**(10), 532–3 (1970).
[44] W. T. Silfvast, *Phys. Rev. Lett.,* **27**(22), 1489–92 (1971).
[45] M-T. I. Soskida and V. S. Shevera, *JETP Lett.,* **22**(11), 269–70 (1975).
[46] R. C. Jensen, G. J. Collins and W. R. Bennett, Jr., *Phys. Rev. Lett.,* **23**(7), 363–7 (1969).
[47] M. K. Dyatlov, E. P. Ostapchenko and V. A. Stepanov, *Opt. Spectrosc.,* **29,** 539–40 (1970).
[48] R. C. Jensen and G. R. Fowles, *Proc. IEEE,* **52**(11), 1350 (1964).
[49] J. J. Rocca and G. J. Collins, *Optoelectronics, Instrumentation and Data Processing,* No. 1, 1–16 (1984).
[50] D. C. Gerstenberger, R. Solanki and G. J. Collins, *IEEE J. Quantum Electron.,* **16**(8), 820–34 (1980).
[51] I. G. Ivanov and M. F. Sem, *Elektron. Tekh., Ser. 4 Elektrovakuumiye i Gazorazryadniye Pribori,* No. 2, 12–16 (1974) (in Russian).
[52] V. V. Zhukov, E. L. Latush and M. F. Sem, *Zh. Prikl. Spektrosk.,* **32**(4), 738–40 (1980) (in Russian).
[53] D. T. Hodges, *Appl. Phys. Lett.,* **18**(10), 454–6 (1971).
[54] W. T. Silfvast, *Appl. Phys. Lett.,* **15**(1), 23–5 (1969).
[55] J. R. McNeil, G. J. Collins, K. B. Persson and D. L. Franzen, *Appl. Phys. Lett.,* **27**(11), 595–8 (1975).
[56] J. Berkowitz and W. A. Chupka, *J. Chem. Phys.,* **48**(12), 5743–4 (1968).

NOTES

1. Note that this technique cannot be directly applied if the metal evaporates in the form of molecules, and then partly dissociates into atoms. The metal atom density cannot then easily be determined. For example, in the case of Se vapours, there are molecules Se_2, Se_5, Se_6, Se_7 and Se_8 which average to $Se_{5.6}$ [56].

4

Metal Vapour Ion Lasers using Recombining Plasmas

4.1 LASER OSCILLATION IN MERCURY, CADMIUM AND ZINC IONS

Recombination of electrons and ions is one of the major pump mechanisms in MVILs. In Table 4.1 are listed the transitions of ion lasers that are pumped by recombination in pulsed PC discharge afterglows. Recombination laser transitions can also be obtained using segmented arc discharges and laser-induced plasmas (sections 4.6 and 4.7).

The first MVIL to be demonstrated, the mercury ion laser [1, 2], oscillates on recombination pumped transitions (as was later established), one of which is amongst the strongest of MVIL laser transitions, $5\ ^2F^o_{7/2} \longrightarrow 6\ ^2D_{5/2}$ at 567.7 nm. The level $5\ ^2F^o_{7/2}$ cannot be populated by charge transfer because it lies 1.1 eV above the potential of He^+. In the cases of the analogous transitions in cadmium and zinc ($4\ ^2F^o \longrightarrow 5\ ^2D$, 537.8 and 533.7 nm Cd II; $4\ ^2F^o \longrightarrow 4\ ^2D$, 492.4 and 491.2 nm Zn II), it was shown in Chapter 3 that in helium discharges they are populated by charge transfer. However, laser transitions on $F \longrightarrow D$ transitions of Zn and Cd, as well as in Hg, can also occur in a mixture of the metal vapour with neon or argon at low pressure, or even in the absence of a buffer gas [2–4]. Such regimes of operation have been studied in detail in [4], where laser oscillation was observed on lines of Hg II (567.7 nm) and Cd II (537.8, 533.7 and 441.6 nm).

The blue Cd line oscillates during the current pulse, which implies excitation via electron impact (section 3.2), while the green laser lines of Hg and Cd are generated in the afterglow (Figure 4.1). The optimum vapour pressures and discharge currents for Cd and Hg are 2.7 Pa, 30 A and 2 Pa, 90 A respectively. Laser oscillation on the Hg line is also observed in He and Ne up to 250 Pa, with similar temporal behaviour (Figure 4.1b). With Ne the gain is higher, which can be explained by the absence (in this case) of parasitic population of the $6\ ^2D_{5/2}$ lower laser level of the 567.7 nm line by charge transfer from He ions.

Table 4.1 Wavelengths of ion laser transitions excited by recombination in the positive columns of pulsed gas discharges.

Ion	λ (nm)	Ion	λ (nm)
Be II	527.2	Cd II	533.7
Mg II	924.4		537.8
	921.8	Sn II	558.9
Al II	622.8		579.9
	623.3		1062.0
	624.3		1074.0
	683.0	Ba II	677.1
	747.1		687.6
	835.4		1109.2
	836.1		1158.1
	836.5		1193.5
	1063.3		1247.8
	1063.6	Hg II[a]	567.7
	1063.9	Tl II	473.7
Ca II	370.6		498.1
	373.7		507.9
	994.0		515.2
Zn II	491.2	Pb II	537.2
	492.4		1162.8
Sr II	416.2		1322.1
	430.5		
	870.0		
	1087.0		
	1123.0		

[a] Hg II laser lines with upper level potentials above that of He^+ (λ = 734.66, 862.2, 855, 739.7, 1058.3, 1117.9 nm) are presumably pumped by recombination as well.

Thus, the pumping mechanism under these conditions begins after the current pulse, when T_e is low, and it is independent of the type of buffer gas used. It is clear that such a mechanism cannot be electron impact [2]. Nor can it be an energy transfer process that depends on the type of buffer gas, such as Penning ionization of Hg atoms by singlet helium atoms, as suggested in [5]. The pumping mechanism in this case is the recombination of doubly charged metal ions in the afterglow,

$$Hg^{2+} + 2e \longrightarrow Hg^{+*} + e\,, \tag{4.1}$$

first proposed by Aleinikov [6]. In order for recombination to be effective, high densities of doubly charged ions, n^{2+}, and electrons, n_e, and a low T_e are required (see section 1.4). Doubly charged ions of Hg, Cd and Zn are produced during the current pulse by electron impact, and a high T_e during that time will lead to effective ionization, which can occur when a low pressure of the buffer gas is used.

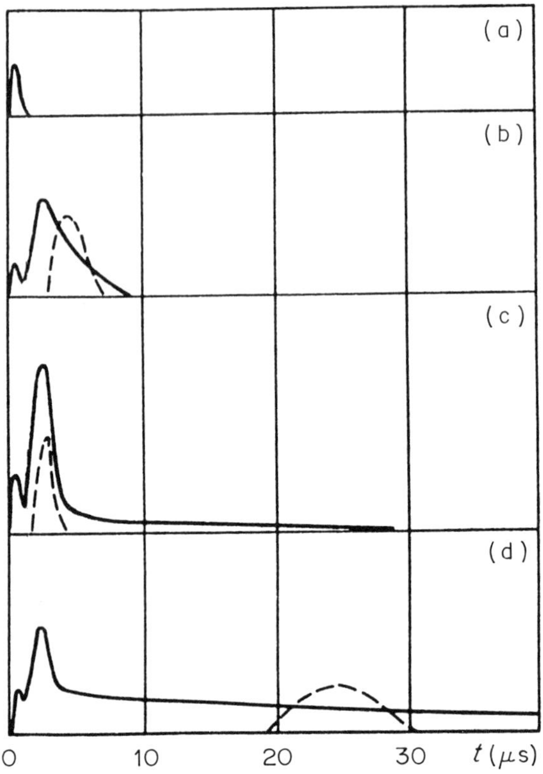

Figure 4.1 Oscillograms of the spontaneous (solid) and stimulated (broken) emission of the green lines of mercury and cadmium. (a) Current pulse. (b) Pure vapour of mercury or cadmium. (c) Hg II 567.7 nm line, He–Hg mixture, p_{He} = 0.2 kPa. (d) Cd II 537.8 nm line, He–Cd mixture, p_{He} = 0.5 kPa.

As the pressure of helium is slowly reduced from 7 kPa to 'zero', the pumping mechanism for the green Cd lines changes from charge transfer to recombination [4]. For $p_{He} \geq 70$ Pa, the form of the spontaneous afterglow emission of Cd II lines is composed of a short-duration recombination spike, superimposed on the longer charge-transfer afterglow (Figure 4.1d). Laser oscillation occurs in the longer afterglow. The maximum gain is 25% m^{-1} at a helium pressure of 0.5 kPa. For buffer-gas pressures below 30 Pa, laser oscillation is observed with He and Ne, and the temporal characteristics of the laser oscillation and spontaneous emission are as in the pure metal vapour (Figure 4.1b).

With the decrease of buffer-gas pressure, the recombination afterglow intensifies due to the more effective production of doubly charged metal ions during the pulse. The gain rises monotonically, reaching a maximum in the pure vapour (100% m^{-1}).

Careful examination of the 567.7 nm Hg II line in the afterglow of a discharge mixture with He or Ne (p = 0.5–1 kPa) shows that, besides the short recombination spike, an extended weak afterglow is present (Figure 4.1c). It is shown in [7] that the extended afterglow is due to the following processes:

$$Hg^{+} + He(m) \longrightarrow Hg^{2+} + He + e\,; \tag{4.2}$$

$$Hg^{+} + Ne(m) \longrightarrow Hg^{+*} + Ne\,; \tag{4.3}$$

i.e. Penning collisions and energy transfer to Hg ions. The contribution of these two processes to the pumping of the levels is not high (up to 5%) compared with recombination, because they are multiple-step processes, and presumably have low cross sections. Such processes are observed in other mixtures also [8].

Hence, laser oscillation on the 567.7 nm line as well as on other Hg II lines, whose upper levels lie above the ionization potential of He, is accounted for by recombination pumping [9], in the same manner as for laser oscillation on the $F \longrightarrow D$ transitions of Cd and Zn at low pressures of the buffer gas or in the pure metal vapour. In helium, at pressures above 100 Pa, laser oscillation on these Cd and Zn lines is due to charge transfer, where the discharge conditions make charge transfer more effective than recombination.

Clearance of the lower laser levels of $F \longrightarrow D$ transitions of Hg, Cd and Zn, in all pumping regimes, is realized by radiative decay. Inversion is favoured by the ratios of upper and lower level lifetimes, due to the fact that the difference in energy between the F and D levels is less than between the D and resonance P levels (see Figs 3.6–3.8). As a result, electron de-excitation, which begins to compete with radiative transitions as n_e is increased, lowers the upper level population earlier (i.e. at lower currents) than the lower laser level population. Hence, laser oscillation on $F \longrightarrow D$ transitions begins to reduce with an increase in current, even though the rate of recombination pumping of the upper level continues to rise.

4.2 LASER OSCILLATION ON ION TRANSITIONS OF Sr AND Ca

4.2.1 The Excitation Mechanism

Strontium and calcium ion lasers are the major examples of recombination pumped systems [10–13]. In Figures 4.2 and 4.3 partial energy-level diagrams are given for Sr II and Ca II, where the laser transitions are shown by the heavier lines. The transitions most suited to recombination pumping are

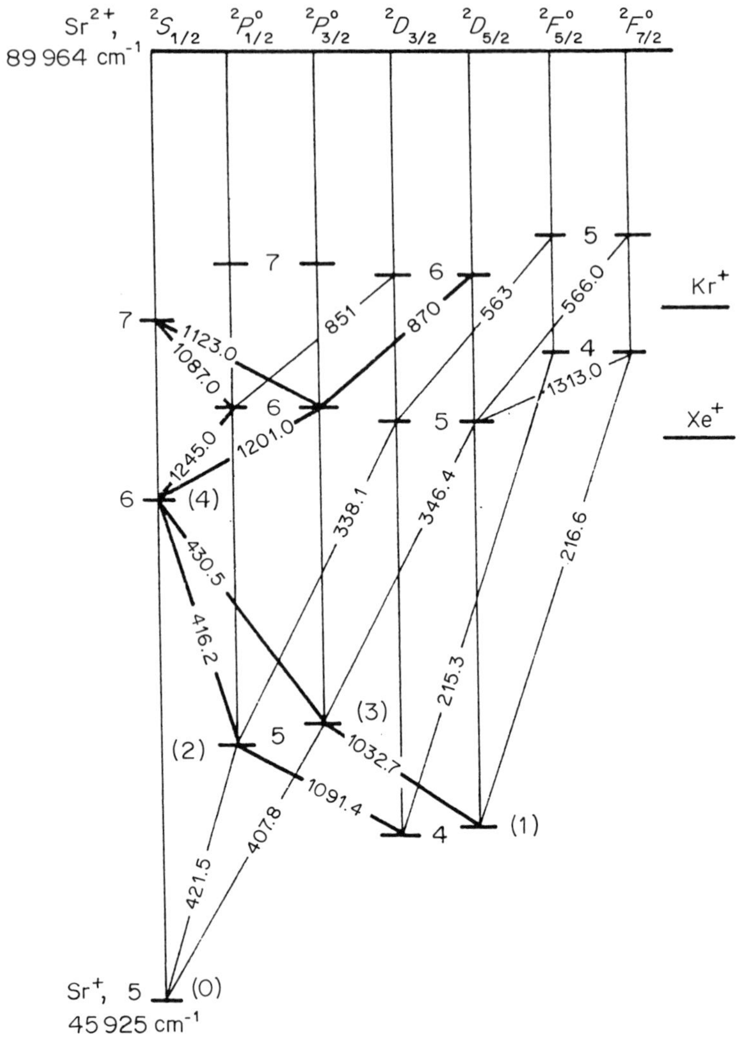

Figure 4.2 Partial energy-level diagram of Sr II.

430.5 and 416.2 nm ($6\,^2S \longrightarrow 5\,^2P$) of Sr II, and 373.7 and 370.6 nm ($5\,^2S \longrightarrow 4\,^2P$) of Ca II. They lie in the short-wavelength region of the spectrum, and exhibit the highest output powers. The similarity in the energy level configurations of Ca II and Sr II leads to similarity in their laser characteristics. Therefore, we will concentrate on Sr II transitions, and

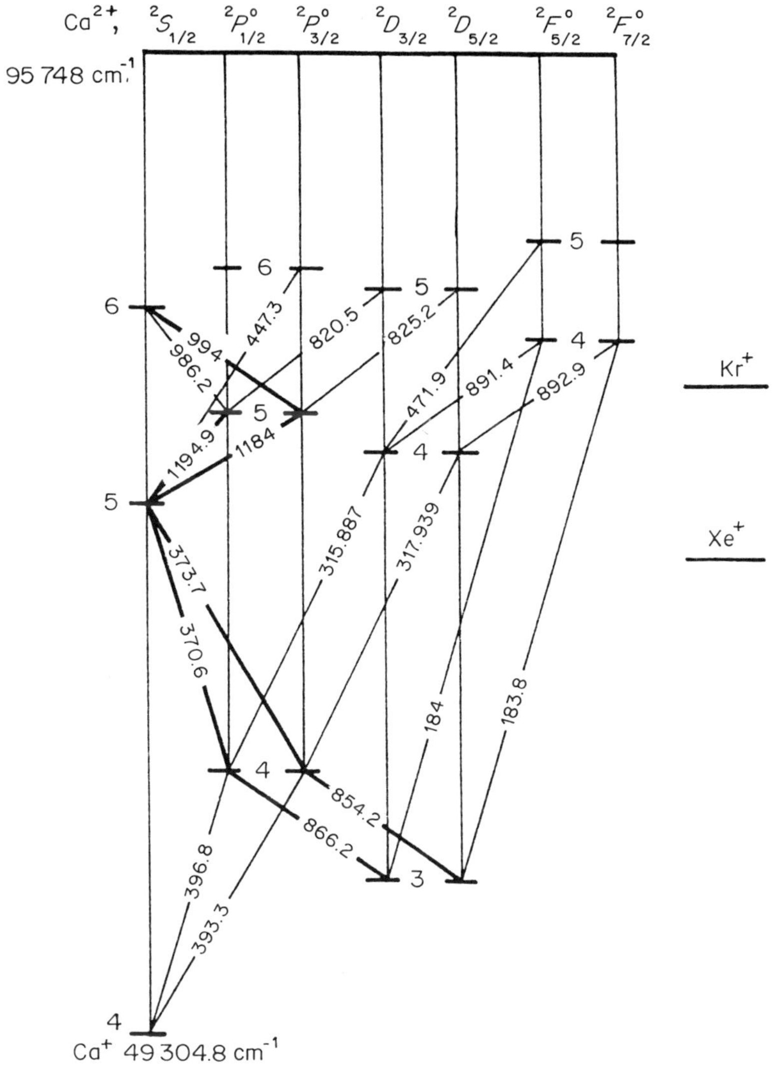

Figure 4.3 Partial energy-level diagram of Ca II.

mention Ca II only occasionally.

Laser oscillation on the transitions listed in Table 4.1 is achieved in mixtures of Sr or Ca vapour with He (mainly) with pulsed discharge excitation. In the first studies [10, 12], the inside diameter and length of the ceramic containment tube were 8 mm and 60 cm, respectively. Rectangular current pulses were produced using a hard-tube pulser circuit (see section 3.1). Their duration varied from 0.35 to 5 μs (usually around 1 μs), and the amplitudes were between 1 and 150 A. The range of metal vapour pressure over which lasing occurred was 0.1–10 Pa. The optimum buffer-gas pressure for the IR lines was about 1 kPa. For the violet Sr II lines and the UV Ca II lines, laser oscillation began at He pressures above 0.3 kPa, and the gain and laser power rose with He pressure. At p_{He} = 4 kPa, the gain coefficient was 28–30 dB.m^{-1} on the 430.5 nm Sr II line, and 10–12 dB m^{-1} on the 373.7 nm Ca II line. The gain on the IR lines was much less at 6–20% m^{-1}. Some of them lase with Ne as a buffer gas, as well as with He.

Laser oscillation on all lines is produced in the discharge afterglow. Typical oscillograms of the spontaneous emission of various lines, and the laser pulses are shown in Figure 4.4. An intensity spike in the early afterglow appears on the metal ion lines (Figure 4.4c) at pulsed currents above 3–5 A. With an increase in current, the amplitude of the spike rises much faster than the spontaneous intensity during the current pulse, and the spike's duration reduces. The duration of the spike also reduces with an increase in buffer-gas pressure. Under the same conditions, the duration of the spike is much less (2–3 times) in a helium afterglow than in a neon afterglow. Raising the metal vapour pressure above optimum causes the duration of the spike to increase many times.

Oscillation on the majority of lines in the regime of strongest laser output is observed during the short-duration afterglow spike (Figure 4.4d). With an increase in current above optimum, the IR laser transitions shift to the long afterglow region (Figure 4.4e). Laser oscillation on the violet and UV lines of Sr II and Ca II lengthens considerably (to 5 μs) if small amounts of hydrogen are added to the discharge (Figure 4.4f).

Below, we shall examine the mechanisms that give rise to the afterglow spectra of ions of alkaline earth metals. The initial spike in the afterglow is observed with all ion lines of second-group elements at high pulsed currents, independent of the type and presence of the buffer gas. This is the result of the recombination of doubly charged ions produced during the current pulse, i.e. as for Hg, Cd and Zn. The short duration of the initial spike compared with the duration of the afterglow of neutral atom lines is accounted for by the much higher recombination rate of multiply charged ions compared with

---

Figure 4.4 Oscillograms of the spontaneous and stimulated emission of strontium ion lines: (a) current pulse; (b) afterglow emission of He I and Sr I lines; (c) 430.5 nm afterglow emission in a He–Sr mixture (broken line is for p_{Sr} above optimum); (d) laser oscillation on 430.5 nm in a He–Sr mixture; (e) laser oscillation on infrared transitions in a He–Sr mixture (broken lines are for increasing current); (f) 430.5 nm afterglow emission in a Ne–Sr mixture; (g) 430.5 nm afterglow emission in a Ne–H_2–Sr mixture; (h) 430.5 nm laser oscillation in a Ne–H_2–Sr mixture.

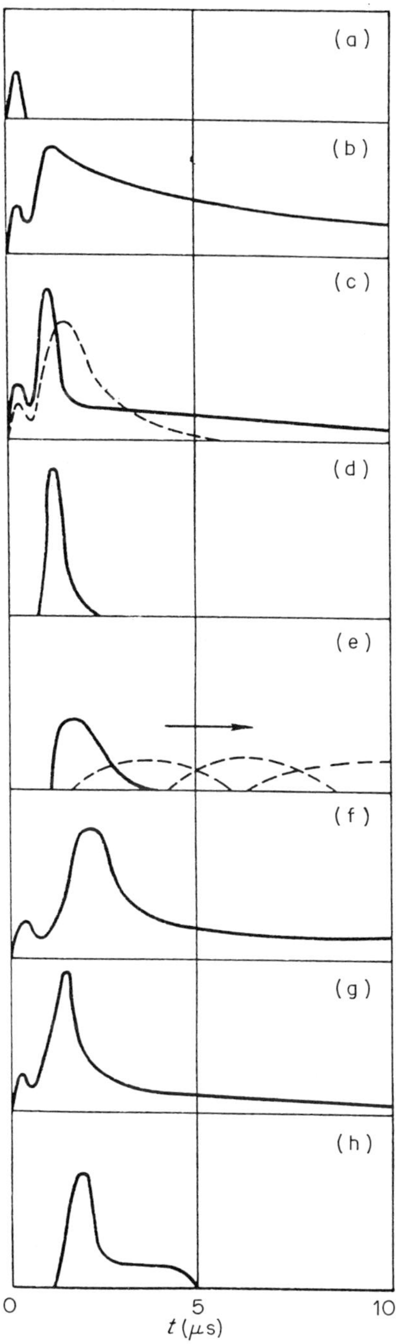
(a)
(b)
(c)
(d)
(e)
(f)
(g)
(h)
0
5
10
t (μs)

singly charged ions. Since the recombination coefficient $\alpha \sim z^3 \ln\sqrt{z^2+1}$, the rate of recombination of doubly charged ions is around 18 times higher than that for singly charged ions.

The potentials for single and double ionization of alkaline earth metal atoms are low, and their sum is less than the ionization potential of He (except for Be). Thus a great concentration of doubly charged ions can be produced during the current pulse even at significant buffer-gas pressures. The high buffer-gas pressures, especially of He, yield rapid cooling of electrons, a condition that is conducive to intense recombination pumping. In the case of the long afterglow that follows the short recombination spike (Figure 4.4c), long-lived excited atoms or ions are responsible. For ion lines of Hg, Zn and Cd, the long afterglow is due to charge transfer, and is observed only on transitions that originate from levels below He^+. The same explanation holds for ion lines of alkaline earth elements when heavy noble gases are used, which have ionization potentials below the double ionization potential of the metal atoms [14, 15]. The long afterglow of the lines of alkaline earth metal ions in a mixture with helium or neon is a result of the following processes [16, 17]:

$$\mathrm{Sr} + \mathrm{He}^+ \longrightarrow \mathrm{Sr}^{2+} + \mathrm{He} + \mathrm{e}\,, \tag{4.4}$$

followed by three-body collisional recombination:

$$\mathrm{Sr}^{2+} + 2\mathrm{e} \longrightarrow \mathrm{Sr}^{+*} + \mathrm{e}\,. \tag{4.5}$$

The recombination mechanism for pumping the ion lines of alkaline-earth metals is proved in experiments on plasma reheating by the application of a second low-amplitude current pulse (Figure 4.5). The afterglow practically disappears during the application of this second pulse—the spike and long afterglow due to metal ion lines, and the recombination afterglow of the helium lines. Corresponding holes appear in the laser output pulses also. Note that the application of such pulses in a charge-transfer-driven afterglow, as in mixtures of Hg, Cd and Zn with He, does not, in general, cause complete removal of the afterglow emission, and on some lines even intensifies the emission (see Figure 3.17); this is reasonable, since with conventional charge transfer excitation, recombination is not required.

4.2.2 Creation of Inversion

The disposition of the laser levels of Sr II (Figure 4.2) is favourable to the creation of inversion by recombination and collisional processes (see section 1.4). Indeed, the levels $7\,^2S$ and $6\,^2D$ are two of the lowest lying in the group $7\,^2S$, $7\,^2P$, $6\,^2D$, $5\,^2F$ and $4\,^2G$, whereas $6\,^2P$ is one of the highest levels in the group $6\,^2S$, $6\,^2P$, $5\,^2D$ and $4\,^2F$. Analogously, the upper level of the most

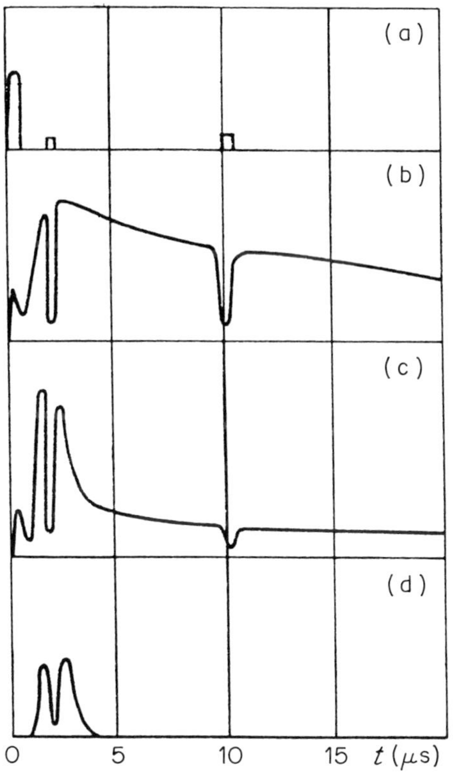

Figure 4.5 The effects of secondary heating current pulses on the afterglow and on laser generation: (a) current pulses; (b) helium line afterglow; (c) afterglow of strontium ion lines; (d) laser oscillation on the Sr II 430.5 nm line.

intense laser transitions, 430.5 and 416.2 nm, is the lowest in the group 6 ^2S, 6 ^2P, 5 ^2D and 4 ^2F, while the lower laser level 5 ^2P is the highest of the pair 5 ^2P and 4 ^2D. Thus, in the presence of electron de-excitation, recombination pumping is funnelled to the upper laser levels, and the lower laser level populations are reduced.

As can be seen from the disposition of the energy levels, inversion on the $S \longrightarrow P$ transitions is not possible by reliance on favourable ratios of radiative probabilities of the upper and lower levels. The probabilities of the radiative transitions on the 6 $S \longrightarrow$ 5 P laser transition and the 5 $P \longrightarrow$ 5 S resonance transition are nearly the same because the respective wavelengths are approximately the same. Moreover, the resonance radiation has strong self absorption which leads to a reduction in the radiative decay of the lower laser levels. However, the level 4 ^2D, lying close to the 5 ^2P level ($\approx$1 eV) causes the latter to be effectively depopulated by electron collisions in the low-temperature recombining plasma. This is the key factor in the production of an inversion on the $6\,^2S \longrightarrow 5\,^2P$ Sr II and $5\,^2S \longrightarrow 4\,^2P$ Ca II transitions in the recombination regime.

Below we shall give a simplified model for the inversion on the strongest laser transition 430.5 nm ($6\,^2S_{1/2} \longrightarrow 5\,^2P_{3/2}$) of Sr II. Resonance trapping of radiation from the lower laser level assuming Doppler broadening is considered, along with electron de-excitation of all levels and electron excitation of the $5\,^2P_{3/2}$ level from the metastable $4\,^2D_{5/2}$ level. We shall also assume that in the optimum regime for lasing, the concentration of Sr ions in the ground state is $n_0 \approx n^+ = 0.5n_e$, and that the pump rate W_r applies only to the upper level, $6\,^2S_{1/2}$. Taking into account the radiative and collisional probabilities under quasi-steady-state conditions, and using the levels designated in brackets in Figure 4.2, we have the following set of balance equations:

$$W_r - (A_{43} + A_{42} + k_{43}n_e + k_{42}n_e)n_4 = 0\,; \tag{4.6}$$

$$\begin{aligned}(A_{43} + k_{43}n_e)n_4 + k_{03}n_e n_0 + k_{13}n_e n_1 - \\ (A_{30}g_0 + A_{31} + k_{31}n_e + k_{30}n_e)n_3 = 0\,;\end{aligned} \tag{4.7}$$

$$(A_{31} + k_{31}n_e)n_3 - (k_{10}n_e + k_{13}n_e)n_1 = 0\,; \tag{4.8}$$

where g_0 is the self-absorption factor [18].

The radiative probabilities are calculated in the Coulomb approximation and the collisional probabilities following [19]. As can be seen from the calculations (Figure 4.6a), there is no inversion at low electron densities up to $n_e \approx 5 \times 10^{13}$ cm^{-3}, due to the unfavourable relation between the radiative lifetimes, and because of self absorption (which leads to an increase in the

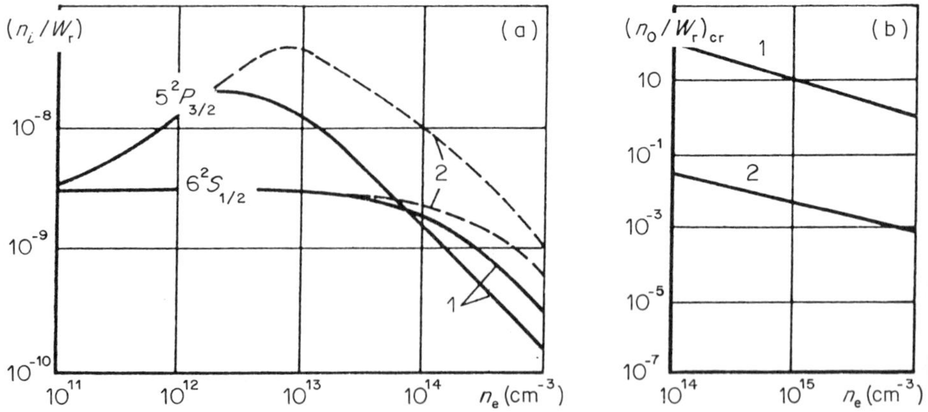

Figure 4.6 Calculations of (a) n_i/W_r for Sr II levels, and (b) $(n_0/W_r)_{cr}$, as functions of electron density. 1, $T_e = 0.124$ eV; 2, $T_e = 1.24$ eV.

$5\,^2P_{3/2}$ level population for $10^{11} \leq n_e \leq 2 \times 10^{12}$ cm^{-3}). Despite the increase in self-absorption with the rise in n_e (for $n_e > 10^{13}$ cm^{-3} the radiation is fully trapped), above $n_e = 5 \times 10^{13}$ cm^{-3}, inversion is produced on the transition $6\,^2S_{1/2} \longrightarrow 5\,^2P_{3/2}$ due to the very effective clearance of the lower laser level by electron de-excitation to the metastable D and the ground S states of Sr^+.

If T_e increases, the inversion disappears due to the reduction in electron de-excitation, and mainly due to the increase in the excitation of the lower laser level $5\,^2P_{3/2}$ by electron impact from the metastable $4\,^2D_{5/2}$ level. If we also consider population of the resonance $5\,^2P$ and metastable $4\,^2D$ states by electron impact from the highly populated $5\,^2S$ Sr II ground state, then the T_e dependence is much stronger. The inversion depends not only on n_e and T_e, but also on the parameter n_0/W_r, e.g. above some critical value, $(n_0/W_r)_{cr}$, the inversion ceases (Figure 4.6b).

The gain on the IR lines is mainly determined by the upper-level pump rate, since in the lower group of terms, long-lived metastable levels are absent. When electron de-excitation increases, the IR laser oscillation shifts to the far afterglow, and the inversion is not so critically dependent on T_e.

Using a more sophisticated model of the kinetic processes in the He–Sr laser, the inversion mechanism described has been confirmed [20, 21]. In those papers, the prospect of applying an external ionizer is discussed, for instance an electron beam or fragments of a nuclear reaction, and the possibility of obtaining steady-state (regarding n_e and T_e) laser oscillation exists. With an RF-modulated e-beam, the laser emission would be modulated with the same frequency, although the density of electrons and ions would actually be the same [21]. This is the result of the sensitivity of the recombination and inversion to T_e.

4.2.3 Electron Temperature and Density in the Plasma Afterglow of Strontium Ion Recombination Lasers

A low T_e is the most important factor for the existence of inversion on the recombination laser transitions of alkaline earth metals, because only this condition ensures both strong recombination pumping and electron de-excitation. According to the analysis in section 2.3.3 of the behaviour of T_e during the discharge afterglow in the mixture of a metal vapour and a noble gas, after the current pulse T_e falls rapidly initially and then reaches a quasi-steady-state value $(T_e)_{eq}$, which is several times higher than the gas temperature.

The rate of fall of T_e and the magnitude of $(T_e)_{eq}$ determine, respectively, the rate of rise and the duration, τ_r, of the recombination of doubly charged ions, e.g. the duration of the recombination radiation spike (Figure 4.4c):

$$\tau_r \approx 1/\alpha n_e^2 = 1/c(Z) n_e^2 (T_e)_{eq}^{-9/2}\,. \tag{4.9}$$

(See equations (1.23) and (1.24).) Using (1.24) and (2.42), in the case of low metal vapour pressure with a domination of helium ions, we find

$$\tau_r \approx 2.14 \times 10^{15} / Z^3 \ln \sqrt{Z^2 + 1}\, n_e^{1/2} N^{3/4}, \quad (4.10)$$

whence it is seen that τ_r decreases with an increase in helium density. For example, for $n_e = 5 \times 10^{14}$ cm^{-3} and $N = 10^{17}$ cm^{-3}, $\tau_r = 2.6$ μs, and for $N = 3 \times 10^{18}$ cm^{-3}, we calculate $\tau_r = 0.2$ μs, which is in good agreement with the duration of the recombination spike and that of the laser oscillation measured under such conditions.

With an increase in Sr vapour pressure, an increase in the amplitude and duration of the recombination spike of Sr II lines is observed experimentally. In this case, the laser power begins to fall long before the peak of the spontaneous emission (Figures 4.4c and 4.7). This occurs especially at relatively low He pressures (<10 kPa). The reduction in the recombination rate of Sr^{2+} is explained by an increase in T_e, which is the result of the substitution of the light He^+ ions by the heavy Sr^+ and Sr^{++} ions, and the higher effective recombination rate of the latter (as shown in section 2.3). Thus the rise in T_e and the increase in the density of the Sr^+ ground state causes the inversion to cease due to the population of the lower laser levels from the ground and metastable Sr^+ states due to electron impact.

From our electrical probe and spectroscopic measurements (see Figure 2.15, and [22]), at a helium pressure of 33 kPa, during laser oscillation on the 430.5 nm and 416.2 nm Sr II lines, we have $T_e = 0.25 \pm 0.1$ eV and $n_e = 2\text{–}5 \times 10^{14}$ cm^{-3}. Some measurements of He–Sr laser plasma parameters made by other investigators can be presented for comparison. In [23] an electron density of $n_e = 3 \times 10^{14}$ cm^{-3} was measured under optimum lasing conditions with $p_{He} = 67$ kPa; in [24, 25] laser interferometric and spectroscopic studies yielded $n_e = 6.8 \pm 0.4 \times 10^{14}$ cm^{-3} with $p_{He} = 33$ kPa and $n_e = 7.5 \pm 0.4 \times 10^{14}$

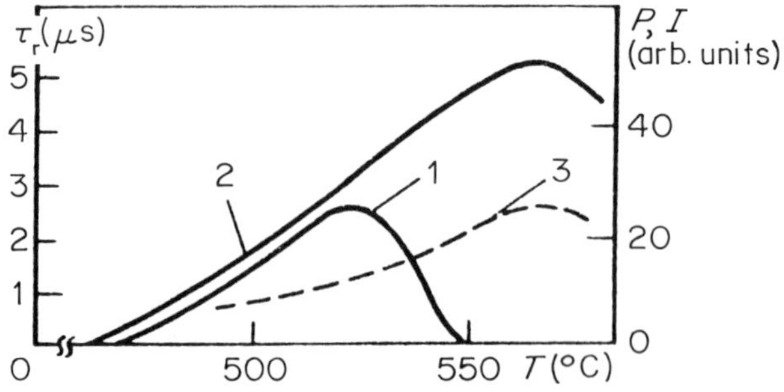

Figure 4.7 Dependences of the laser power (1), amplitude (2) and length (3) of the 430.5 nm Sr II recombination spike on laser tube temperature; $p_{He} = 9.3$ kPa.

cm^{-3} with p_{He} = 50 kPa. The electron temperature deduced from these data turned out to be 0.25–0.35 eV for p_{He} = 33 kPa and 0.39 eV for p_{He} = 50 kPa.

4.2.4 The Role of the Buffer Gas

The type and pressure of the buffer gas are of great importance for the attainment of low T_e in the early afterglow. Helium has unique properties in this regard. Because helium is the lightest noble gas, electrons transfer a relatively large part of their energy in elastic collisions with atoms and ions of that element. Moreover, the cross-section for elastic electron collisions with He at $T_e \approx 1$ eV is about 5×10^{-16} cm^2 (much higher than for other noble gases). The high excitation and ionization potentials of He atoms are conducive to the production of a large density of doubly charged metal ions during the current pulse. The energy of He ions can be utilized for the production of doubly charged ions of alkaline earth metals in the afterglow by charge transfer with ionization (equation (4.4)). Finally, the low ionization energy of metastable He atoms ($\approx$5 eV) is responsible for their effective ionization during the current pulse. Thus, immediately following the current pulse, it follows from section 2.3 that T_e can be lower than $(T_e)_{eq}$, i.e. there is a minimum in T_e during the recombination of doubly charged metal ions. For those reasons, a large concentration of doubly charged metal ions is produced, and fast and effective recombination and electron de-excitation follow. Hence, laser oscillation on the violet and UV transitions of Sr II and Ca II is observed in mixtures with helium. With an increase in p_{He} up to 80–100 kPa, the optimum pressure of Sr vapour increases also, and there is an increase in laser output power (Figures 4.8 and 4.9). The tube voltage increase that accompanies the increase in p_{He}

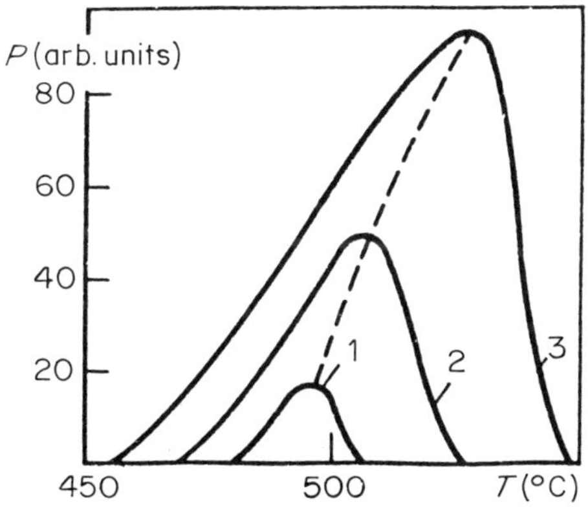

Figure 4.8 Dependence of laser power of the Sr II 430.5 nm line on laser tube temperature with d = 1.1 cm, l = 50 cm and helium pressure: 0.67 kPa (1), 2.7 kPa (2), 9.4 kPa (3).

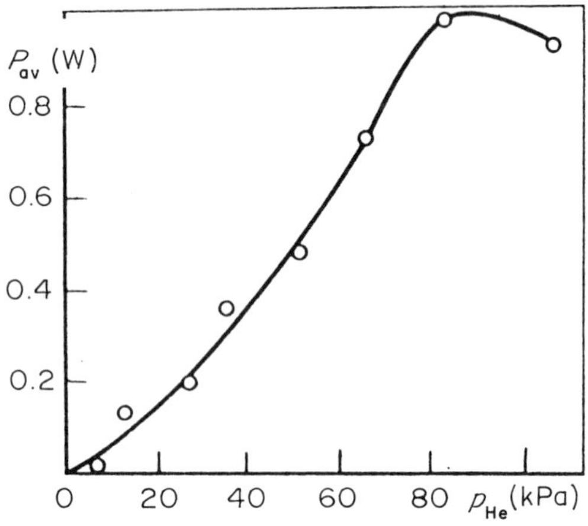

Figure 4.9 Dependence of average output power on the Sr II 430.5 nm line on helium pressure with d = 1.1 cm, l = 50 cm, and using Blumlein circuit excitation (see Fig. 3.2c) with C_s = 1.1 nF and f = 5 kHz.

leads to more electrical input energy deposited per unit length of the discharge, again yielding higher laser power. These trends persist up to p_{He} = 200 kPa as the studies [26, 27] of transversely excited He–Sr and He–Ca mixtures show, and up to p_{He} = 400 kPa with a short longitudinal discharge [22, 28, 29].

Use of the lighter isotope ^3He, with which the electrons are better cooled by elastic collisions, causes the laser oscillation to begin at lower helium pressures (0.1–0.2 kPa) and the laser power to increase at higher p_{He} by 15–20%.

Hydrogen is an interesting candidate for a buffer gas in recombination lasers. However, it has been found that at p_{H_2} ≥100 Pa, the intensity of the recombination afterglow of the metal ion lines is much less than in He, and for p_{H_2} ≥ 1 kPa, it is absent due to the lower (compared with He) ionization potential of hydrogen, and possibly the occurrence of plasma-chemical processes that produce hydrides of the alkaline earth metals.

In mixtures of metal vapour with hydrogen, laser oscillation is found only on the IR transitions of Sr II and Ca II, being optimum at p_{H_2} = 30–80 Pa [30]. However, lasing on the violet and UV transitions of Sr II and Ca II is not obtained in hydrogen mixtures [11, 30]. This is explained by the fact that n_0/W is higher than the critical value for inversion (Figure 4.6b), since the lower ionization potential of H_2 does not favour production of sufficient Sr^{2+} density, and the recombination pumping of the laser levels is low.

However, hydrogen can be used as an additive to mixtures of Sr with noble gases. In this way, a large concentration of H^+ is produced, which can additionally cool the electrons via elastic collisions. Thus, the addition of small amounts of hydrogen (0.1–10 Pa) to a Ne–Sr mixture does not reduce the density of Sr^{2+}, but decreases T_e, and enhances the recombination of Sr^{2+}, i.e. the duration of the recombination spike in the line afterglow shortens (Figure 4.4g) and lasing on the 430.5 nm line appears. Also, when H_2 is added to the mixtures of He–Sr and Ne–Sr, laser oscillation is generated in the long afterglow, and the duration of the laser pulse can reach 6–12 μs (Figure 4.4h). For laser oscillation in the long afterglow, the optimum pressure of the metal vapour is higher than in the short recombination spike (Figure 4.10). Thus, the addition of H_2 decreases T_e in the afterglow, and permits operation at

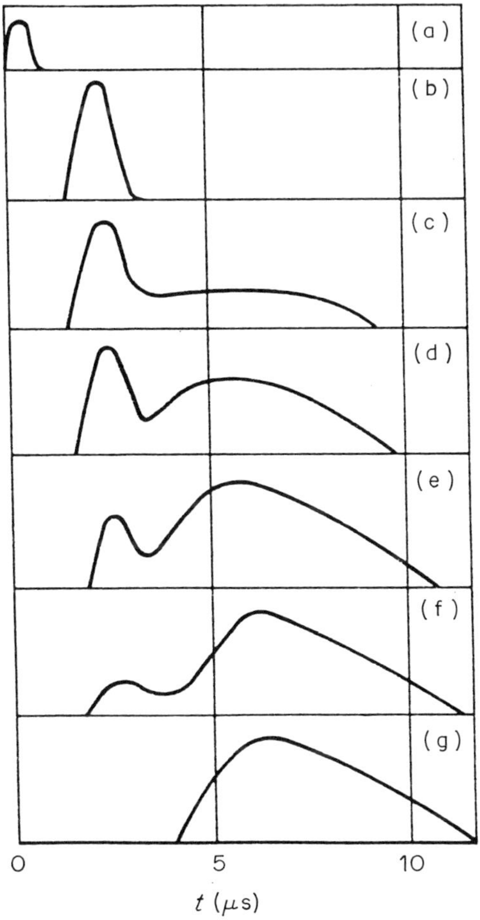

Figure 4.10 Oscillograms of the current pulse (a), and the laser intensity at 430.5 nm (Sr II) in helium with a small admixture of hydrogen ($p_{H_2} \approx 1$ Pa) with various strontium vapour pressures: (b) 0.67 Pa, (c) 0.94 Pa, (d) 1.2 Pa, (e) 1.33 Pa, (f) 2.7 Pa and (g) 6.7 Pa.

higher metal vapour pressures, when the density of doubly charged metal ions increases due to the process in equation (4.4), and parasitic population of the lower laser levels due to electron impact from the ground and metastable Sr^+ states does not yet quench the inversion.

In [31–33] two regimes of operation of the 430.5 nm Sr II line were also observed, the 'short-pulse' and the 'long-pulse' modes of oscillation. A silica laser tube was employed in those experiments (further details are given in section 4.5). At low vapour pressures, laser oscillation began immediately after the current pulse, during the short recombination spike. With an increase in pressure and current, it appeared in the long afterglow (Figure 4.11), similar to the behaviour of mixtures with hydrogen (Figure 4.10). It is very likely that in the experiments [31, 32] that hydrogen was evolved as a

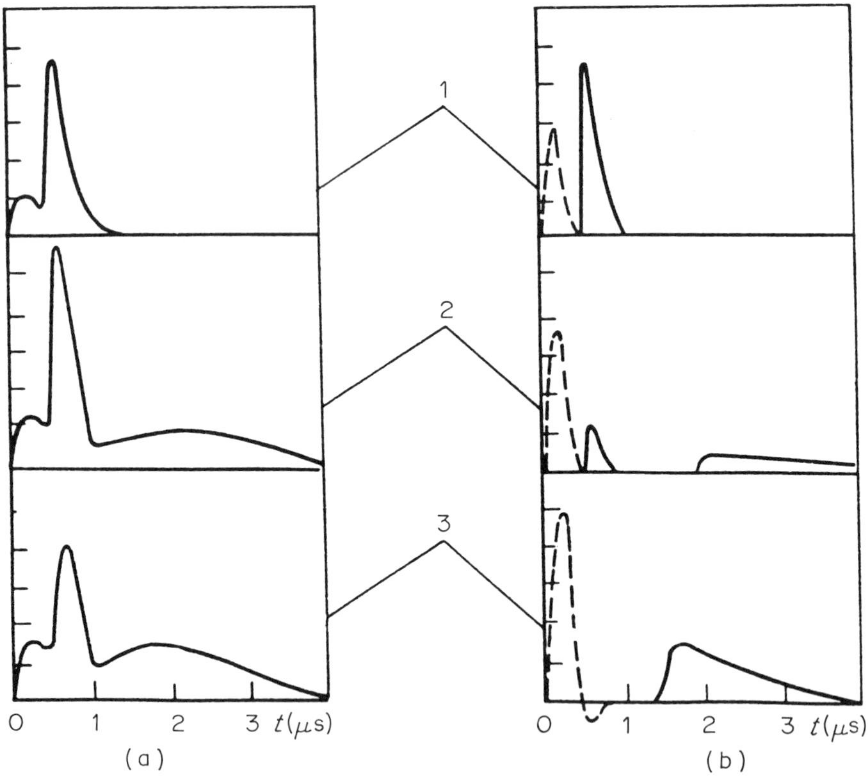

Figure 4.11 Oscillograms of (a) spontaneous and (b) stimulated emission of the Sr II 430.5 nm line [31], with p_{He} = 12 kPa, d = 1.3 cm, l = 40 cm, and tube wall temperatures of (1) 595 °C, (2) 620 °C and (3) 660 °C; the broken lines are the current pulses (100 A/division); the laser intensity is calibrated to 100 W/division.

result of reaction of Sr vapour with the silica tube. Another reason could be the much larger energy deposited in the discharge and the higher vapour pressure in the long-pulse regime. Then, full single and double ionization of the strontium would be expected [22, 34, 35], which is conducive to the production of a high He^+ density.

As a result of a slower decrease in T_e, laser oscillation during the initial recombination spike was reduced. In the late afterglow, Sr atoms were produced again due to Sr^+ recombination and thus the Sr^{2+} density rose due to process (4.4). Also, at higher He pressures, a process analogous to charge transfer with ionization may proceed involving molecular He ions [36]:

$$Sr + He_2^+ \longrightarrow Sr^{2+} + 2He + e\,, \tag{4.11}$$

which is seen in the rise of the intensity of the spontaneous emission of Sr II lines (Figure 4.11a). If T_e decreases sufficiently, the lasing starts again later in the afterglow due to charge transfer with ionization (equations (4.4) and (4.11)) followed by recombination (4.5).

4.2.5 Maximum Pulse Recurrence Frequency

The excitation mechanism of Sr II and Ca II lasers favours high-frequency operation. The only requirement is that each current pulse does not overlap the recombination spike, or the laser pulse, of the preceding pulse (Figures 4.4c,d). Thus, the inter-pulse period can be as short as 1 μs (for the short pulse regime), which has been shown using double pulse excitation [10, 11, 13, 26, 32, 37], i.e. the upper limit to PRF is around 1 MHz. Moreover, experiments show that for pulse separations of less than 10–15 μs, the laser power of the second pulse is 20–40% higher than that of the first pulse, with equal input energy deposition. This observation can be explained by the formation of doubly charged ions from singly charged ions which have not yet recombined after the first current pulse; thus, the efficiency of the second pulse is higher.

The possibility of very high PRF operation without detriment to laser power or efficiency is an advantageous difference between recombination lasers and self-terminating lasers, the latter of which require the number density of the lower laser level to be sufficiently low at the onset of each current pulse for oscillation to occur. However, at high steady-state PRFs, problems of extraction of heat from the discharge arise.

4.3 LASER OSCILLATION ON TRANSITIONS OF OTHER IONS

We shall now outline the features of laser oscillation on ion transitions of the other elements listed in Table 4.1. Laser tubes of 40–60 cm active length and 7–8 mm diameter were used; excitation current pulses of 0.35–5 μs duration were used, with between 1 and 150 A amplitude.

Laser oscillation on Ba II transitions (Figure 4.12) is obtained in discharge mixtures of He, Ne or H_2 with barium vapour (0.7–7 Pa) in the afterglow [13, 15]. During the recombination spike, IR lasing is obtained in the presence of noble gases (0.2–2 kPa). Visible lasing (677.1 and 687.6 nm) is obtained in the long afterglow with 15–50 μs delay from the current pulse, which increases with a rise in current. As was shown in [15], the pumping of Ba II, like Sr II, proceeds via recombination of Ba^{2+} ions, produced in the current pulse and by charge transfer (as in equation (4.4)). Inversion is produced by electron de-excitation and radiative depopulation of ion levels. The latter is the main pump route for the red laser lines.

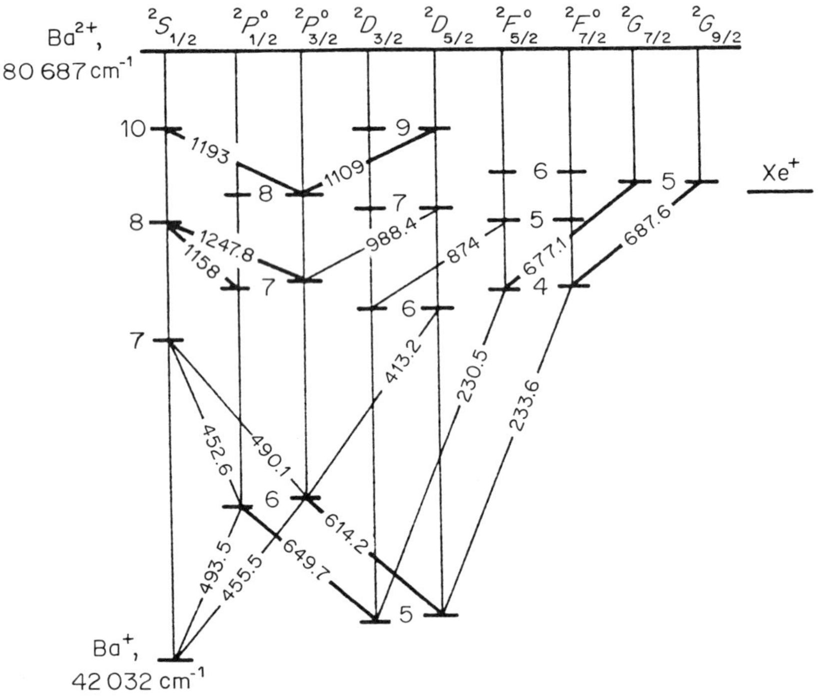

Figure 4.12 Partial energy-level diagram of Ba II.

In a mixture with H_2, laser oscillation is observed at low gas pressures (<1–2 Pa), i.e. almost in pure Ba vapour. From Figure 4.13(c) it can be seen that pumping is realized through the recombination of Ba^{2+} ions produced during the current pulse.

Lasing on the Ba II transitions 490.1 and 452.6 nm ($7\ {}^2S_{1/2} \longrightarrow 6\ {}^2P_{3/2,1/2}$), which are analogous to the violet and UV transitions of Sr II and Ca II, is not

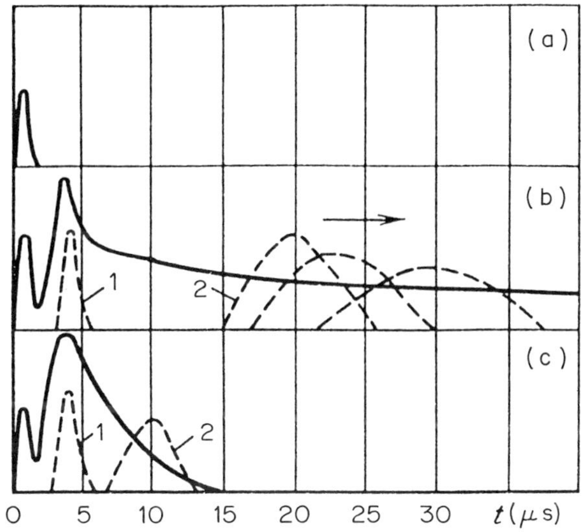

Figure 4.13 (a) The current pulse. (b) The spontaneous and stimulated emission from Ba II transitions, excited in a mixture with helium or neon and (c) with hydrogen. 1, infrared laser oscillation; 2, visible laser oscillation. The arrow shows the shift in position of the laser pulse as the current is increased.

found. This is due to the disposition of the metastable 5 D level of Ba II—it lies lower than the corresponding levels of Sr II and Ca II and is close to the 6 S ground state of Ba^+. Thus, electron de-excitation of the 6 P level to the 5 D level is low, and inversion on the transition 7 $S \longrightarrow 6$ P is not produced. This is confirmed by calculation of the radiative and collisional probabilities [11, 13].

Laser oscillation has been observed on two IR transitions of Mg II in a mixture of Mg vapour and He or Ne (0.5 kPa) (Figure 4.14) [13]. In the presence of Ne, laser oscillation is produced in the early afterglow during the short spike, for pulsed currents of ~ 2 A amplitude. Laser emission is a maximum about 4 A, and ceases at 8 A. In mixtures with He, lasing exists in the current range 13–70 A. In both cases, with the increase in current above the optimum, the laser pulse shifts to the long afterglow. The shift in the He–Mg mixture is particularly large: 30–50 μs.

The potential of Ne^+ is less than that of Mg^{2+}, and thus the long afterglow in the Ne–Mg mixture is due to conventional charge-transfer excitation:

$$Mg + Ne^+ \longrightarrow Mg^{+*} + Ne\,. \qquad (4.12)$$

The shift of laser oscillation in the afterglow with an increase in current is caused by electron de-excitation, since the inversion is mainly supported by

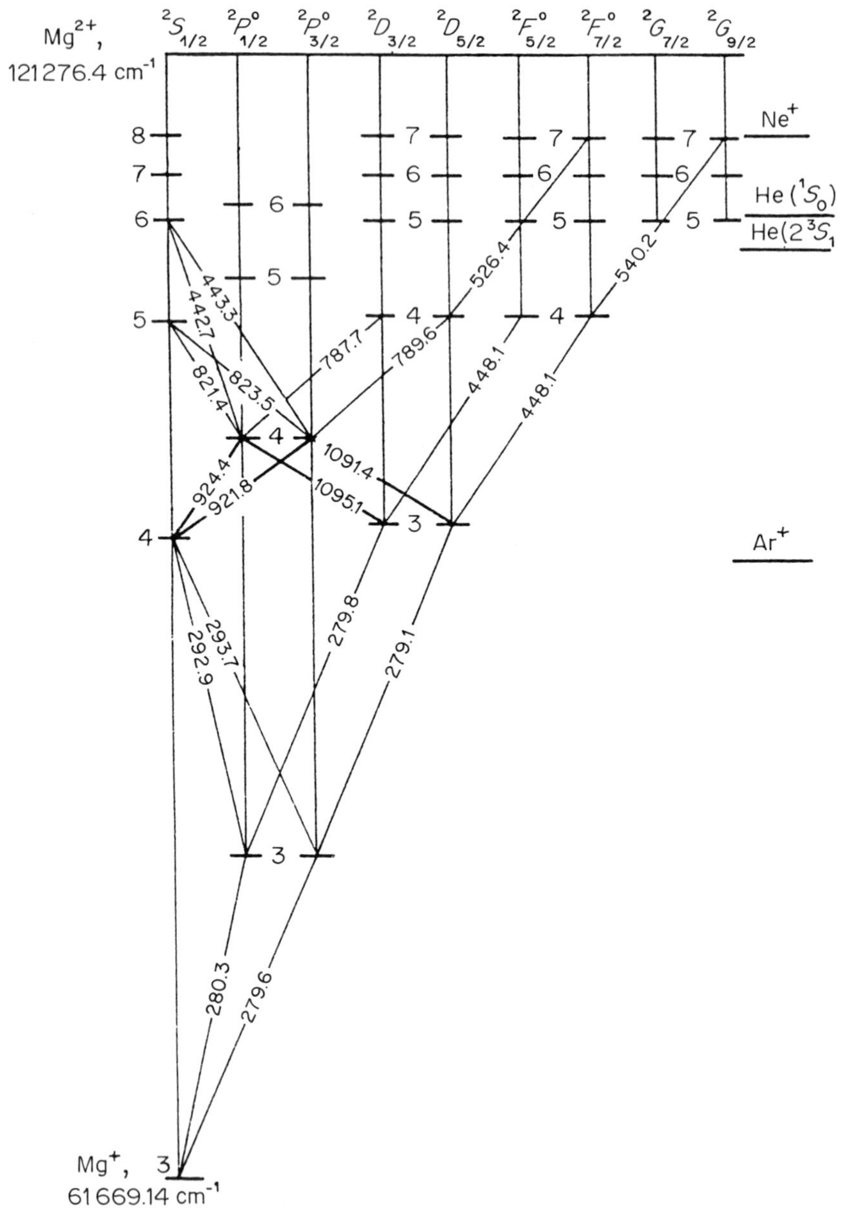

Figure 4.14 Partial energy-level diagram of Mg II.

radiative decay from the level 4 $^2S_{1/2}$ of Mg II. In He mixtures, charge transfer with ionization, of the type exemplified by equation (4.4), is responsible for the long afterglow [16].

In Be, laser oscillation has been achieved on transitions of the Be II ion: 467.5 nm (4 $^2F \longrightarrow 3\ ^2D$), 527.2 nm (4 $^2S \longrightarrow 3\ ^2P$) and 1209.9 nm (3 $^2P \longrightarrow 3\ ^2S$) [38]. For the first two lines, a mixture of Be with He (0.5 kPa) was used, whereas a Ne-Be mixture was used for the third line [39]. The ionization potentials of He and Ne are less than the potential of double ionization of Be, and they lie near the Be II levels with $n = 4$ (for He^+) and $n = 3$ (for Ne^+). Thus, in these mixtures, besides pumping due to recombination of Be^{2+} ions produced during the current pulse, effective population of Be II levels occurs via charge transfer:

$$Be + He^+ \longrightarrow Be^{+*}(n = 4) + He\ ; \tag{4.13}$$

$$Be + Ne^+ \longrightarrow Be^{+*}(n = 3) + Ne\ . \tag{4.14}$$

The lines 467.5 and 1209.6 nm of Be II lase during the long afterglow at relatively low currents ($\approx$5 A), and are excited via charge transfer from He and Ne ions, respectively. Lasing on the 527.2 nm (4 $S \longrightarrow 3\ P$) transition begins at higher currents during the recombination spike in the early afterglow (similar to Figure 4.13b, curve 1). Pumping of this line is due to a combination of recombination and charge transfer in approximately equal measures.

As seen from Figure 4.15, the 527.2 nm transition is most suited to laser oscillation in the recombination regime. Its upper level is the lowest in the Be II level group with $n = 4$, and the lower level is one of the upper levels of the group with $n = 3$. Thus the pumping channels to the upper laser level, and clearance of the lower laser level is effected by electron de-excitation. From that point of view, the 177.6 nm (3 $^2S \longrightarrow 2\ ^2P$) transition should also be considered. Although there is no intermediate level between the lower 2 ^2P level and the ion ground state 2 ^2S in Be II, as there is in Sr II and Ca II, the energy difference between 2 ^2P and 2 ^2S is half that between the 3 ^2S and 2 ^2P levels. Therefore, a population inversion could be expected on the UV 177.6 nm transition in the recombination regime due to electron de-excitation of the lower level, which is predicted by calculation [40].

Laser oscillation is observed on Al II transitions with He or Ne buffer gases (Figure 4.16) [11, 15]. All laser lines oscillate in the discharge afterglow during the short recombination spike, which is delayed 1–2 μs from the current pulse. The most powerful is the 747.1 nm line. The gain coefficient increases to 80–90% m^{-1} with a rise in helium pressure to 4 kPa. The good characteristics of this laser line originate from a favourable disposition of the laser levels; the upper level is one of the lowest in the group 5 $^1P_1$, 4 ^1D, 5 $^1S_0$ and 4 $^1F_3$, and the lower laser level is the highest in the group 3 $^1D_2$, 4 $^1P_1$ and 4 $^1S_0$.

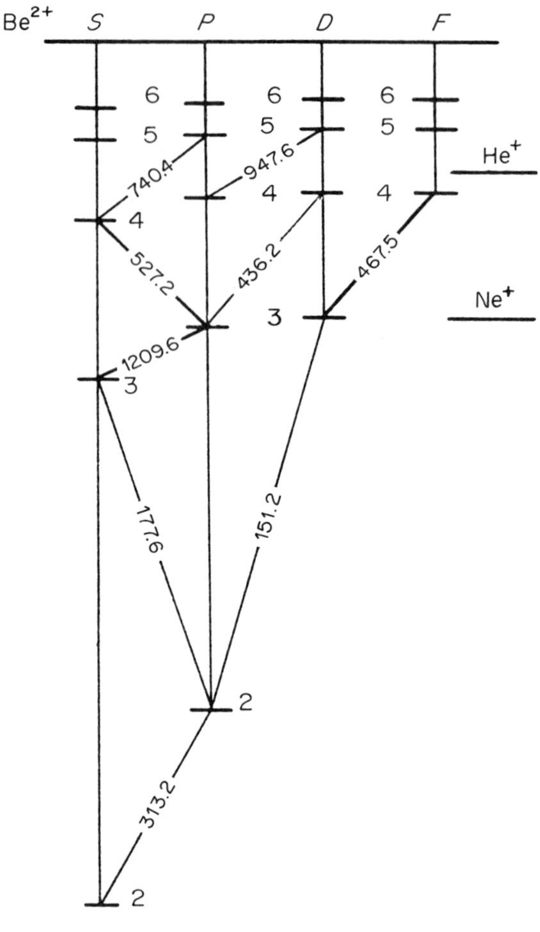

Figure 4.15 Partial energy-level diagram of Be II.

Lasing in Tl II ions is also obtained in mixtures of the metal vapour with He or Ne [41, 42]. In mixtures with Ne, the upper laser levels are pumped entirely by charge transfer, and lasing appears on yellow-red lines with excitation potentials below Ne^+ (see Table 3.3). In He mixtures, lasing is produced on other Tl II lines in the green spectral region (see Table 4.1), with upper levels near the energy of He^+, and it has a complex temporal behaviour (Figure 4.17). At helium pressures of 25–250 Pa, the laser pulse occurs in the early afterglow, and coincides with the recombination spike. Upon increase of the He pressure, the intensity of the long charge-transfer pumped afterglow increases, the intensity of the initial spike reduces and a

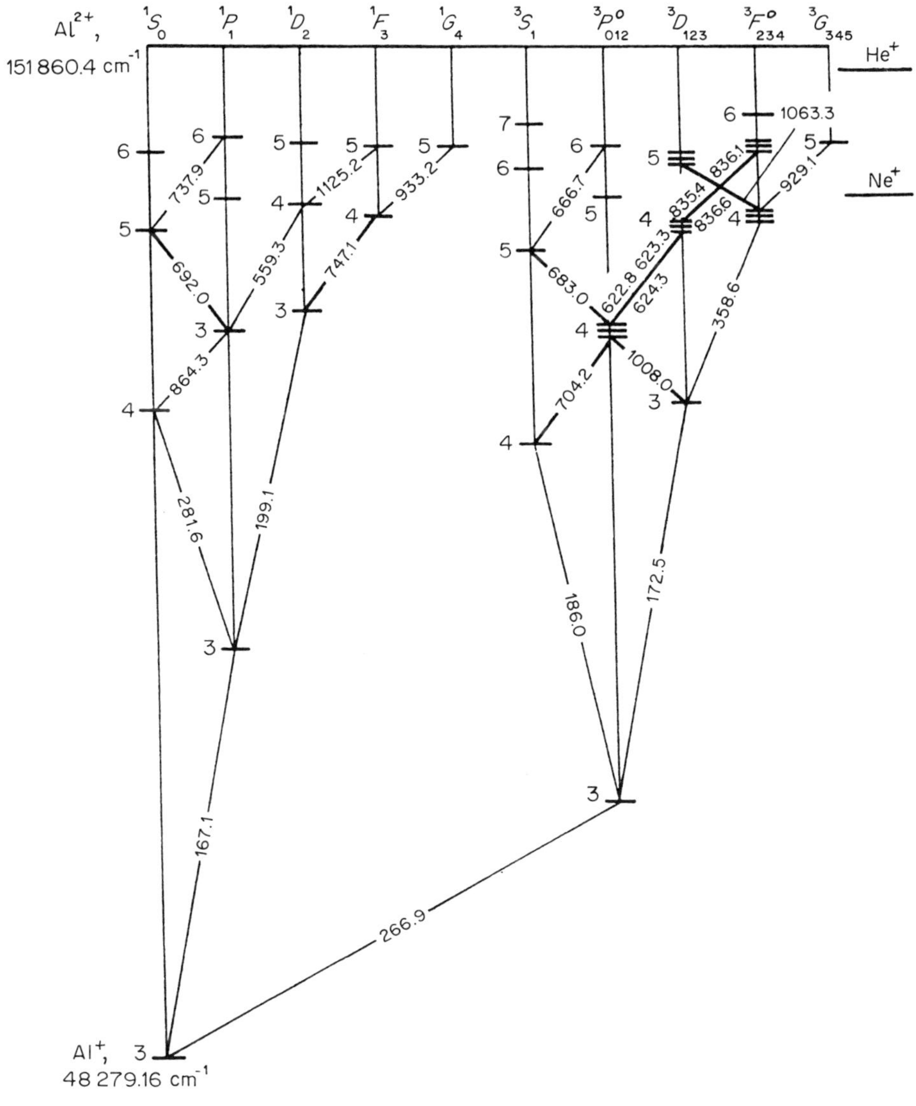

Figure 4.16 Partial energy-level diagram of Al II.

laser pulse appears in the long afterglow. When the current is increased, the second pulse moves further into the afterglow, while the first pulse does not—the behaviour of the second pulse is typical of charge-transfer pumped lines, and is caused by electron de-excitation.

For $p_{He} > 0.7$ kPa, only the second pulse is present. Thus, in He–Tl

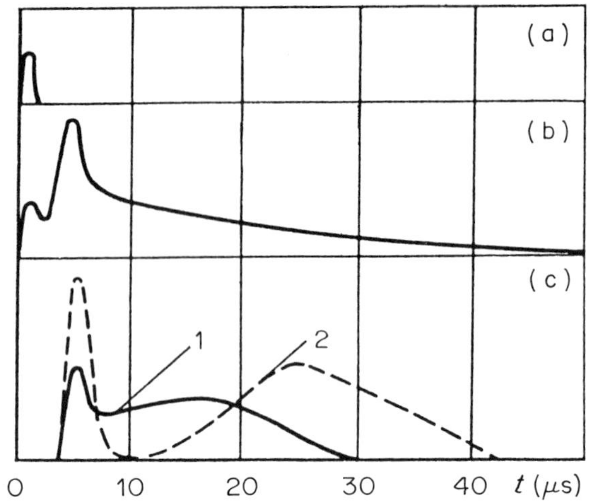

Figure 4.17 (a) The current pulse. (b) Spontaneous emission and (c) laser emission from the Tl II 515.2 nm line in a He–Tl mixture with two different discharge currents: 1, 15 A; 2, 40 A.

mixtures, two regimes of pumping the Tl II laser levels have been realized, recombination and charge transfer pumping. This view is supported by experiments where a small reheating pulse is superimposed on the first laser pulse, which is sharply reduced in power, and the recombination spike is suppressed. Reheating pulses impressed on the second laser pulse and the long afterglow do not induce changes in their forms. Note that the only Tl II transitions observed in the long afterglow originate from levels energy below that of He^+.

Similar temporal behaviour is observed in the cases of the recombination transitions of Sn II and Pb II [11, 43, 44]. The laser levels are pumped by recombination of doubly charged metal ions created in the current pulse by electron impact, and in the afterglow by charge transfer with ionization. In Ne mixtures, charge transfer with excitation is present. Radiative decay and inelastic collisions with slow electrons yield inversion. The inversions are favoured by the dispositions of the various energy levels, which satisfactorily meet the requirements for collisional recombination kinetics (see section 1.4).

We should note that from the point of view of collisional recombination kinetics, charge transfer can be considered as a process that intensifies the recombination flux, beginning from particular levels around the ionization potential of the participating noble gas. Charge transfer with ionization can also be treated as charge transfer into a continuum. For charge transfer with excitation, a low T_e is not required, since subsequent recombination of doubly charged ions is not necessary. This makes CW operation possible on lines pumped by charge transfer.

4.4 ALTERNATIVE MECHANISMS OF ION-LASER PUMPING IN RELAXING PLASMAS

Despite the fact that collisional recombination is one of the most fundamental and well-studied plasma processes, from both experimental and theoretical points of view, there was originally some doubt that it could be used to pump laser transitions ([1], p. 21; [18], p. 160) because it is the reverse of excitation and ionization processes. In section 4.1, it was remarked that the process proposed by [5] for the excitation of the green and some of the other Hg II laser lines, which originate from above the potential of He^+, was stepwise Penning ionization:

$$Hg^*(6\,^3P) + He^*(2\,^1S_0) \longrightarrow Hg^{+*}(5\,^2F) + He + e\,. \qquad (4.15)$$

However, analysis of process (4.15) does not explain why there should be minima in intensities of the ion lines after the current pulse, nor why all the ion lines, including those that originate from far above the upper level of the green Hg II laser line (i.e. beyond the reach of reaction (4.15)) exhibit the short afterglow spike. Moreover, as can be seen from results discussed in section 4.1, such a form of the afterglow is observed not only in mixtures with helium, but also with neon and argon, and even in pure metal vapour, where process (4.15) cannot apply. However, this form of afterglow can be explained easily with recombination of doubly charged ions [9], without assumptions (such as the (doubtful) recourse to different electron temperatures across the bore of the laser tube [2]).

In order to explain laser oscillation on ion transitions of Sr, Ca and many other metals, the alternative mechanism of charge transfer from doubly charged ions to metastable atoms has been proposed [45], and in later works [46, 47] insisted, e.g.

$$Sr^{2+} + Sr^*(He^*) \longrightarrow Sr^{+*} + Sr^+(He^+)\,. \qquad (4.16)$$

None of the particles that take part in this process is in the atomic ground state, which allows us to designate reaction (4.16) as stepwise charge transfer, based on the classification proposed in section 1.3. Therefore, to obtain a sufficient pumping rate for this process, its cross-section would have to be $\approx 10^{-13}$ cm^2, which is a very large value. Even if such a large cross-section were used, as calculations show [48], the pump rate of Sr levels by that process is insufficient due to the low density of excited state populations.

Moreover, the two ions produced as products of the process (4.16) experience strong Coulomb repulsion. Thus, not only will the upper laser level and other near-resonant levels be populated as proposed in [45], but also much lower Sr II levels. This is the result of the intersection of each term of the $Sr^{2+}Sr^*$ complex with the many ion–ion terms of the $Sr^{+*}Sr^+$ complex. Further, for ion separations of 0.5–1 nm, inter-crossing of the more populated $5\ ^1S_0$ Sr I ground state with the $5\ ^2P$ Sr II lower laser level should occur, causing significant population of the latter, which reduces the population inversion. Due to Coulomb repulsion, process (4.16) should liberate large amounts of kinetic energy, and lead to strong Doppler broadening of Sr II lines, which is not observed experimentally. The time dependence of Sr II, Sr I and He I lines cannot be explained using (4.16). If we assume that there is no electron-ion recombination in the early afterglow, then the ion density should decrease due to reactions such as (4.16), and the metastable density will also fall due to electron de-excitation, e.g. approximately exponentially. The same exponential fall must be observed on Sr II lines as well, in a similar manner to the behaviour of Hg, Zn, Cd and other ion lines that are pumped by charge transfer. Thus the complex intensity dependence of Sr II lines (Figure 4.4) in the early afterglow cannot be explained only by processes such as (4.16). It is also difficult to explain why the same behaviour is observed on all Sr II lines almost up to the ionization potential of Sr^+. It cannot be explained by a shift of the 'bottleneck' position of recombination flux through excited Sr II levels [45]. For all reasonable parameters of the plasma afterglow (T_e = 0.1–0.6 eV), the bottleneck (corresponding to $3T_e/2$ below the potential of Sr^{2+} [18]) is far above the upper laser level of the 416.2 and 430.5 nm Sr II lines, which is 5.3 eV below Sr^{2+}. Thus, the depopulation of this level must be downwards, and the temporal behaviour of the line intensities must correspond to the pumping of their upper levels. Thus, charge transfer of the (4.16) type cannot be the main mechanism of pumping of the laser levels of Sr ions and other recombination laser transitions.

At helium pressures of around 100 kPa, double charge transfer from molecular helium ions (equation (4.11)) can make significant contributions to the production of doubly charged alkaline earth metal ions. This process will be important when atomic helium ions are converted to molecular ones within about 1 μs of the current pulse by

$$He^+ + 2He \longrightarrow He_2^+ + He\,. \tag{4.17}$$

The conversion rate is $k = 1.1 \times 10^{-31}(300/T_g)$ cm^6 s^{-1}. Hence, the time for conversion of He^+ to He_2^+ at $p_{He} = 13.3$ kPa is $\approx$10 μs, which is much longer than the current pulse and laser pulses. However, since the rate of conversion is proportional to $(p_{He})^2$, at $p_{He} = 125$ kPa the time for conversion falls to 0.1 μs.

It seems unlikely that molecular ions such as $HeSr^+$ or $HeSr^{2+}$ [45] have any effect on the populations of Sr II, since the binding energies of those molecules are much less than 1 eV [49], and they are not observed in experiments with He–Sr lasers (where the gas temperature is 800–900 K).

Although a minor role for the stepwise charge-transfer reaction (4.16) follows from the above general considerations, the investigations [45–47] have stimulated more careful plasma diagnostics and studies of various level populations and long-lived particle densities in the He–Sr laser afterglow [22, 24, 25, 31, 34, 35, 50, 51], as well as mathematical modelling of this laser [22, 51–53]. In particular, T_e and n_e have been measured by different techniques and independent scientific groups, and have turned out to be on average 0.3 ± 0.1 eV and 5 (±3) × 10^{14} cm^{-3} during the time period of the laser pulse (see section 4.2). A value for T_e of 0.6 eV has mostly been used for estimates in [45–47], so that the recombination rate is greatly underestimated, due to its $T_e^{-9/2}$ proportionality (see equation (1.24)). However, measured strontium and helium metastable densities have been found to be rather small during the laser pulse: 2–5 (× 10^{11}) cm^{-3} and 3–5 (× 10^{12}) cm^{-3} for Sr(*m*) and He(*m*), respectively [22, 34, 35, 51]. Corresponding calculations using these experimentally obtained data show that the fraction of the laser levels pumped by stepwise charge-transfer (4.16) cannot exceed several percent [22, 51]. This conclusion is corroborated by the results of computer simulation [51–53], which agree with a recombination pump mechanism for He–Sr lasers, and with the observed temporal spontaneous emission and laser output characteristics. These facts together permit us to conclude that electron–ion recombination is indeed the dominant mechanism for upper level excitation in gas-discharge He–Sr and He–Ca violet and UV ion lasers. However, deeper studies of the active media of these interesting systems are still worthwhile and continue to date [36, 54–59].

4.5 ENGINEERING OF STRONTIUM AND CALCIUM ION RECOMBINATION LASERS

4.5.1 Longitudinal Discharge Lasers

Among recombination pumped lasers, the best transitions are those at 430.5 nm (Sr II) and 373.7 nm (Ca II). As found in the first studies [10, 11, 13], these laser lines are useful for their short wavelengths, useful average output powers of 0.5–1 W, high PRFs, and reasonable (for ion lasers) efficiencies of 0.1–0.15%.

Relatively low temperatures (500–800 °C) are necessary in order to obtain laser oscillation with Sr and Ca vapours. Therefore the self-heating regime, which is applied successfully to copper vapour lasers, is even more applicable to these lasers, because there is no need for additional thermal insulation around the discharge tube, and its construction can be simple—either a bare BeO tube, or such tubing placed within a silica or glass jacket (see Figures 3.1a,c) [60]. Cold cylindrical electrodes, usually made of Ni or Ta, are inserted 10 to 20 mm into the BeO ceramic tube. It is advisable that BeO ceramic be used for its resistance against chemical attack by alkaline earth metal vapour. However, in [31–33] it was found that fused silica can also be used, provided the active zone is first passivated by heating it for a few hours in the presence of the metal vapour in order to form a film (probably Si) that is resistant to further attack by the vapour.

For optimum operation of He–Sr and He–Ca lasers, the pressure of He must be high (up to 150 kPa). Thus the longitudinal discharge requires high tube voltages, and rapid falls in current ($\leq 0.1\ \mu s$). The Blumlein excitation circuit is the most suitable (Figure 3.2c)—it can produce a tube voltage four times higher than that supplied at the rectifier. Thus, a power supply of 3–8 kV can be used for tubes 30–50 cm long, and with 40–80 kPa He pressure. For a discharge tube of 15 mm inside diameter, the storage capacitance is typically 1–2 nF. The typical PRF in the self-heating regime is 5–10 kHz. The warm-up time to laser oscillation is 5–10 min.

The optimum optical cavity for He–Sr and He–Ca lasers consists of a high reflector and an output coupler of 30% transmission. The lasers operate stably with plane mirrors, which ensures a good mode structure in the output beam, and the opportunity for focusing the beam to a spot of 10–100 μm diameter. The typical pulse length is 150–250 ns, and the average laser power is 0.5–1.5 W at helium pressures of 70–80 kPa [11, 60–63].

With a non-selective cavity, laser oscillation on the second components of the doublets (416.2 nm (Sr II) and 370.6 nm (Ca II)) is usually suppressed by competition of transitions for the upper level population. With appropriate mirrors, simultaneous oscillation at 430.5 and 416.2 nm can be obtained. The laser power of the latter is 20–30% of the total power. With a dispersive prism or diffraction grating, the laser power at 416.2 nm can reach 60% of the 430.5 nm laser power, and oscillation at 370.6 nm can produce 20–30% of the 373.7 nm laser power.

For helium pressures of around 1 atmosphere, the 430.5 nm Sr II and 373.7 nm Ca II lines are strongly collisionally broadened ($\Delta\nu \approx 12$ GHz). Since the broadening is homogeneous, the spectral width of the laser line is quite narrow (0.3–1.5 GHz) [60]. This allows a Fabry–Perot étalon to be placed in the cavity, to permit the laser output to be tuned over a 12 GHz range [64]. This is useful for pumping dimer lasers [65]. In some cases, i.e. overheating of the active zone and diaphragming of the beam inside the cavity, mode self-locking is observed [64].

Further scaling of the laser power of He–Sr and He–Ca lasers can be realized by increasing the active volume. The peak laser power is proportional to the square of the tube bore, which is a result of the volumetric character of recombination. The energy input must be correspondingly increased also.

The active volume can be enlarged by making the tube longer. However, the high helium pressure proves to be disadvantageous when discharge tubes longer than 50 cm are employed—very high voltages are required. In that case the discharge tube can be sectioned. Thus, a two-section He–Sr laser, with a total discharge length of 90 cm, a 15 mm bore and operation with a helium pressure of 73 kPa has been demonstrated; it has an output power of 3 W produced at 0.12% efficiency [66]. From a similar He–Ca laser tube, 1.35 W output at 373.7 nm was obtained, although the operation regime in that case had not been fully optimized.

In [31–33], investigations were conducted on the optimization of a He–Sr laser excited with rarely recurrent pulses in silica tubes with external heating. This regime allows very good laser output characteristics to be obtained in both the short recombination spike and in the long afterglow (the short and long pulse regimes, see section 4.2) at elevated energy input to the active medium.

In a tube of 13 mm bore and 40 cm active length, with p_{He} = 15–20 kPa, p_{Sr} = 4–8 Pa and a current density j = 260 A cm^{-2} , laser oscillation in the short-pulse (180 ns) regime is obtained with a laser pulse energy $\mathcal{E}$ = 180 μJ, and a peak laser power of P_{pk} = 1 kW. At a higher vapour pressure of 30–60 Pa, and j = 450 A cm^{-2}, long-pulse mode ($\approx$1 μs) laser output is obtained with $\mathcal{E}$ = 250 μJ and P_{pk} = 250 W [31, 67]. Scaling of the active volume to 1000 cm^3 yields short-pulse mode (300 ns) output pulses of 6 mJ energy and 20 kW peak power [33, 67].

In self-heated lasers, the optimum pulse recurrence frequency is essentially determined by the rate of heat extraction from the discharge tube, and by the optimum vapour pressure. Thus, by enhancing the rate of heat transfer from the tube, it is possible to increase the PRF, and the average laser output power.

We shall now consider the transfer of heat from a bare discharge tube, fabricated from BeO ceramic (Figure 4.18a). This type of ceramic has a high thermal conductivity ($\approx$60 W m^{-1} K^{-1}) in the temperature range 673–1073 K. Thus the difference in temperature between the inside and outside tube walls is a few degrees, and the inside wall temperature depends on free convection and radiation from the outside wall.

In [68], an expression is given for the thermal power extracted from a bare BeO tube by means of convection, Q_1, and radiation, Q_2:

$$Q = Q_1 + Q_2 = l[0.0276(T_t + T_a)^{0.2} (T_t - T_a)^{1.25} d_{od}^{0.75} + 1.78 \times 10^{-9} d_{od} \varepsilon T_t^4] , \qquad (4.18)$$

where l is the length of tube (in m); d_{od} is the outside diameter (in cm); T_t and

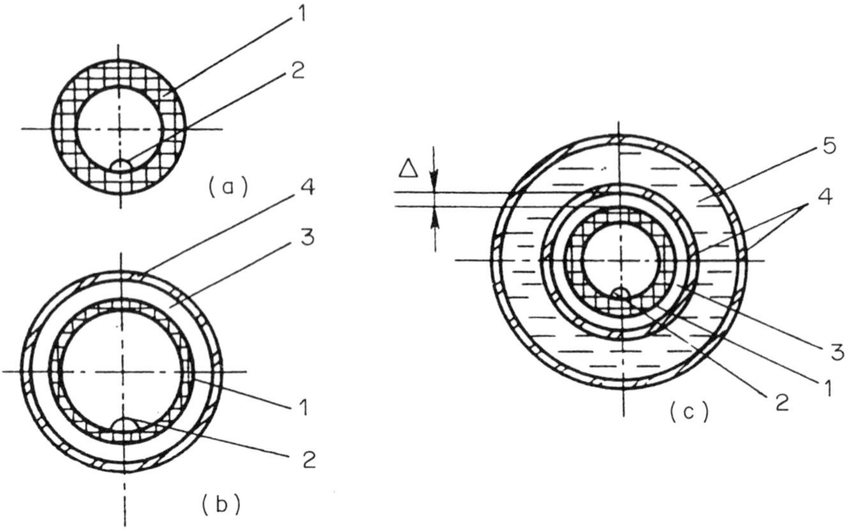

Figure 4.18 Schematic cross-sections of He–Sr laser tubes: (a) bare ceramic tube; (b) in quartz envelope; (c) in water jacket. 1, ceramic tube; 2, strontium; 3, helium layer; 4, quartz; 5, water.

T_a are the temperatures of the tube and the ambient air (in K); ε is the emissivity of the tube surface. Based on the efficiencies of He–Sr and He–Ca lasers (0.12% and 0.08%, respectively), and using the formula (4.18), it is possible to estimate the average laser power attainable in the self-heating regime. For a typical laser tube (l = 0.5 m, d_{od} = 2 cm, ε_{BeO} = 0.5, T_a = 300 K, T_t = 873 K, which corresponds to the optimum Sr vapour pressure at a He pressure of 80 kPa), we obtain $Q = Q_1 + Q_2 = 285 + 570 = 855$ W. For a He–Ca laser under similar conditions (but with T_t = 923 K), $Q = Q_1 + Q_2 = 320 + 710 = 1030$ W. Then, taking the laser efficiencies into account, we find that the average laser power of the He–Sr laser is 1–1.2 W, and of the He–Ca laser is about 0.8 W, which coincides with the experimental values.

It can be inferred that radiation is the major heat loss mechanism, which can be raised by increasing the emissivity of the outside surface of the laser tube (by up to a factor of two). The experimental proof was described in [68] and confirmed this point: as the emissivity for the BeO ceramic surface was raised up to 0.9, the optimum PRF along with output power rose by a factor of 1.6. A similar experiment in [69] gave an increase in optimum input power of 15%; output laser power rose by about 40%, and the efficiency increased by 25%. In addition, due to the high thermal conductivity of the BeO tube, the outside tube diameter is of great importance, and it is desirable to increase it, while keeping the inside tube diameter constant. The outside surface of the tube can be finned, and thus 2–3 times intensification of heat removal from the tube can be attained.

Calculations of heat transfer from the ceramic tubes mounted in silica jackets (Figure 4.18b) show that in such a construction the heat transfer is practically the same as in a bare ceramic tube (due to the high thermal conductivity of He), and the attainable average laser powers in the two cases are approximately the same.

The high thermal conductivity of He can be exploited in tubes with water cooling of the outside surface. Direct contact of the water with the BeO tube is not possible because its thermal conductivity is very high, and too high an input power would be required in order to establish the necessary 500°C temperature difference across the ceramic tube wall. Thus, instead the silica jacket is water cooled (Figure 4.18c), permitting the He gas layer to moderate the removal of heat from the ceramic laser tube. On the basis of the attainable input power and PRF, we can set the width of the He layer to be appropriate for watercooling, with a ceramic wall temperature corresponding to the optimum vapour pressure. The dependence of the input power per unit length of tube with watercooling on the width of the helium layer, Δ, is given in Figure 4.19. Note that Q increases with a reduction in Δ, being very sensitive for $\Delta < 1$ mm. For uniform heating of the laser active zone, the helium layer width must be maintained very precisely at Δ along the tube length. Simple considerations indicate that it is not possible to scale the laser power by continuously increasing the PRF, and to rely on the increase in heat extraction from the outside surface of the laser tube in order to maintain the

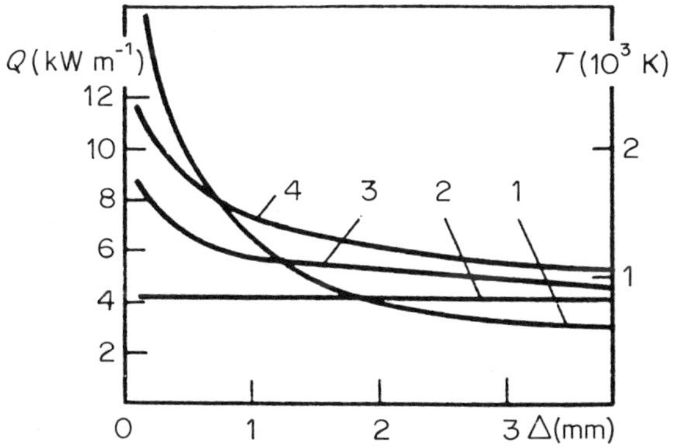

Figure 4.19 Thermophysical parameters of the water-cooled laser tube versus the thickness, Δ, of the helium layer. 1, Heat transfer power per unit length, 2, ceramic wall temperature, 3, average gas temperature inside the ceramic tube, 4, gas temperature on the tube axis.

tube operating temperature, because the output power is limited ultimately by overheating of the axial zone of the gas discharge. As calculations show, the critical gas temperature for destruction of the inversion in a He–Sr laser is about 2300 K. Therefore, we see that, from Figure 4.19, there exists scope for increasing the laser output power by raising the PRF together with intensified heat removal.

Experiments were reported in [70] using a He–Sr laser tube (bore = 1.1 cm, l = 30 cm) with watercooling and a He layer of width 0.5 mm. The PRF was extended to 29 kHz, and the average laser power at 430.5 nm reached 3.9 W or 130 mW cm^{-3}. To date, these are the best characteristics reported for self-heated He–Sr lasers, and are comparable to those of copper vapour lasers. Our estimates [70] indicate that laser powers of 13 W m^{-1} are achievable for these lasers, but recent studies reveal the existence of radial non-uniformities in particle densities which, obviously, would grow with any increase in PRF, especially in large-bore tubes [34, 36, 52, 53]. This might prevent the attainment of high average laser output powers from large-bore water-cooled tubes. Therefore, a proposal to utilize rectangular discharge channels [36] looks quite attractive, and indeed was confirmed by the experiments in [71] where, with a rectangular-bore BeO tube average and specific output powers of P_{av} = 1.82 W and P_{sp} = 16 mW cm^{-3}, respectively, were reached on the 430.5 nm line with a PRF of 6.7 kHz; P_{sp} turned out to be 4 times that of a comparable cylindrical bore tube. Enhancement of He–Sr laser output power by means of forced-air cooling of the tube was demonstrated in [72].

In order to prevent overheating at elevated PRFs, the metal can be placed away from the discharge zone in special cavities (Figure 4.20). In the presence of increased input power and operating PRF, this construction also gives more stable operation. The discharge does not contact the metal pieces and produce sudden changes in metal vapour pressure. In order to ensure

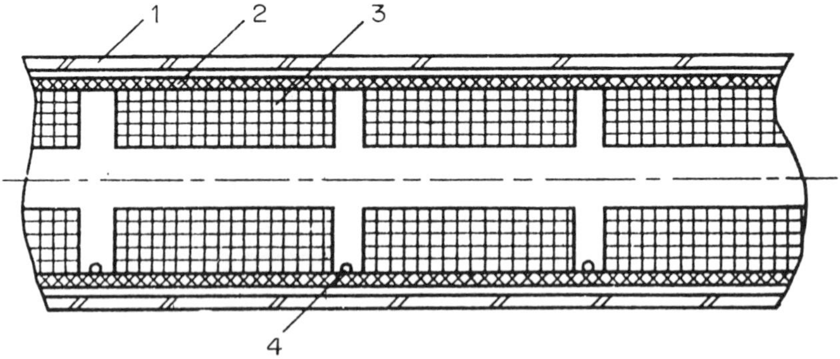

Figure 4.20 Longitudinal cross-section of a segmented discharge tube. 1, quartz sleeve; 2, ceramic tube; 3, ceramic insert; 4, lasant metal.

stability of laser output, it is necessary that the ceramic inserts make good contact with the main tube for good thermal transfer and to prevent the discharge from passing between the inserts and the main tube. With such a tube (segment dimensions: length 2.5 cm, outer diameter 2 cm, inner diameter 1 cm; full active length of the tube 45 cm), $P_{av} = 2$ W was obtained from a He–Sr laser operating at a PRF of 6.5 kHz [22, 28].

An average laser power of 2.05 W was reported in [73] from a He–Sr laser tube of 27 mm bore and 28 cm active length with $p_{He} = 90$ kPa; thus the output power per unit length reached 7.3 W m^{-1}. Distinctive features of the laser were: first, the use of a slow flow of helium through the tube, allowing the active medium to be more uniform and clean; second, employment of a current excitation scheme with a 'clipping' thyratron, which made the trailing edge of the current pulse steeper and damped undesirable current oscillations in afterglow. The flow of helium also supplied strontium vapour to the tube by entrainment from a special separately heated zone outside the discharge zone. Such a technique allows the excitation parameters and the strontium vapour pressure to be independently adjusted, and thus it is easier to obtain 'optimum' output characteristics.

A similar technique was employed in [74] where strontium pieces were placed in a special container providing good mixing of the metal vapour with the helium buffer gas, and prevented strontium oxide dust from entering the gas flow. The average output power in this experiment amounted to 1.7 W at 430.5 nm from a laser tube of 15.5 mm bore and 70 cm active length, with $p_{He} = 66$ kPa and a PRF of 6 kHz. The device exhibited good stability and reproducibility of laser characteristics.

For some applications, simultaneous oscillation at several different wavelengths is required. Therefore, the feasibility of dual oscillation in the same discharge tube on the recombination laser lines 430.5 nm (Sr II) and 373.7 nm (Ca II) were investigated [22, 28]. Pieces of Sr and Ca were placed in spatially separated internal zones of the discharge tube (1.1 cm bore, full length 30 cm, length with Sr 10 cm, length with Ca 20 cm), so that Sr (the metal with the lower ionization energy and higher vapour pressure) was situated at the cathode end, and Ca was placed on the anode side of the discharge tube. (With placement *vice versa*, Sr vapour would enter the Ca zone due to cataphoresis and suppress the Ca II laser oscillations.) As a result, simultaneous lasing at 430.5 nm and 373.7 nm was obtained with a total average output power of 0.3 W. It is interesting to note that in this tube, a heating regime close to optimum for each zone (Sr and Ca) was maintained automatically to some extent, despite the different temperatures needed for maximum laser power (50° higher for calcium): if the strontium zone of the discharge becomes overheated, its resistance falls due to the increase in metal vapour pressure, and a larger proportion of the electrical power is deposited in the calcium zone, thus heating it up.

In addition to recombination laser lines, rather strong self-terminating laser transitions excited by electron impact can be produced in Sr and Ca ion and neutral spectra. Oscillations on a number of lines with the same excitation pulse over a wide range of helium pressures and PRFs are possible in the discharge-heated regime of operation. A total power on all lines of 0.52 W was obtained in our experiments [22, 28] from a laser tube of 1.1 cm bore and active length 50 cm, with $p_{He} = 60$ kPa and a PRF of 6 kHz. The power was distributed between the lines as follows: 70%, 430.5 nm Sr II; 16%, 6.45 μm Sr I; 8%, 3.05 μm Sr I; and 3%, 1.03, 1.09 μm Sr II. Similar self-terminating laser transitions are also present in Ca I,II spectra in the He–Ca laser. Thus, by combining strontium and calcium in one tube, laser action on a number of lines ranging from 0.37 to 6.45 μm can be obtained under the same discharge conditions.

The output powers of Sr–He and Ca–He lasers rapidly grow with an increase in helium pressure, and no signs of saturation have been observed latterly up to a pressure of 200 kPa. Most recently [22, 29], we staged an experiment with a small-scale longitudinal-discharge tube in order to determine the pressure of helium above which output power would begin to decrease. The tube had metallic ends with rigidly mounted quartz windows at Brewster's angle. The discharge was excited in a BeO tube of 0.8 cm bore and 20 cm length. Unfortunately, a part of the tube aperture was blocked with a damaged electrode, resulting in an actual optical active volume of only 1.4 cm^{-3}. A maximum output power was reached at a pressure of $p_{He} = 400$ kPa, and from 100 kPa up to 400 kPa of helium the power output increased by a factor of four. Specific output characteristics from this tube reached record values for self-heated longitudinal He–Sr lasers: $P_{av} = 44.1$ mW cm^{-3}, $P_{pk} = 180$ W cm^{-3} and pulse energy $\mathcal{E} = 13.8$ μJ cm^{-3} at a PRF of 3.2 kHz.

Of the latest investigations, the work in [55] should be mentioned, where excitation of a small-scale strontium laser tube was performed using semiconductor switches. With careful matching of the plasma tube to the electrical discharge circuit, current ringing was practically eliminated. As a result, a rather good specific output power was reached, $P_{sp} = 13.7$ mW cm^{-3}, with a total average power of 118 mW. The characteristics of various types of longitudinal He–Sr and He–Ca lasers are given in Table 4.2.

4.5.2 Transverse-Discharge and e-Beam Excited Lasers

For optimum operation of He–Sr and He–Ca lasers, the helium pressure should be 1 atmosphere or more. In longitudinal discharges at such high pressures, very high voltages are required, and it is preferable to use transversely excited discharge excitation.

Oscillation was first reported at 430.5 nm in a transverse-discharge laser in [75] under non-optimized conditions. Transverse-discharge-excited recombination lasers were studied in more detail in [26, 27, 32]. In [27], a single-section tube of stainless steel of 80 mm diameter and length 70 cm was employed with ceramic inserts. Cylindrical electrodes of 200 mm length and diameters 6.4, 12.5 and 25.4 mm were spaced 25 mm from each other (for 25.4 mm diameter, the spacing was 5 mm). The best results were obtained with electrodes of 12.5 mm diameter. In [26, 32], an external silica tube of 40 cm length and 64 mm diameter was used, with four pairs of stainless-steel cylindrical electrodes (5 mm diameter, 60 mm length, 25 mm spacing). In both cases, an excitation circuit with a peaking capacitor was used (Figure 3.2b). The storage capacitance was varied in the range 15–70 nF, and the peaking capacitance 10–35 nF. The optimum ratio between storage and peaking capacitance was 1.5. The maximum applied voltage was 30 kV, and the discharge current reached 10 kA. The tube was externally heated. The best results were obtained in the single-pulse regime, or with a recurrence frequency of 1–5 Hz. However, in [27], a transversely excited laser was operated with a higher PRF, although not exceeding 500 Hz.

Uniformity of the transverse discharge is obtained without pre-ionization due to the low ionization potentials of the alkaline earth elements. The absence of means of pre-ionization greatly simplifies the construction of the tube. A drawback of the transverse discharge is the difficulty of matching the tube to the excitation circuit, which usually results in ringing of electric energy in the circuit and discharge, and can even prevent lasing in the afterglow. However, by suitable modification of the circuit, laser oscillation was obtained in the space between two successive current oscillations of equal amplitude, with a separation of 0.5 μs. This indicates that very high PRFs are feasible under certain conditions.

Laser pulse energy rises very quickly with an increase in helium pressure (Figure 4.21). As in longitudinal-discharge excited lasers, the optimum Sr concentration and discharge current increase with pressure (Figure 4.22). The fall in power noted in [27] for $p_{He} > 150$ kPa was probably due to the upper limit in discharge current in the circuit (500 A cm^{-2}). The best results obtained in He–Sr and He–Ca lasers with transverse excitation are given in Table 4.2, where it can be seen that the specific output power is an order of magnitude higher than in longitudinally excited lasers of the same type.

Hollow-cathode discharges and so-called ‘open’ discharges [76] can also be considered as transverse ones. Their characteristic feature is a highly non-Maxwellian EEDF containing a group of fast electrons, generated by ion bombardment of the cathode surface and acceleration of ejected electrons in the cathode potential fall (U = 200–300 V); and the coexistence of a far more populated group of secondary slow electrons (ε = 0.2–1 eV). This is the reason why in such kinds of discharges it is considered possible to have the

Table 4.2 Characteristic parameters of metal ion plasma recombination lasers, He–Sr (430.5 nm) and He–Ca (373.3 nm).

Laser	d (cm)	l (cm)	V_{active} (cm³)	PRF (kHz)	P_{av} (W)	P_{sp} (mW cm⁻³)	$\mathcal{E}$ (mJ)	$\mathcal{E}_{sp}$ (mJ cm⁻³)	P_{pk} (kW)	Effic. (%)	p_{He} (kPa)	Remarks
						Self-heated longitudinally excited lasers						
He–Sr	1.5	40	70.7	5	1.2	17	0.24	3.4	1.2	0.12	75	The parameters given are for the optimum regime [11, 22, 28].
He–Ca	1.5	40	70.7	5	0.7	10	0.14	2	0.7	0.07	75	
He–Sr	2.5	40	160	0.59	0.75	4.7	1.3	8.1	5	–	40	Admixture of 5% Ne [36].
He–Sr	2.5	40	160	0.25	0.3	1.88	1.1	7.1	8.1	–	40	Admixture of 15% Ne [36].
He–Ca	2.5	40	160	0.365	0.37	2.3	1.2	7	7	0.03	100	[54]
He–Sr	1.5	50	88	5.6	1.92	21.8	0.34	3.9	1.7	0.09	63	Sr is evaporated outside discharge zone and transported by He flow. Trailing edge of current clipped using thyratron [73].
He–Sr	2.7	28	160	5.6	2.05	12.8	0.37	2.29	2.5	0.108	90	Sr is evaporated outside discharge zone and transported by He flow. Trailing edge of current clipped using thyratron [73].
He–Sr	1.55	70	132	5.8	1.7	12.9	0.293	2.22	1.2	0.074	66	Slow flow of He mixed with Sr generated from an outside oven. Dust removal. Stable and long-lived operation [74].
He–Sr	1.0	45	35.3	6.5	2	57	0.31	8.7	1.55	0.12	75	Segmented laser tube: 15 segments, l = 2.5 cm, o.d. = 2.0 cm, i.d. = 1 cm, 0.7 cm segment spacing. Sleeved in BeO tube.
He–Sr	1.5	90	159	7	3	19	0.43	2.7	2.15	0.12	73	Two-section laser tube [66].
He–Ca	1.5	90	159	7	1.32	8.5	0.19	1.2	0.95	0.054	73	Two-section laser tube [22, 28].
He–Sr	1.1	30	28.5	29	3.9	130	0.13	4.6	0.65	0.14	67	Water-cooled through He layer [68].
He–Sr	0.8	20	1.4	3.2	0.062	44.1	0.019	13.8	0.25	0.055	400	The highest He pressure [22, 29].

					Externally heated longitudinally excited lasers							
He–Sr	1.3	22	30	0.01	–	–	0.18	6	1	0.05	20	Short-pulse regime ($\tau = 180$ ns) [31, 67].
He–Sr	1.3	22	30	0.01	–	–	0.25	8	0.25	0.04	16	Long-pulse regime ($\tau = 1$ μs) [31, 67].
He–Sr	4	80	1000	0.001–0.1	–	–	6	6	20	0.1	15–20	Short-pulse regime ($\tau = 300$ ns) [33].
He–Ca	4	80	750	0.001–0.1	–	–	3.2	4.2	12	0.05	40	Short-pulse regime ($\tau = 250$ ns) [118].
					Transversely excited lasers							
He–Sr	2.5[a]	25	20	0.001	–	–	1.2	50	12	0.07	170	[26]
He–Sr	2.5[a]	20	40	0.001	–	–	2.1	53	20	0.08	140	[27]
He–Ca	2.5[a]	25	20	0.001	–	–	0.75	35	7.5	0.06	150	[26]
He–Sr	2.5[a]	20	40	0.200	0.09	2.3	0.45	11	2.25	0.1	110	[27]
He–Sr	0.45	50	8	1000–3000	20	2500	20	2515	0.02	0.2	50	e-Beam excitation in an 'open discharge'. Quasi-CW oscillation with $\tau_{las} = 1$ ms. Mixture He-Kr-Sr: 50 kPa, 1.3 kPa, 1–50 Pa. Additional Sr II level excitation via charge transfer from Kr^{+} ions [76].

[a] Distance between electrodes.

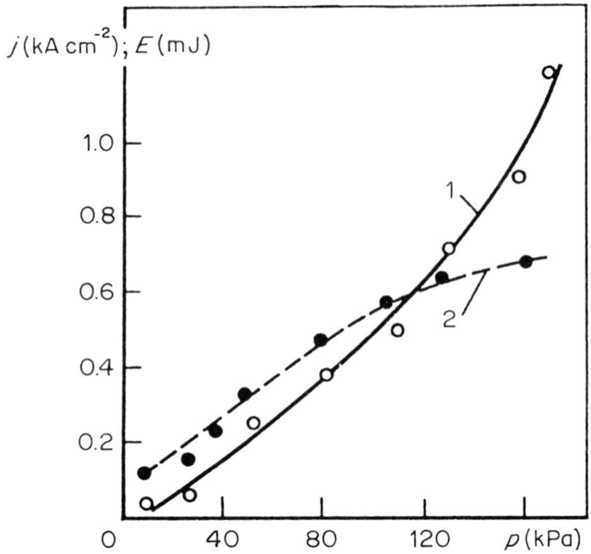

Figure 4.21 Dependences of (1) laser pulse energy, and (2) optimum discharge current density, on helium pressure in a He–Sr laser with transverse discharge excitation [26]. The active laser volume is 20 cm^3.

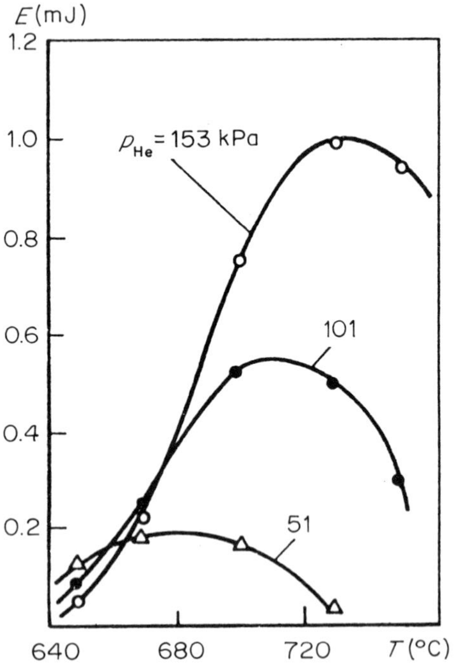

Figure 4.22 Dependence of laser pulse energy on tube temperature in a transversely excited He–Sr laser [26].

simultaneous occurrence of fast ionization and recombination processes and, consequently, to have a quasi-CW regime for recombination lasing.

Our studies [77, 78] have demonstrated the possibility of employing a hollow-cathode (HC) discharge for laser action on the IR lines of Sr II: 1123, 1087, 1201 and 1245 nm (see Figure 4.2); and of Ca II: 1184 and 1195 nm (Figure 4.3) via charge-transfer excitation by krypton ions. The laser oscillations occurred in both CW and pulsed regimes, with high small-signal gains. However, lasing at 430.5 nm was not obtained then because the current densities j and the helium pressures were too low. Attempts to increase j beyond 6 $A\,cm^{-2}$ led to arcing and the spoiling of discharge uniformity.

In later work [79], the difficulty of discharge non-uniformity at high current densities in the presence of high gas pressures in a HC discharge was overcome by the aid of pre-ionization. A 25 cm long, slot hollow cathode was used with transverse slot dimensions of 4 $\times$ 1.5 mm^2. It was found to be important to keep the gap between the cathode cylinder and the anode narrow (less than 0.5 mm) in order to allow discharges to be run at buffer-gas pressures of up to 15 kPa, without a discharge appearing outside the slot. The pre-ionization was arranged with a corona discharge between the cathode and an auxiliary wire electrode confined in a quartz capillary tube mounted inside the cathode. As a result, the attainable current density reached $j = 9\ A\,cm^{-2}$ before arcing started, and that enabled lasing on the 430.5 nm Sr II line in the recombination regime, although the output laser power and energy were low.

Better output characteristics have been obtained in 'open' discharges of coaxial design. A high-energy electron beam appears in the gap between the cylinder cathode and anode, the latter being made of molybdenum mesh. The distance between the mesh and the outer electrode was 0.5 mm, and the active length was 9 cm. The molybdenum mesh with a cell size of 0.5 $\times$ 0.5 mm^2 had a transparency coefficient of 80%. Two operation regimes of this tube were investigated depending on the polarity of the applied voltage: the 'open' discharge, i.e. a positive high voltage on the mesh, and the reverse electron beam (REB) with a negative high voltage on the mesh. The latter is very much like a hollow-cathode discharge. Laser action at 430.5 nm was attained in both regimes, although in the REB regime, laser output intensity was an order of magnitude higher under the same discharge conditions. In a mixture of He–Kr–Sr, simultaneous lasing on several ion lines pumped by different mechanisms was observed: self-terminating electron-impact excited emission at 1033 and 1091 nm, charge-transfer excitation by Kr^+ ions at 1087, 1123 and 1201 nm, and recombination pumped emission at 430.5 nm.

Far more impressive results on the Sr II 430.5 nm line in an 'open' discharge were reported in [76]. Quasi-CW laser action of 20 kW average output power was realized in a mixture of He–Kr–Sr with corresponding partial pressures of 50 kPa, 1.3 kPa and 0.01–0.05 kPa, under excitation with a burst of current pulses (PRF in the burst 1–3 MHz, burst duration 1 ms); the laser tube dimensions were 4.5 mm bore, 50 cm active length and a gap of 0.2 mm. A laser efficiency (based on discharge input power) of 0.2% was reached (see Table 4.2).

In general, as has been demonstrated in a number of studies, summarized in [80, 81], pumping of active media with a hard ionizer (an electron or proton beam of keV energies) is a universal and effective method for the creation of powerful metal vapour and gas recombination lasers, especially at high pressures of the gas mixture, although such kinds of lasers are much more complicated than gas discharge ones.

Recently, laser oscillation was for the first time obtained on the 430.5 nm Sr II line with a pulsed microwave-pumped discharge in self-heated mode, at helium pressures between 50 and 90 kPa [82]. Although the output parameters were just above threshold, further improvements of coupling of microwave energy to the plasma tube would probably enhance the laser output characteristics.

4.5.3 Applications

The lasers discussed above emit in a short-wavelength spectral region and therefore are of interest for a variety of applications; for instance, as optical pump sources for laser active media in order to transform their light into other spectral regions.

In [65], we realized quasi-resonant optical pumping of thallium vapours in a mixture with helium using a He–Ca recombination laser, resulting in laser oscillation on the green 535 nm Tl I line in a superluminescent mode.

Another possibility of converting wavelengths of recombination lasers in gaseous media is to use optical pumping of dimer molecules. By this method, conversion of 430.5 nm He–Sr laser radiation was achieved down to yellow-red wavelengths in a Te vapour cell [83]. Lasing was observed at 17 wavelengths of Te molecules, the most intense being at 642, 641.66, 631.95, 632.46 and 624.45 nm.

As a pump source for pulsed dye lasers, He–Sr and He–Ca lasers are of interest not only due to the high-energy photons they generate, but also because of the relatively long pulse-width (200–300 ns), which is an order of magnitude larger than that of traditional pump sources such as nitrogen and copper vapour lasers. Consequently, dye laser output radiation of higher quality can be obtained, i.e. with linewidths of 10's of MHz, which is close to the monochromaticity of CW dye lasers. In fact, we have successfully tested a dye laser with a He–Sr recombination laser pump [84]. Two versions of dye laser set-up were realized: one with a dye jet and the other with a dye cell.

He–Sr and He–Ca lasers may also be suitable for pumping solid-state tunable lasers. For example, the 430.5 nm strontium laser line corresponds to one of the absorption bands of an alexandrite, which can generate frequency tunable light centred on a wavelength of 760 nm with a tuning range of 100 nm.

He–Ca laser emission (373.7 nm) is of special interest in semiconductor technology, for it closely coincides with the absorption maxima of the most efficient photoresists. The He–Sr laser can be employed for the preparation and cleaning of semiconductor substrates.

These lasers have good prospects in microtechnology for thin-film processing, thermal trimming of integrated circuit elements and the repair of defects in integrated circuits and masks.

A sufficiently narrow linewidth (0.5–1.5 GHz) and adequate coherence length (10–30 cm) together with the short wavelengths of these lasers make them convenient for holographic applications: in holographic art and optical memory and for the manufacture of embossed holograms. In the last case, the valuable advantage of the He–Ca (373.7 nm) laser is that its wavelength corresponds to the maximum of sensitivity of the commonly used photoresists. As a result, the use of a 0.5 W He–Ca laser is equivalent to the use of an argon-ion laser delivering more than 100 W total output power.

The application of a strontium laser as one of the components in a laser projection television system is also possible [85]. Such a system has already been tested with a 10 W copper vapour laser [86]. The approach to image formation in it demands the laser to have a PRF of about 15 kHz in order to match laser operation to a standard TV signal. This PRF is attainable by strontium lasers. Thus, a blue 430.5 nm Sr II line in conjunction with green 511 nm Cu I and red 628 nm Au I laser lines would provide a high-quality colour image on a relatively large screen ($\approx$10 m^2). However, for suitable performance, a He–Sr laser with an output power of 5–10 W must be developed. Metal vapour lasers (mainly copper vapour lasers) are now actively used as brightness amplifiers in so-called 'projection microscopes'. A He–Sr laser does not have small-signal gain as high as that of a copper vapour laser, but the gain is still high enough for that type of application by making use of a multi-pass amplifier arrangement. This has been demonstrated in experiments [87–89]: a magnified image of an object of appropriate brightness, contrast and resolution was formed with a He–Sr amplifier and viewed on a screen of about 1 m^2 in size.

Various applications of strontium and calcium recombination lasers in medicine can be forecast for the UV irradiation of blood, for phototherapy of neonatal jaundice, for fluorescence diagnostics and for the photodynamic therapy (PDT) of cancer.

In the case of PDT, we can give a little more detail. The 416.2 nm Sr II laser wavelength almost exactly matches one of the absorption maxima of haematoporphyrin derivative (HPD), the key substance used for treatment in PDT. Hence the HPD can be made to fluoresce efficiently using this type of laser, which is important for diagnosis and monitoring of cancers. Thus far, rather expensive and inefficient Kr-ion lasers are usually used for excitation of the fluorescence. Hence, the He–Sr laser could be a good alternative to the Kr-ion laser in cancer diagnostics. For the treatment of cancer by PDT, absorption of HPD in the red region (610–635 nm) of its absorption spectrum is normally used, supplied usually from an argon-ion laser pumped dye laser. However, red laser dyes can also be pumped by a He–Sr laser. Consequently, a medical installation based on a He–Sr laser might be much simpler and more cost-effective than one employing krypton-ion and argon-ion lasers.

The list of He–Sr and He–Ca laser applications will, of course, be added to when these simple and relatively cheap sources of short-wavelength coherent radiation appear in the marketplace. Therefore, their development, especially towards greater reliability and higher output powers, is certainly worthwhile and should be continued.

4.6 PUMP MECHANISMS IN SEGMENTED-PLASMA-EXCITED RECOMBINATION LASERS

In [90–92], a recombination laser with novel excitation was reported: i.e. the segmented-plasma-excited recombination (SPER) laser (Figure 4.23). Its operation is as follows. On a long electrically insulating substrate is mounted a series of up to 65 metal strips, spaced by 1–2 mm, each of 1 mm thickness, 2 mm width and 10–20 mm length. A capacitor of 10 nF charged to 10–20 kV is discharged across the strips. Pulsed arc discharges are produced in the gaps between the strips, with current pulse durations of 2–6 μs, and peak currents of 50–800 A. The discharge causes metal to be evaporated, ionized and expanded hemispherically into the buffer gas surrounding it (usually helium at 0.1–0.5 kPa). The speed of expansion is about 5×10^5 cm s^{-1} The plasma cools by collisions of electrons with helium atoms, and due to the adiabatic

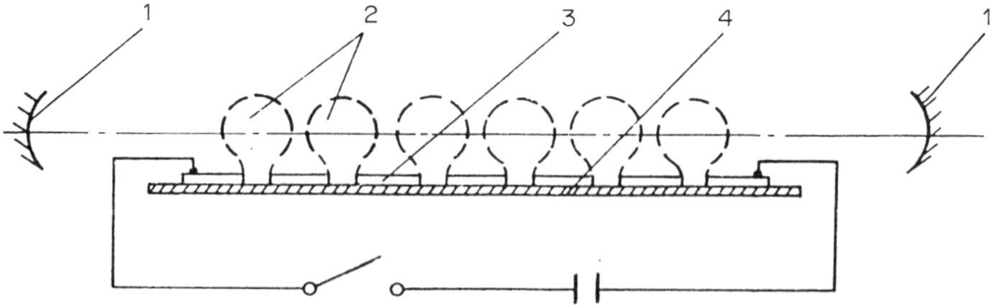

Figure 4.23 Schematic diagram of a segmented-plasma-excited recombination laser [90]. 1, Mirrors; 2, expanding plasma; 3, metal strips; 4, dielectric substrate.

expansion. Helium is necessary in order to prevent the very fast expansion, which would rapidly decrease the plasma density. The axis of the optical cavity is approx. 7 mm above the metal strips. An attractive feature of the SPER laser is that it is simple, with no external heating, and can easily produce intense recombining plasmas of non-volatile metals.

Laser oscillation has been obtained in SPER lasers on 60 new transitions of neutral and ion spectra of Ag, C, Cd, In, Mg, Pb, Sn, Zn, Ca, Cu, Ba, Li and Al [90–92] (see Appendix 4). Some of these transitions also oscillate in conventional discharge afterglows. Laser oscillation is observed in the expanding afterglow during the peaks of the spontaneous emission of all lines. The delay and duration of the laser pulse of a neutral line are 20–40 μs, those of a singly charged ion are around 6.5 μs, and for a doubly charged ion they are 3.5–4 μs, which is in agreement with the expected dependence of the rate of recombination on ion charge. A rapid fall in the current pulse is required for laser oscillation on the ion lines.

It was found that the metals evaporate with different rates in SPER lasers, depending on the helium pressure (0.1–1 kPa). Stable arcing is produced in each gap with Li, Al, Ca and Cu of standard strip size at $p_{He} \geq 5$ kPa, while at lower pressures, the discharge appears partly in the helium, bypassing the strips. The best results at low He pressures are achieved with thin foils of 0.1–0.2 mm thickness and separations of 0.2 mm (Li, Ca) or 0.5 mm (Al, Cu) [92]. The improved production of arcs and foil evaporation is due to the lower heat transfer in comparison with larger strips. This causes an increase in plasma density and in the rate of expansion, since less energy is wasted on heating. This is particularly important for elements of high heat capacity and thermal conductivity, such as copper and aluminium. A useful parameter for determining the vapour production in SPER lasers is $M = (\Theta c \rho T^2)^{-1}$, where T is the temperature required to produce 13 Pa over the metal surface, Θ is the thermal conductivity, c is the specific heat capacity and ρ is the density of the metal [92]. The elements can be arranged in order of decreasing M: Cd ($M = 1.0$), Na (0.99), Pb (0.57), In (0.41), Zn (0.39), Li (0.33), Ca (0.27), Sn (0.11), Ni (0.03), Al (0.02), Cu (0.01). One of the reasons for the weak thermal evaporation of the alkali and alkaline earth elements in the arc discharge is their chemical activity, which leads to the formation of an oxide layer on the surface. Investigations of the plasma parameters and excitation of the metal lines of SPER lasers are discussed in [93–105].

Among the lines obtained in [90] that are of particular interest are those transitions that belong to isoelectronic ion series (isoelectronic scaling), e.g. 4 $^2F \longrightarrow 5\,^2D$, to which belong 1.838, 1.832 μm (Ag I), 537.8, 533.7 nm (Cd II) and 300.8, 298.3 nm (In III). Laser oscillation on those lines indicates that the dispositions of energy levels are analogous, and meet the requirements for population inversion in the recombination regime, *i.e.* recombination pumping of the upper levels and radiative decay of the lower levels. This series can be continued to heavier multiply charged ions, which can be expected to be the source of very short wavelength laser oscillation. Generally, nearly all the new recombination laser transitions in the SPER laser are between more or less narrow groups of levels, i.e. they are in accordance with the requirements set out in section 1.4 for recombination lasers.

The SPER laser can be scaled and improved. In [90], two parallel series of metal strips were used, which produced plasmas expanding towards each other, and the laser intensity at 1.433 μm (Cd I) increased by a factor of 5 as a result. When glass plates were placed either side of the discharge channel in order to confine the plasma expansion, the laser power increased by 4 times. It is clear that by placing strips of different metals in series or parallel, simultaneous lasing on a number of different lines of different elements may be obtained.

The peak laser power of SPER lasers is typically 0.5–5 W, but could be as high as 1 kW, as predictions in [91] show.

4.7 LASER-INDUCED-PLASMA RECOMBINATION LASERS

The development of pulsed neodymium–yttrium-aluminium-garnet (Nd:YAG) and CO_2 lasers of sufficient peak power to create plasmas in gases and over the surfaces of solids when focused has given rise to a new field in physics, that of laser-induced plasmas. Such plasmas are very promising as active media for new laser transitions, especially of the recombination type. A short laser pump pulse can be focused to a small area, thus causing evaporation of the irradiated matter, creation of a plasma containing multiply charged ions and rapid expansion and adiabatic cooling conducive to intense recombination.

A laser-induced plasma was first reported as an active medium for MVILs in [106]. A Cd target was placed in a vacuum and irradiated with a focused CO_2 laser beam of 10 MW peak power and 80 ns duration. In the Cd vapour, a laser pulse of 1 μs duration (delayed by 200 ns from the pump pulse) was observed on the 537.8 nm line of Cd II. The gain was estimated to be 0.01 cm^{-1}. The axis of the cavity was 5 mm above the target. It was proposed that the upper laser level was excited by electron impact. However, as was shown in the experiments of [107,108], the pump mechanism is more likely to be recombination of doubly charged ions, because the laser pulse was generated after the excitation pulse, and was characterized by a decay time typical for recombination.

Further experiments with laser-induced Cd vapour plasmas [107,108] have shown that the presence of helium is important for lasing in the recombination regime. Helium accelerates the cooling of electrons, reduces the rate of expansion of the plasma, and confines the region of recombination. These three factors combine to sharply increase the intensity of Cd I and Cd II recombination lines.

By using plates to confine the plasma expansion in the transverse direction [109], the laser pulse energy of the 1.43 μm Cd I recombination line increases by two orders of magnitude under the same pumping conditions. The pulse duration increases from 5 to 45 μs, and the amplitude by 22 times. First, the confining plates impede the rate of plasma expansion and the rate of decrease in electron density, which, however, does not hinder the degree of cooling of the electrons, since the plasma is produced in helium (0.5 kPa). Second, the plasma expands longitudinally, i.e. in the direction of the output beam. Laser oscillation is obtained between the plates, and above them. Maximum laser power is generated at the top of the plates, where plasma cooling is enhanced by the adiabatic expansion.

Control of the plasma expansion using opaque plates in [110] yielded, for the first time, laser oscillation in a recombining laser-induced plasma on Be ion lines. Nd:YAG laser pulses (20 J, 10 ns) were focused on a Be target (20 mm long and 0.2 mm wide) with an incident power density of 5×10^{10} W cm^{-2}. A shield with a slit for the pump beam was placed at a distance of 6–12 mm from the target. Emission of the expanding plasma near the shield was enhanced, and laser oscillation was obtained on the Be II transitions 467 nm ($4\,^2F \longrightarrow 3\,^2D$), 436 nm ($4\,^2D \longrightarrow 3\,^2P$) and 527 nm ($4\,^2S \longrightarrow 3\,^2P$). By employing roof-shaped shields, the plasma is confined to a small region, and laser power increases further. The effect of the obstructing shields can be explained in the following way [111]. Where the plasma flows around the shields, shock waves are developed. The plasma concentration just beyond the wave front increases greatly, but the temperature changes slowly, e.g. the shock wave heats the heavy particles but not the electrons. In the case of adiabatic compression, and the resultant increase in T_e, n_e and n^+, the rate of recombination is given by $W_r \sim T_e^{-s} n_e^2 n^+$, where $s = 9/2$ for three-body recombination and $s < 9/2$ for collisional-radiative recombination. If n_e and n^+ rose K times adiabatically, T_e would increase by $K^{(5/3-1)} = K^{2/3}$ times, and the recombination flux would change by $K^{(9-2s)/3}$ times. Thus, if three-body recombination is the main pump process, W_r would not change, but if collisional-radiative recombination prevails (e.g. $s < 9/2$) W_r would, in general, increase (more details in [112]).

A great number of laser lines of singly charged ions of various metals in laser-induced recombining plasmas was reported in [108] (see Appendix 4). A CO_2 laser was used with 2 J pulse energy. The output of that laser was split into eight beams, each focused onto the target, so that a linear sequence of spherically expanding laser plasmas was produced. Expansion of the plasmas took place in helium at a pressure of 0.1–0.3 kPa. The output laser pulses were ≈5 μs long, and delayed by 5 μs from the pump pulse.

Nd:YAG (1064 nm, 700 mJ, 10 ns), KrF (248 nm, 600 mJ, 20 ns; 10 mJ, 360 fs) lasers were employed in [113] for the creation of expanding metal vapour laser plasmas. The outputs of these lasers were divided into nine beams and plasma confinement for the first stage of expansion was used. As a result, a number of new recombination laser lines were obtained both in neutral and in ion spectra of Mg, Zn, Cd, Sn and Pb.

By employing a more powerful Nd:YAG laser delivering 20 J pulses of 3 ns duration, whose output was focused by a special objective to five spots of 0.4 mm diameter, and spaced 8 mm from each other, laser oscillation was obtained on the 253 nm (4f ⟶ 5g) transition of Be IV [114]. The duration of the laser pulse was 60 ns (full width at half-maximum). Studies of this laser were continued in [115, 116].

The experiments in [108, 113–116] have demonstrated the capabilities of laser-induced plasmas as sources of ion laser transitions in the recombination regime. However, the efficiency of conversion from laser pump energy to laser output energy is very low. Nevertheless, it is one of the most promising methods for producing laser oscillation in the vacuum UV and soft X-ray spectral regions, and the experience gained from obtaining laser oscillation in low-charged ions in laser-induced recombining plasmas is certainly valuable. This has been confirmed by the latest successes in X-ray laser realization, mostly on laser-induced plasmas, including laser plasmas using recombination pump mechanisms for obtaining oscillation [117].

REFERENCES

[1] A. V. Yeletski and B. M. Smirnov, *Physical Processes in Gas Lasers,* Moscow: Energoatomizdat, (1985) (in Russian).

[2] A. L. Bloom, W. E. Bell and F. O. Lopez, *Phys. Rev.,* **135**(3A), 578–9 (1964).

[3] M. F. Sem and V. S. Mikhalevskii, *Zh. Prikl. Spektrosk.,* **6**(5), 668–9 (1967) (in Russian); *Elektron. Tekh., Ser. 1 Elektronika SVCh,* No. 9, 98–102 (1969) (in Russian).

[4] V. V. Zhukov, I. G. Ivanov and M. F. Sem, *Zh. Prikl. Spektrosk.,* **26**(3), 544–7 (1977) (in Russian).

[5] H. Ninomiya, H. Osumi and S. Horiguchi, *J. Appl. Phys.,* **51**(12), 6091–4 (1980); **52**(4), 2699-704 (1981); **52**(6), 3889-91 (1981); **52**(5), 3229–32 (1981).

[6] V. S. Aleinikov, *Opt. Spectrosc.,* **28,** 15–17 (1970).

[7] E. L. Latush, *Izv. Ser.-Kav. Nauch Tsentra Vissh. Shkoli. Ser. Yestyestv. Nauk,* No. 2, 101–2 (1973) (in Russian).

[8] A. A. Gudkov, V. F. Kravchenko and V. F. Papakin, *JETP Lett.,* **17**(1), 11–13 (1973).

[9] D. A. Korogodin, E. L. Latush, M. F. Sém and G. D. Chebotarev, *Sov. J. Quantum Electron.,* **20**(6), 634–8 (1990).

[10] E. L. Latush and M. F. Sem, *Sov. Phys.-JETP,* **37**(6), 1017–18 (1973).

[11] V. V. Zhukov, V. S. Kucherov, E. L. Latush and M. F. Sém, *Sov. J. Quantum Electron.*, **7**(6), 708–14 (1977).
[12] V. V. Zhukov, E. L. Latush, V. S. Mikhalevskii and M. F. Sém, *Sov. J. Quantum Electron.*, **7**(6), 704–8 (1977).
[13] E. L. Latush and M. F. Sém, *Sov. J. Quantum Electron.*, **3**(3), 216–19 (1973).
[14] V. V. Zhukov, E. L. Latush and M. F. Sém, *Zh. Prikl. Spektrosk.*, **32**(4), 738–40 (1980) (in Russian).
[15] V. V. Zhukov, E. L. Latush and M. F. Sém, *Izv. Ser.-Kav. Nauch Tsentra Vissh. Shkoli. Ser. Yestyestv. Nauk.*, No. 11, 38–42 (1977) (in Russian).
[16] E. L. Latush and M. F. Sém, *JETP Lett.*, **15**(11), 457–9 (1972).
[17] V. V. Zhukov, E. L. Latush and M. F. Sém, *Sov. Phys. J.*, **20**(7), 957–9 (1978).
[18] L. M. Biberman, V. S. Vorobev and I. T. Yakubov, *Nonequilibrium Kinetics of Low Temperature Plasmas,* Moscow: Nauka (1982) (in Russian).
[19] L. A. Vainshteyn, I. I. Sobelman and E. D. Yukov, *Excitation of Atoms and Spectral Line Broadening,* Moscow: Nauka (1979) (in Russian); *Cross Sections for Electronic Excitation of Atoms and Ions,* Moscow: Nauka (1973) (in Russian).
[20] S. M. Babenko and S. I. Yakovlenko, *Analysis of Kinetic Processes in the He-Sr Laser,* Preprint No. 3192, Moscow: Institute of Atomic Energy (1979) (in Russian).
[21] A. L. Belov, Yu I. Syts'ko and S. I. Yakovlenko, *Sov. J. Quantum Electron.*, **14**(8), 1083–7 (1984).
[22] E. L. Latush, M. F. Sem, Yu V. Koptev, L. M. Bukshpun and S. N. Atamas', *Metal Vapor Lasers and Their Applications: CIS Selected Papers,* ed. G. G. Petrash, *Proc. SPIE,* **2110,** 106–27 (1993).
[23] V. M. Klimkin, V. E. Prokop'ev and L. V. Fadin, *Sov. J. Quantum Electron.*, **9**(3), 357–9 (1979).
[24] D. G. Loveland and C. E. Webb, *J. Phys. D.*, **25,** 597–604 (1992).
[25] D. G. Loveland, A. F. Zerrouk and C. E., Webb *Plasma Sources Sci. Technol.*, **1**(2), 141–6 (1992).
[26] M. S. Butler and J. A. Piper, *Appl. Phys. Lett.*, **42**(12), 1008–10 (1983); **43**(9), 823–5 (1983).
[27] M. Brandt, *Appl. Phys. Lett.*, **42**(2), 127–9 (1983); *IEEE J. Quantum Electron.* **20**(9), 1006–7 (1984).
[28] E. L. Latush, M. F. Sem, L. M. Bukshpun, Yu V. Koptev and S. N. Atamas', *Opt. Spektrosk.*, **72**(5), 672–80 (1992) (in Russian).
[29] S. N. Atamas', E. L. Latush and M. F. Sem, *J. Russ. Laser Res.*, **15**(1), 66–8 (1994).
[30] É. K. Karabut, V. F. Kravchenko, V. S. Mikhalevskii and A. P. Shelepo, *Sov. J. Quantum Electron.*, **5**(11), 1371–2 (1975).
[31] M. S. Butler and J. A. Piper, *Appl. Phys. Lett.*, **45**(7), 707–9 (1984).
[32] J. A. Piper and M. S. Butler, *CLEO '84, Anaheim, CA, USA, 19–22 June 1984*, Washington, DC: OSA/IEEE, *Tech. Dig.*, 218, (1984).
[33] M. S. Butler and J. A. Piper, *IEEE J. Quantum Electron.*, **21**(10), 1563–6 (1985).
[34] R. Künnemeyer, C. W. McLucas, D. J. W. Brown and A. I. McIntosh, *IEEE J. Quantum Electron.*, **23**(11), 2028–32 (1987).
[35] Yu V. Koptev, E. L. Latush and M. F. Sem, *Opt. Spektrosk.*, **72**(5), 681–4 (1992) (in Russian).
[36] C. E. Little and J. A. Piper, *IEEE J. Quantum Electron.*, **26**(5), 903–10 (1990).
[37] M. K. Dyatlov, E. P. Ostapchenko and S. B. Spasin, *Zh. Prikl. Spektrosk.*, **26**(4), 613–17 (1977) (in Russian).

[38] *Spectroscopy of Gas-Discharge Plasmas: Collected Papers,* ed. S. E. Frisha, Leningrad: Nauka (1970) (in Russian).
[39] V. V. Zhukov, V. G. Il'yushko, E. L. Latush and M. F. Sém, *Sov. J. Quantum Electron.,* **5**(7), 757–60 (1975).
[40] L. I. Gudzenko and S. I. Yakovlenko, *Plasma Lasers,* Moscow: Atomizdat (1978) (in Russian).
[41] I. G. Ivanov and M. F. Sem, *Elektron. Tekh., Ser. 4 Elektrovakuumiye i Gazorazryadniye Pribori,* No. 2, 12–16 (1974) (in Russian).
[42] I. G. Ivanov and M. F. Sem, *Zh. Prikl. Spektrosk.,* **19**(2), 358–60 (1973) (in Russian).
[43] V. V. Zhukov, E. L. Latush, V. S. Michalevskii and M. F. Sém, *Sov. J. Quantum Electron.,* **5**(4), 468–9 (1975).
[44] W. T. Silfvast, L. H. Szeto and O. R. Wood II., *Appl. Phys. Lett.,* **33**(11), 936–8 (1978).
[45] B. I. Grinchenko, *Sov. Phys.-Tech. Phys.,* **27**(9), 1163–4 (1982); Preprint No. 5-155, Moscow: Institute of High Temperatures, Acad. Sci. USSR (1985) (in Russian).
[46] B. I. Grinchenko, *Teplofiz. Vys. Temp.,* **27**(3), 621 (1989) (in Russian).
[47] B. I. Grinchenko, *High Temp.,* **31**(2), 189–96 (1993).
[48] T. U. Arslanvekov, V. I. Derzhiev, M. E. Talis *et al., Summaries of Papers. Population Inversion and Oscillation on Atomic and Molecular Transitions,* Part 1, ed. O. V. Ravodina and I. I. Murav'ev, pp. 260–1, Tomsk: Institute of Atmospheric Optics (Sept. 1986) (in Russian).
[49] B. M. Smirnov, *Complex Ions,* Moscow: Nauka (1983) (in Russian).
[50] C. W. McLucas and A. I. McIntosh, *J. Phys. D.,* **20**(5), 591–6 (1987).
[51] E. L. Latush, Yu V. Koptev, M. F. Sém, G. D. Chebotarev and D. A. Korogodin, *Sov. J. Quantum Electron.,* **21**(12), 1314–20 (1991).
[52] R. J. Carman, *IEEE J. Quantum Electron.,* **26**(9), 1588–1608 (1990).
[53] R. J. Carman, *J. Phys. D.,* **24,** 1803–10 (1991).
[54] C. E. Little and J. A. Piper, *Opt. Commun.,* **68**(4), 282–6 (1988).
[55] Y. Takeda, A. Iwata, S. Ueguri and K. Fujii, *IEEE J. Quantum Electron.,* **30**(5), 1176–80 (1994).
[56] D. Orchard and C. E. Webb, *CLEO '90, Anaheim, CA, USA, 21–25 May 1990,* OSA. Technical Digest, pp. 108–9, Washington DC: Optical Society of America (1990).
[57] V. S. Pavlenko, *Sov. J. Quantum Electron.,* **22**(11), 1002–6 (1992).
[58] V. S. Pavlenko and V. G. Egorov, *Quantum Electronics,* **23**(11), 969–71 (1993).
[59] B. J. Tarte and J. A. Piper, *CLEO '88, Anaheim, CA, USA, 25–29 April 1988,* OSA. Technical Digest, pp. 276–8, Washington DC: Optical Society of America (1988).
[60] V. V. Zhukov, V. S. Kucherov, E. L. Latush, M. F. Sem and G. N. Tolmachev, *Pis'ma Zh. Tekh. Fiz.,* **2**(12), 550–3 (1976) (in Russian).
[61] V. S. Efimov, V. V. Kyun, Yu P. Mikhailov *et al., Elektron. Prom.,* No. 5–6 127 (1981).
[62] C. W. McLucas and A. I. McIntosh, *J. Phys. D.,* **19**(7), 1189–95 (1986).
[63] C. E. Little and J. A. Piper, *Opt. Commun.,* **68**(4), 282–6 (1988).
[64] L. M. Bukshpun, V. V. Zhukov, E. L. Latush and M. F. Sém, *Sov. J. Quantum Electron.,* **11**(6), 804–5 (1981).
[65] S. N. Atamas', Yu V. Koptev and E. L. Latush, *Sov. J. Quantum Electron.,* **15**(2), 288–9 (1985).

[66] L. M. Bukshpun, S. N. Atamas, V. V. Zhukov *et al., Izv. Vyssh. Uchebn. Zaved., Fiz,*. No. 6, 105–7 (1983) (in Russian).
[67] C. E. Little and J. A. Piper, *Metal Vapor, Deep Blue and Ultraviolet Lasers,* ed J. J. Kim, R. Kimball, P. J. Wisoff, *Proc. SPIE,* **1041,** 167–74 (1989).
[68] L. M. Bukshpun, E. L. Latush and M. F. Sem, *Teplofiz. Vys. Temp.,* **24**(2), 402–5 (1986).
[69] T. R. Pugsley and C. E. Little III, *European and X Natl. Quantum Electronics Conf., Heriot-Watt University, Edinburgh, 27–30 Aug. 1991, Techn. Dig.* 60 (1991).
[70] L. M. Bukshpun, E. L. Latush and M. F. Sém, *Sov. J. Quantum Electron.,* **18**(9), 1098–100 (1988).
[71] R. M. Hentschel, R. J. Carman and J. A. Piper, *CLEO '92, Anaheim, CA, USA, 10–15 May 1992*, OSA. Technical Digest, pp. 544, 546–7, Washington DC: Optical Society of America (1992).
[72] J. W. Bethel and C. E. Little, *Opt. Commun.,* **84**(5,6), 317–22 (1991).
[73] P. A. Bokhan and D. É. Zakrevskii, *Sov. J. Quantum Electron.,* **21**(8), 838–9 (1991).
[74] D. G. Loveland, D. A. Orchard, A. F. Zerrouk and C. E. Webb, *Meas. Sci. Technol.,* **2**(11), 1083–7 (1991).
[75] A. M. Bogus, V. L. Dzhikiya and A. A. Chernov, *Sov. J. Quantum Electron.,* **8**(2), 259–60 (1978).
[76] P. A. Bokhan and A. R. Sorokin, *Opt. Quantum Electron.,* **23**(4), S523–38 (1991).
[77] V. V. Zhukov, E. L. Latush and M. F. Sem, *Zh. Prikl. Spektrosk.,* **32**(4), 738–40 (1980) (in Russian).
[78] E. L. Latush, R. Solanki and G. J. Collins, *Phys. Lett. A,* **73**(5,6), 387–8 (1979).
[79] Yu V. Koptev, E. L. Latush and M. F. Sém, *Sov. J. Quantum Electron.,* **20**(4), 346–8 (1990).
[80] S. I. Yakovlenko (ed.), *Plasma Lasers in Visible and Near-UV Regions, Proc. General Physics Inst. of USSR Academy of Sciences,* **21,** (1989) (in Russian) .
[81] A. V. Karelin and S. I. Yakovlenko, *Quantum Electronics,* **23**(7), 545–63 (1993).
[82] J. W. Bethel, A. Maitland, P. M. Beecham, T. C. Bell, S. Webb and C. E. Little, *Opt. Quantum Electron.,* **26**(12), 1079–87 (1994).
[83] S. N. Atamas', Yu V. Koptev and E. L. Latush, *Sov. J. Quantum Electron.,* **15**(2), 288–9 (1985).
[84] S. P. Zinchenko, E. L. Latush, M. F. Sém and R. A. Tikidzhi-Khambur'yan, *Sov. J. Quantum Electron.,* **22**(9), 798–9 (1992).
[85] H. W. Messenger, *Laser Focus World,* **26**(4), 87–92 (1990).
[86] R. R. Lewis, G. A. Naylor, N. Salked, A. J. Kearsley and C. E., Webb *New Developments and Applications in Gas Lasers,* ed. L. R. Carlson, *Proc. SPIE,* **737,** 10–16 (1987).
[87] S. P. Zinchenko, G. G. Petrash and M. F. Sem, *Sov. J. Quantum Electron.,* **23**(7), 585–7 (1993).
[88] D. V. Vlasov, P. I. Ivashkin, A. A. Isaev, M. A. Kazaryan, T. I. Kuznetsova and V. V. Chykov, *Phys. Scr.,* **48**(4), 461–3 (1993).
[89] Yu P. Vasiliev, Yu D. Dumarevsky, K. I. Zemskov, M. A. Kazaryan, L. V. Medvedeva, G. A. Petrovicheva, V. V. Chvykov , D. V. Vlasov and P. I. Ivashkin, *Phys. Scr.,* **51,** 92–3 (1995).
[90] W. T. Silfvast, L. H. Szeto and O. R. Wood II, *Appl. Phys. Lett.,* **36**(8), 615–7 (1980); **39**(3), 212–14 (1981).

[91] W. T. Silfvast and O. R. Wood II, *Opt. Lett.*, **7**(1), 34–6 (1982).
[92] J. J. Macklin, O. R. Wood II and W. T. Silfvast, *IEEE J. Quantum Electron.*, **18**(11), 1832–5 (1982).
[93] A. A. Babin, I. I. Murav'ev, L. D. Shatova and A. M. Yancharina, *Sov. Phys. J.*, **29**(7), 531–4 (1986).
[94] V. D. Burlakov, I. I. Murav'ev, A. N. Soldatov, E. V. Chernikova and A. M. Yancharina, *J. Appl. Spectrosc.*, **46**(1), 19–22 (1987).
[95] V. V. Apollonov, S. I. Derzhavin, A. M. Prokhorov and A. A. Sirotkin, *Sov. Tech. Phys. Lett.*, **14**(11), 434–5 (1988); *Sov. J. Quantum Electron.*, **19**(5), 618–20 (1989).
[96] V. V. Apollonov and A. A. Sirotkin, *Sov. J. Quantum Electron.*, **19**(8), 1029–30 (1989).
[97] V. V. Apollonov and A. A. Sirotkin, *Sov. J. Quantum Electron.*, **19**(10), 1319–20 (1989).
[98] V. V. Apollonov, A. V. Ermachenko and A. A. Sirotkin, *Sov. Tech. Phys. Lett.*, **15**(2), 125–6 (1989) .
[99] R. K. Thareja and A. Khare, *Opt. Lett.*, **12**(1), 28–9 (1987).
[100] A. Khare, V. Kumar and R. K. Thareja, *Z. Phys. D.*, **6**(1), 67–71 (1987).
[101] A. Khare and R. K. Thareja, *J. Appl. Phys.*, **62**(9), 3981–4 (1987).
[102] A. Khare and R. K. Thareja, *J. Appl. Phys.*, **63**(1), 253–4 (1988).
[103] A. Khare and R. K. Thareja, *Appl. Opt.*, **27**(5), 834–5 (1988).
[104] A. Khare and R. K. Thareja, *IEEE J. Quantum Electron.*, **24**(12), 2525–9 (1988).
[105] I. I. Murav'ev, T. M. Gorbunova and A. M. Yancharina, *Sov. Phys. J.*, **31**(4), 299–302 (1988).
[106] K. Toyoda, M. Kobiyama and S. Namba, *Jpn. J. Appl. Phys.*, **15**(10), 2033–4 (1976).
[107] J. M. Green, W. T. Silfvast and O. R. Wood II, *J. Appl. Phys.*, **48**(7), 2753–61 (1977).
[108] O. R. Wood II, J. J. Macklin and W. T. Silfvast, *Appl. Phys. Lett.*, **44**(12), 1123–5 (1984).
[109] W. T. Silfvast, L. H. Szeto and O. R. Wood II, *Appl. Phys. Lett.*, **36**(7), 500–2 (1980).
[110] V. A. Boiko, B. A. Bryunetkin, V. F. Bunkin *et al.*, *Sov. J. Quantum Electron.*, **13**(5), 565–6 (1983); **15**(11), 1543–4 (1985).
[111] V. A. Boiko, B. A. Bryunetkin and V. F. Bunkin, *Sov. Tech. Phys. Lett.*, **9**(6), 289–91 (1983); *Sov. J. Quantum Electron.*, **14**(7), 901–4 (1984).
[112] V. I. Derzhiev, S. A. Maiorov and S. I. Yakovlenko, *Sov. Phys.-Tech. Phys.*, **30**(2), 221–3 (1985).
[113] C. Momma, M. Hube, A. Tünnermann, K. Mossavi and B. Wellegehausen, *Appl. Phys. B*, **54**(3), 234–8 (1992).
[114] B. A. Bryunetkin, V. I. Derzhiev, V. M. Dyakin, S. A. Maiorov and S. I. Yakovlenko *Sov. Tech. Phys. Lett.* **12**(5) 252–3 (1986).
[115] B. A. Bryunetkin, V. M. Dyakin, I. Yu Skobelev, A. Ya Faenov and S. Ya Khakhalin, *Sov. J. Quantum Electron.*, **19**(3), 271–2 (1989).
[116] B. A. Bryunetkin, V. M. Dyakin, G. A. Koldashov, I. Yu Skobelev and A. Ya Faenov, *Sov. J. Quantum Electron.*, **20**(10), 1248–50 (1990).
[117] C. H. Skinner, *Phys. Fluids, B*, **3**(8), 2420–9 (1991).
[118] M. S. Butler and J. A. Piper, *Opt. Lett.*, **12**(3), 166–8 (1987).

5

Positive-Column Excitation of CW Metal Vapour Ion Lasers

5.1 TUBE CONSTRUCTION

Although the positive-column (PC) discharge is not the best type of discharge for the excitation of MVILs (see section 2.6), it is the most widely exploited in present-day systems. This has mainly been due to clever engineering, resulting in the development of laser tubes no more complicated than those of He–Ne lasers. The earliest example of a PC-discharge-excited laser made use of cataphoresis to seed the PC discharge with metal vapour: the so-called 'cataphoretic' MVLs. Positive-column lasers produce a great number of laser lines from the ultraviolet through to the infrared regions of the spectrum, as can be seen in Table 5.1, making them useful in many scientific and technological applications. Until ultraviolet hollow-cathode lasers using the vapours of Cu, Ag and other refractory metals were developed (see Chapter 6), cataphoretic He–Cd lasers were the sources of the shortest-wavelength CW laser radiation (325.0 and 441.6 nm). Nevertheless, the cataphoretic He–Cd laser still provides one of the shortest wavelengths available commercially (together with UV noble-gas ion lasers).

Cataphoretic laser tubes have a major advantage over externally heated laser tubes: there are no major falls in temperature at the ends of the gain medium, which would prevent utilization of the full laser gain medium and cause losses of the lasant by diffusion, and thereby contamination of the laser windows [1, 2].

In most cataphoretic laser designs [3–6], the active zone has a diameter of 1–4 mm and is heated by the discharge current; the metal source oven is near the anode and a condenser is provided near the cathode (Figures 5.1 and 5.2). Diffusion of metal vapour to the laser windows (or internal mirrors) can be prevented by the use of an additional anode [5]. The length of the additional retarding region must be sufficient (see equation (2.27)). The metal vapour is usually produced by using an external heater, or self-heating by the discharge; cathode sputtering [7] or dissociation of chemical

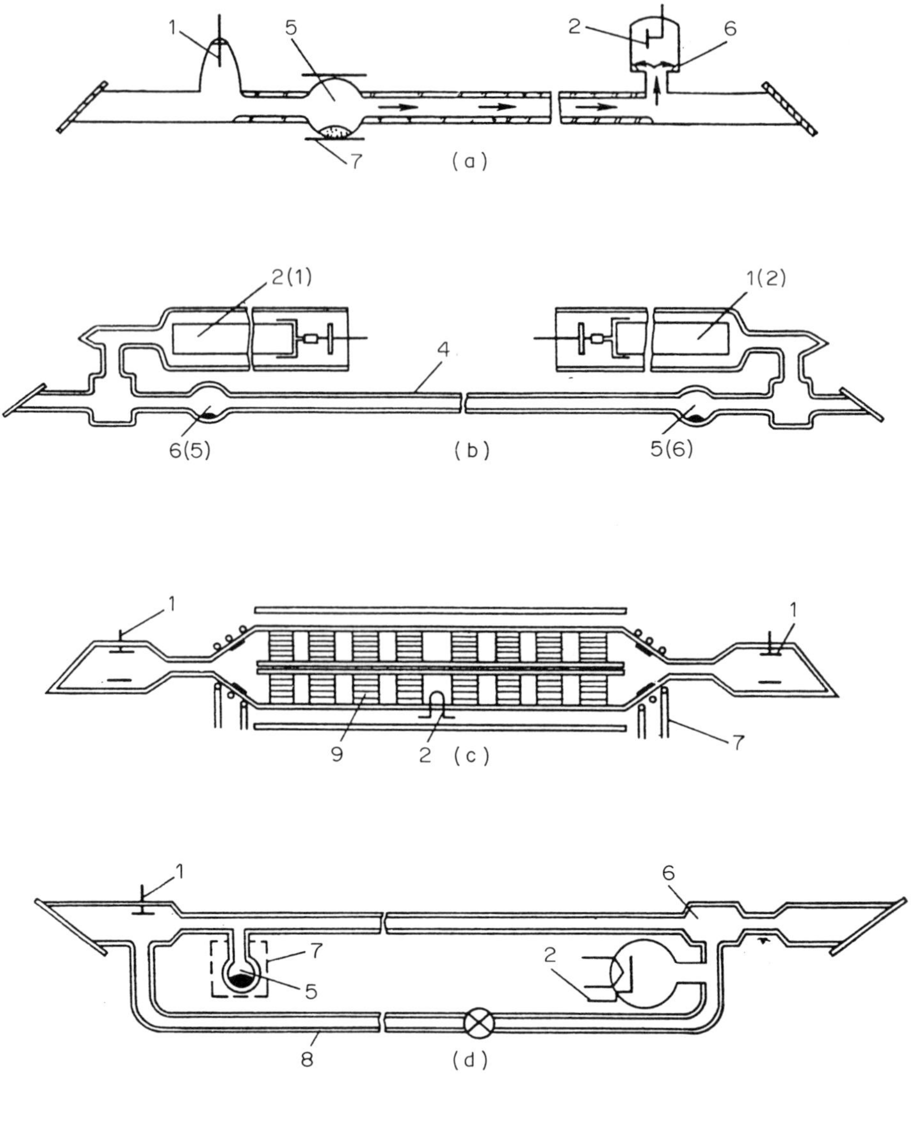

Figure 5.1 Discharge tubes of cataphoretic MVILs. 1, Anode; 2, cathode; 3, auxiliary (confinement) anode; 4, capillary; 5, metal reservoir; 6, condenser; 7, reservoir heater; 8, bypass channel; 9, diffusion return channel for metal vapours; 10, metal segments; 11, additional channels for admission of metal vapours.

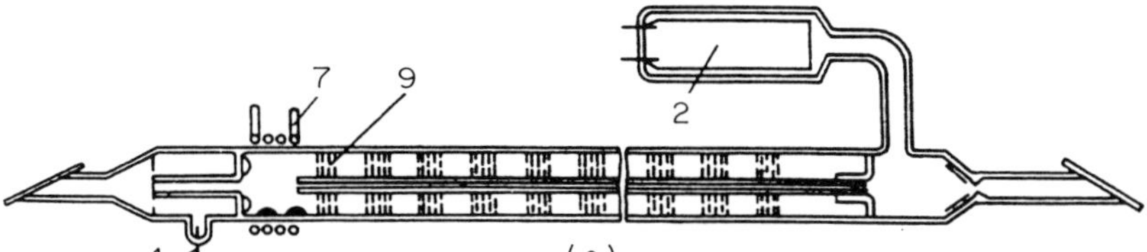

(e)

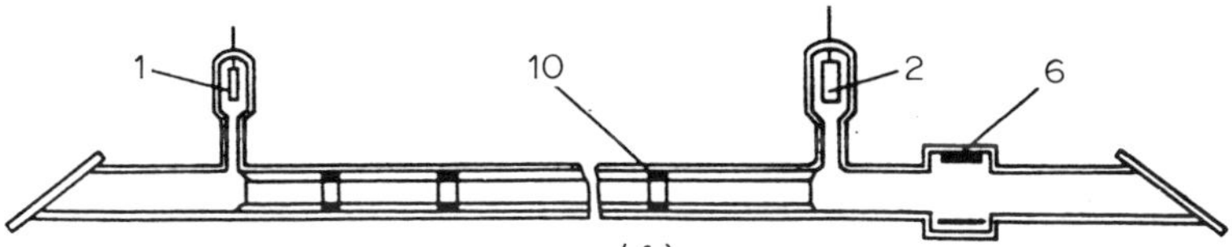

(f)

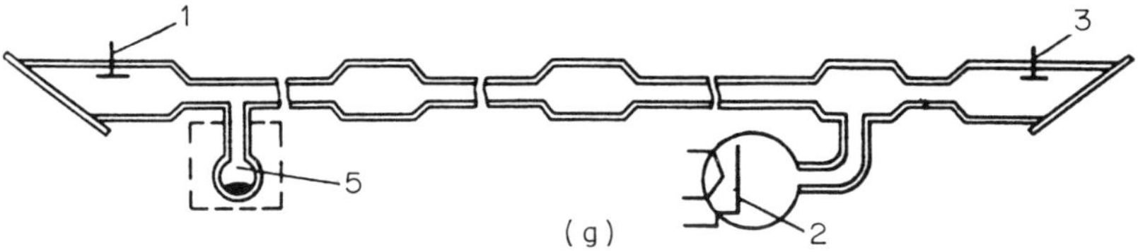

(g)

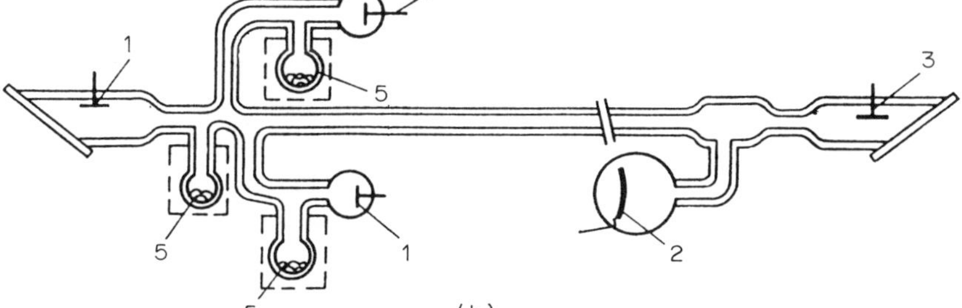

(h)

Table 5.1 CW ion lasers employing the positive column, and which use cataphoretic seeding of metal vapour. T_{oven} and T_{cond} are the temperatures of the metal source and the condenser, respectively.

Ion	Active	Laser wavelengths (nm)	Number of lines	T_{oven} (°C)	T_{cond} (°C)
Cd II	He–Cd	325.0–806.7 (325.0; 441.6)[a]	5(2)[a]	240	360
Zn II	He–Zn	492.4–758.8 (589.4; 747.8)	4(2)	310	450
Mg II	He–Mg	921.8–1095.2	4	400	550
Se II	He–Se	446.7–1258.67 (497.6–530.5)	49(6)	220	330
Te II	He–Te	484.28–873.38 (570.8)	19	350	475
	Ne–Te	544.98–937.85 (575.5)	20		
I II	He–I	540.7–880.4 (540.7–658.6)	9(5)	20	30
	{He–CdI_2}			260	
P II	He–P	602.4–604.3	3	170	240
As II	He–As	549.8–617.0	4	260	345
Sb II	He(Ne)–Sb	612.9–699.6	2	500	680
Bi II	He(Ne)–Bi	571.9	1	630	840
Hg II	He–Hg	615.0; 794.5 (615.0; 794.5)	2(2)	30	100
Tl II	Ne–Tl	595.0	1	570	770
Sn II	He–Sn[b]	645.3; 684.4	2	1180	1510
Eu II	He–Eu[b]	664.5–1477	6	570	770

[a]Wavelengths of practical use are placed in brackets. Lasing at 441.6 nm has also been observed with Ne, Ar and other buffer gases.
[b]Cataphoretic seeding was not used in these cases [38, 65].

compounds vaporized in an oven [8] are sometimes used to introduce the metal.

As mentioned earlier (see section 2.2), the density of metal atoms close to the tube wall, n_w, is about two orders of magnitude higher than the density along the axis of the tube $n(0)$, which is of the same order as the density in the oven, n_{oven} due to strong radial cataphoresis. Condensation of metal on the tube wall will not occur if the tube wall temperature $T_w \geq T_{dew}$ for a given n_w (T_{dew} is the temperature of formation of condensate for a given vapour pressure near the wall). Typical values of T_{dew} are given in Table 5.1.

However, when there is considerable overheating of the tube wall, i.e. when $T_w >> T_{dew}$, which is typical for volatile elements such as mercury, waves (pulses) of vapour propagate along the tube in an unstable manner with a time period of the order of 1 second [9] and prevent laser oscillation. The wave formation can be explained by the following. Any accidental rise of Hg vapour density in the capillary tube cannot be compensated for by the loss of atoms to the wall due to condensation. The longitudinal drift velocity of the vapour increases with increasing vapour density (see Figure 2.8), which forms a density wave with a shock front and a more gently falling trailing

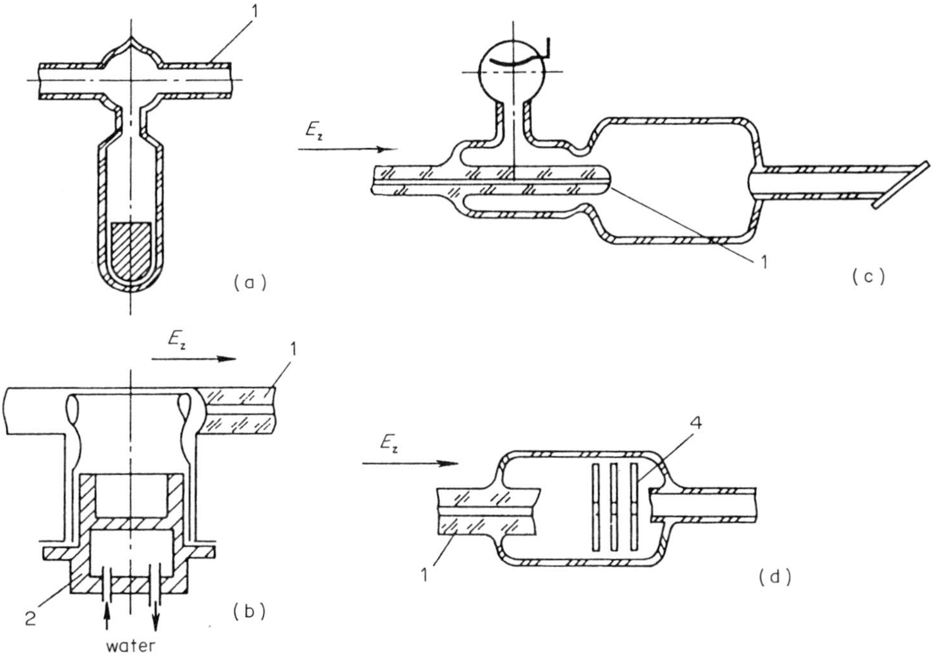

Figure 5.2 Details of the constructions of (a), (b) the ovens and (c), (d) the condensers of cataphoretic MVILs. (a) A side-arm heater connected to the capillary. (b) Oven based on cathodic sputtering [7]. (c) Condenser with reduced heating of the cooling zone [14]. (d) Single design heater/cooler with diaphragms to improve the capture of metal condensate [15]. 1, Capillary; 2, sputtering cathode; 3, metal; 4, diaphragms.

edge. The avalanche-like increase in the vapour density wave as it progresses towards the cathode is enforced by detachment of metal atoms from the tube surface as a result of its bombardment by heavy metal ions. The density of the wave increases until either that part of the capillary tube is clean, or the density of metal atoms reaches the point where condensation onto the wall can take place.

In the first, simple tubes, the metal vapour was transported in one direction (Figure 5.1a). Exhaustion of the metal lasant (sometimes very expensive single isotopes) was high (usually more than 1 g per 1000 h), and the buffer gas was strongly sorbed in the condenser. The highest rate of sorption is observed in mixtures of cadmium vapour and helium [10, 11]. This is due to the binding of helium within the outermost porous Cd layer, and its overcoating with newly condensed atoms (adsorption), and the deeper and faster implantation of helium ions, accelerated into the film of Cd by the radial field (absorption). The sorption depends strongly on temperature and current density in the condensation zone, and can be tens of pascals per hour. The sorbed helium can be completely recovered if the Cd film is heated to the melting point of Cd (320°C).

The problems of high metal expenditure and helium sorption were solved by the use of a symmetric tube design, which is now widely used in the commercial production of cataphoretic lasers (Figure 5.1b) [12, 13]. In this construction, the cathode and anode (respectively, the oven and condenser) are alternated for a period of time (tens of hours) sufficiently short to prevent sorption from affecting the laser characteristics. In that case, the lifetime of the tube is 1/Y times higher than the duration of one cycle (where Y is the relative loss of metal atoms in a cycle); the loss of helium is also $1/Y$ times lower. For example, if the initial quantity of lasant is sufficient for 50 h lifetime without cycling, then the fill lifetime for the tube with cycling for $Y = 1\%$ will be 5000 h. The absorption of gas atoms due to ion bombardment can be prevented by the use of a special design of condenser (Figures 5.2c and d): the conjunction of the capillary tube and condenser must be such that the plasma does not touch the place of condensation [14], or the condenser temperature must be as close as possible to the oven temperature [10]. The condenser must also be efficient at 'capturing' the metal vapour. Since the mean free path of metal atoms to the area of condensation in the condenser is proportional to the diffusion coefficient (i.e. $1/p$), as the pressure is raised the chance of capture of the vapour at the wall increases, and the size of the condenser can be reduced. Capture of the metal atoms can be enhanced by the use of metal diaphragms [15] (Figure 5.2d).

The adverse losses of metal and sorption of gas are also absent in a tube design that uses a diffusional gas return path, as in Figure 5.1(c) [16], which incorporates periodic comb structures (to prevent electrical breakdown in the return channel). To stop condensation in the return channel, it must be heated. Furthermore, a pressure gradient must exist from the cathode end of the capillary towards the oven, i.e. the cathode end temperature should be higher than that of the oven. However, as was pointed out in [16], this situation leads to a greater level of noise in the laser output intensity.

For cataphoretic MVILs operating at high current density and low buffer gas pressure, electrophoresis of the buffer gas is usual (see section 2.2), which produces a significant difference in the pressure between the cathode and anode ends of the capillary. This causes not only a deterioration of the laser conditions but also a gradient in E_z, and a gradient in wall temperature along the capillary tube which leads to insufficient heating of the cathode end of the capillary. The gas pressure can be equalized by connecting a bypass channel between the anode and cathode chambers [17, 18]. The nett velocity of the vapour flux is reduced due to the reverse flux of gas in the capillary from cathode to anode, since collisions with buffer-gas atoms reduce the drift of metal ions—this leads to a reduction in oven temperature and in the loss of the lasant.

Two types of output radiation noise are observed: low frequency and high frequency noise, which both arise due to fluctuations in vapour density in the capillary (see section 5.5). Thus, different designs of tube have been proposed: perforated capillary tubes [19], segmented capillaries [20] and a variable cross-section capillary [9] (Figure 5.1e–g). The first design comprises a capillary tube with equidistant holes, which is placed within a thermal enclosure; fine comb baffles, which separate the capillary tube from the envelope, also prevent electrical breakdown down its outside. This design ensures a local circulation of the metal vapour, which smooths out the axial distribution of the vapour density. In cadmium vapour lasers, this construction gives 1.1–1.5 times higher laser power, and 5 times less radiation noise in comparison with a laser tube with a diffusion return channel. In the discharge tube with a segmented capillary tube (Figure 5.1f), there is no special oven; the capillary tube is assembled from a number of glass segments of 6–8 cm length, separated by rings of the lasant (cadmium of natural isotopic composition). The length of each ring is approximately 4 mm. The segmented capillary tube is placed within a glass envelope. Stabilization of the vapour pressure and the laser output power are attained by stabilization of the discharge current. It is clear that such a construction can be used with lasants that operate at temperatures below the melting point of the metal (see Appendix 2), i.e. they are solid. Division of the capillary tube into short sections with stabilizing reservoirs of larger cross-section between them (Figure 5.1g) provides for volatile lasants, and acts against waves of vapour density. The same effect is observed when the capillary is made of ceramic [21].

Temperature stabilization of the oven is not sufficient to maintain the vapour density constant in the capillary, since, according to equation (2.29), it depends on the discharge current, the gas pressure, the temperature and quality of the tube walls, the amount of lasant in the oven, etc. Currently the most popular technique used to stabilize the metal vapour density is to use the dependence of the vapour density, n, on electric field E_z (see Figure 2.3a). An analogue automatic regulator can maintain the tube voltage $U = E_z l$ constant by adjusting the temperature of the oven. If the density is higher than optimum, i.e. $n > n^{opt}$, then $E_z < E_z^{opt}$ and the oven temperature is turned off; if $n < n^{opt}$, then $E_z > E_z^{opt}$ and the heater is switched on again. The accuracy of the input signal processing determines the fluctuations in the oven temperature and in the laser power. Since the slope of $E_z(n)$ is a function of the buffer-gas pressure, and it is small for lasers that operate at high pressures, it is usual to use the dynamic discharge resistance rather than E_z (see Figure 2.3b).

5.2 ENERGY CHARACTERISTICS AND PUMPING MECHANISMS

The laser characteristics and optimum discharge conditions of MVILs excited using positive-column discharges are presented in Table 5.2. The unsaturated gain coefficient g_0 is usually determined by insertion of calibrated losses into the cavity and measuring g_0 above the inherent cavity losses β_0.

The cavity losses β_0 can be easily determined by the following method: the investigated tube should have two additional ovens and two anodes (Figure 5.1h), which permit the alternating introduction of vapours of natural and single isotopes of an auxiliary metal. The calibrated loss of the single isotope, β^1, and that of the natural mixture β^* of the auxiliary metal, which lead to termination of the laser oscillation, are measured on the isotopically split line of the probe metal. Taking into account that $g_0 = \beta + \beta_0$ and the relation $g_0{}^1/g_0{}^* = 1/X$ (X is the isotopic factor for the measured line [22]), we find that

$$\beta_0 = (X\beta^1 - \beta^*) / (1 - X) . \tag{5.1}$$

As probes, the lines 441.6 nm Cd II ($X = 0.28$), 589.4 nm, 747.8 nm Zn II ($X \approx 0.5$) and 615 nm Hg II ($X \approx 0.3$) can be used, which cover the visible spectrum.

The most thoroughly investigated MVIL with PC excitation is the He–Cd laser. The interest in this laser is due to many factors: first, its UV and visible lines are of practical importance; stable transport of the vapour can be realized in tubes made from simple glass; the optimum discharge current is not high. The highest laser power on the 441.6 nm Cd II line is achieved in long (approx. 1 m) tubes and is 200–250 mW when a single Cd isotope is employed. The maximum laser power per unit volume is obtained in small-bore tubes; up to 68 mW cm^{-3} in a perforated capillary tube of 1 mm bore and 150 mm length.

Since the transition $5s^2\ {}^2D^*{}_{5/2} \longrightarrow 5\ {}^2P^o{}_{3/2}$ (441.6 nm) involves two electrons, the radiative lifetime of the upper laser level in the absence of any perturbation is quite long ($\tau_2 \approx 10^{-6}$ s). It can be measured by the spontaneous emission observed through the side of the tube, whilst rapidly modulating the cavity Q factor (for example, with an acousto-optic modulator), without perturbation of the plasma parameters and the pumping rate, but altering the population of the upper laser level [23]. Extrapolating τ_2 to the zero current point, when electron collisions are negligible, yields $\tau_2 =$ 670 ns [23]. In [24], the radiative lifetime was determined from the width of the magnetic resonance curve (in this case the half-width at half-height corresponds to the decay rate of the upper laser level), and it was found to be half as much (approx. 350 ns). It was found that the decay rate increases slowly with increasing current and helium pressure, and with decreasing

Table 5.2. Some similarity relations and energy characteristics of CW MVILs with PC discharge excitation (d = 1–4 mm).

λ (nm)		Buffer gas	$p_{He}d$ (kPa cm)	E_zd (V)	i/d (A cm^{-1})	P_{in}/l (W cm^{-1})	g_0d (10^{-4})	P (mW) (P_{sp}, mW cm^{-3})
Cd II	325.0	He	0.11–0.14 [3, 49], 0.16 [20]	6–8 [49, 66]	0.4–0.43	2.4–3.5	1.56[a] [20]	25 (3) 20 (3,1) [3] 4.5 (13.2) [66]
	441.6	He	0.11–0.14 [3, 49], 0.16 [20]	6–8 [49, 66]	0.4–0.43 [49, 66]	2.4–3.5	3.75[a] [20], 1.4 [52]	250 (35) [57], 200 (30.9) [3], 8 (68) [66]
Zn II	589.4	He	0.12–0.14 [6]	6–8	0.4–0.43	2.4–3.5	2.8[a] [6]	7 (0.6) [37], 4 (4.6) [12]
	747.8	He	0.12–0.14	6–8	0.4–0.43	2.4–3.5	1.9[a] [6]	15 (17) [12]
Se II	497.5–530.5	He	0.32 [40], 0.35–0.43 [49]	8–9	1.33 [41], 1.7 [40]	11.3–14.4	3.3 [6], 2.7 [49], 2.75 [40]	32 (13) [6], 250 (12) [41], 28 (32) [12]
Te II	570.8	He	0.2 [6, 47]	6–7.5 [48]	1.5	12	1.3 [47], 1.5 [6]	10 (1) [6, 47]
	575.5	Ne	0.08 [6, 47]	5.0	1.5	7.5	0.75 [47]	
Hg II	615.0	He	0.2 [9, 44]	7.5	1.4 [44]	10.2–11	1.4, 4.7[a] [9]	10 (4) [9]
	794.5	He	0.2 [9, 44]	7.5	1.3 [9]	10.2–11	2.4 [9]	8 (3.2) [9]
I II	540.7–658.5	He	0.2–0.28 [6, 26, 67]	7.6	1.2–1.4 [26]	9.8–11.3	1.63 [6]	30 (6), 40 (3.9) [26]
Tl II	595.0	Ne	0.1 [68]	5.1	2.3	11.8	0.5	–
Sb II	699.6	He (Ne)						
Bi II	571.9	He (Ne)	0.05 [17]	7	3.2	22.4	0.5	–

[a] An isotopically enriched lasant was used.

cadmium vapour density. This can be explained by electron-impact deactivation and the behaviour of electron density which is proportional to the pressure of the gas mixture components and the discharge current in the positive column (see section 2.2).

The investigation of the pumping mechanisms of the 441.6 nm and 325 nm lines in the CW regime resulted in the development of cataphoretic lasers. The major excitation mechanisms in He–Cd mixtures are considered to be Penning ionization for population of the upper laser levels, and fast radiative decay for the lower laser levels [1, 25]. The laser emission produced with buffer gases other than helium (Ne, Kr, Xe, etc.) [27] and in 'pure' metal vapours [28] mean that there are other mechanisms also, such as electron-impact excitation from the atomic ground state or ion ground state (5s $^2S_{1/2}$), which is the most universally applicable.

Experiments in which helium metastable atoms were optically pumped into high excited states $n\ ^1P$, $n\ ^3P$, etc. ($n \geq 2$), where the resonance transitions of He are blocked, revealed that, in addition to the helium metastables, other excited states of helium take part in the Penning ionization of the ^2D Cd II levels [29]. It was found that the spontaneous and laser emissions both increase (5% and 20%, respectively) due to the increase in the Penning ionization rate because the probability of the Penning process is higher, the higher the energy defect [30]. The pump rate W_2 of the upper level can be found (according to [29]) by producing 100% intra-cavity modulation of the 441.6 nm line and observing the modulation of the spontaneous emission on the same transition, since W_2 is dependent on the upper laser level populations with (n_2) and without (n_2^0) laser oscillation according to

$$W_2 = \rho\beta_0 / (1 - \Gamma_2)\,, \tag{5.2}$$

where $\Gamma_2 = n_2^0 / n_2$ is the coefficient of population modulation, and ρ is the flux density of laser radiation. Under optimum conditions for the 441.6 nm line (see Table 5.2), $\Gamma_2 = 0.7$ with $\rho = 1.4 \times 10^8$ erg cm^{-2}s^{-1}. From (5.2), we can obtain $W_2 = 7.5 \times 10^{16}$ cm^{-3}s^{-1}, which coincides with the rate of pumping by Penning ionization (see Figure 2.6) and is 2 to 4 times higher than the pump rate due to electron excitation.

As could be seen in Figure 2.6, excitation by electron impact is the dominant process in pumping the 441.6 nm line for $n_e > 3 \times 10^{12}$ cm^{-3}, i.e. for pump rates much higher than the optimum one, and when inversion is prevented due to efficient excitation of the lower laser level. In mixtures of Cd with gases other than He, pumping is produced by electron impact, and is thus lower. It is not possible to raise the inversion by increasing n_e because of the much faster increase in population of the lower laser level, leading to a reduction of the gain and the laser power compared with that from helium mixtures. In only one case, when 'pure' cadmium vapour is excited, is the gain a little greater due to the higher (by 2) optimum density of Cd vapour required to achieve practically the same values of n_e and T_e as in He mixtures [28]. The He–Cd mixture, in general, is one of the few systems where various mechanisms act to pump one laser transition. The reason for this is that the upper laser level is the ground state of the ion in the Beutler term system [31] (thus the upper level has a large cross-section for electron impact) and decays radiatively (see also section 3.3.5). The pumping mechanism for the 441.6 nm transition is still under examination (see for example [32, 33]), although discussions are now focused on defining the relative contributions of Penning ionization and electron-impact excitation.

Experiments with modulation of the laser have been useful in other ways. For the lower laser level, $\Gamma_1 = 1.064$ (i.e. the stimulated emission increases its population by only 6.4%). Using a relation between n_1, n_2 and the decay rate γ[34],

$$n_1 - n_1^0 = (\gamma_2 - \gamma_{21})(n_2^0 - n_2) \,/\, \gamma_1 \,, \tag{5.3}$$

the lifetime of the upper laser level in the discharge is found to be $\approx 0.25 \times 10^{-7}$ s. This lifetime is an order of magnitude less than the radiative lifetime and indicates the existence of transitions (due to electrons etc.) that bypass the laser transition. At the same time, there is no modulation on the 325 nm transition from the $^2D^*_{3/2}$ level with Q-modulated 441.6 nm oscillation, which means that there is only weak mixing of the levels $^2D^*_{5/2}$ and $^2D^*_{3/2}$ in comparison with their decay routes in the plasma. In [35], experiments on the influence of the laser field in the cavity on the acceleration of ions in the $^2D^*_{5/2}$ state was reported. It was found that the lifetimes of ions in the $^2D^*_{3/2}$ and $^2D^*_{5/2}$ states are essentially limited by ambipolar diffusion as well. The typical value measured was 10^{-7} s. The radial velocity of ions is 1.2×10^5 cm s^{-1}, which is 3 times higher than the thermal velocity and 6 times higher than the longitudinal drift velocity, which occurs when $E_r > E_z$. Such a high radial velocity (compared with the thermal velocity) causes deformation and shifting of the laser lines at 441.6 nm and 325 nm [36].

The excitation mechanism of the $^2D^*_{3/2}$ level (325 nm line) is the same as for the 441.6 nm line, but the laser power on the 325 nm line is lower than that of the violet one. This is the result of the lower gain ($\lambda^3(442)/\lambda^3(325) = 2.5$) [20], a lower saturation parameter and, as a rule, the lower quality of UV optical components. The population of the other Cd II lines, which oscillate only in He mixtures, is produced as in the other types of discharge, via charge transfer of He ions to Cd atoms directly, or followed by cascade transitions. However, the pumping is not as effective as it is in hollow-cathode discharges, and on the green $F \longrightarrow D$ lines in the steady-state PC discharge laser oscillation is absent.

The Zn ion has a similar disposition of energy levels as the Cd ion, and the pumping mechanisms and optimum discharge conditions are similar.

However, the laser powers of the 747.8 nm and 589.4 nm Zn II lines (analogous to the 441.6 and 325 nm lines of Cd II) are lower, which is due mainly to the lower quantum energies. Some differences from the Cd II energy level structure allow lasing on the $F \longrightarrow D$ transition of Zn II at 492.4 nm, but with very low power [2, 6, 37–39].

For He–Se mixtures, CW laser oscillation was obtained independently in [6, 40, 41] on 49 ion transitions in the range 0.44 to 1.2 μm. The most powerful laser lines are the six in the green spectrum (0.49–0.53 μm). Their total power is about the power of the Cd II 441.6 nm line (see Table 5.2). In the cataphoretic selenium MVIL, the optimum discharge current density and the value of *pd* (gas pressure × discharge tube bore) are 3 to 4 times higher than in a similar Cd II 441.6 nm laser (populated by Penning ionization), and are typical for processes involving ions of the buffer gas. The energies of the upper laser levels of Se II relative to the atomic ground state (23.5–24.7 eV) coincide with the energy of He^+ (24.56 eV) to within several kT. Hence, we can infer that pumping of the laser transitions of Se II in the CW regime is the same as in the pulsed regime, *i.e.* by charge transfer of helium ions to selenium atoms. The vapour of selenium consists of molecules containing various numbers of atoms—from 2 to 8 (Se_2, Se_5, Se_6, Se_7, Se_8), depending on the gas temperature. The average number of atoms in a molecule of selenium at 196°C is 5.6 [42]. Presumably, the complete dissociation of Se molecules under steady-state excitation is the main reason for the larger number of laser lines, compared with the pulsed regime.

The effect of acceleration of the Se II ions in the upper laser level and the resultant Doppler line deformation is weak because the lifetimes of the upper levels are of the order of tens of nanoseconds [43], and the excitation is produced from the atomic ground state. Note that the decay rates of the upper laser levels do not depend on He pressure, Se vapour pressure or the discharge current [43] since the frequencies of the radiative transitions are higher than those of the electron collisions (about 0.5 MHz) and those of ambipolar diffusion of ions to the tube wall (about 1 MHz).

As in the case of Cd, it is possible to realize long-life cataphoretic laser tubes. This makes the Se laser an attractive source of CW radiation in the green spectrum with some advantages compared with other ion lasers (e.g. in comparison with the Ar^+ laser, its efficiency is 3–5 times higher, and it is cheaper and simpler to construct).

In mercury, CW laser oscillation is observed on both components of the doublet $7\,^2P_{3/2} \longrightarrow 7\,^2S_{1/2}$ with λ = 615 nm and 794.5 nm only in a mixture with He (up to 3 kPa) [9, 44]. Excitation by electron impact cannot be the main process in either CW or pulsed Hg lasers because it is insensitive to the type of buffer gas, and would be most intense at low pressures when T_e is high. Also, Penning ionization of Hg metastable atoms via He metastables cannot be the excitation mechanism for the $7\,^2P$ Hg II levels (see equation

(3.2)), because the intensities of the 615 and 794.5 nm lines increase with current to 4 $A\,cm^{-2}$, while the He metastable populations are saturated with current before 1.5 $A\,cm^{-2}$. The population of the upper laser level increases linearly, which implies that pumping is due to charged particles—energy transfer from He^+ to Hg atoms. The partial cross-sections for charge transfer to the levels 7 $^2P_{3/2}$ and 7 $^2P_{1/2}$ can be derived from the measurements of gain at 615 and 794.5 nm, and are 2.6×10^{-15} and 0.36×10^{-15} cm^2, respectively [9, 45]. Thus, the largest fraction of the energy of He^+ ions that take part in charge transfer goes to the excitation of the quartet terms, 12 of them having energies between 23.5 and 24.5 eV [31]. When an even isotope of mercury is employed, the gain coefficient increases by a factor of 3.3, which yields higher laser output power.

The major role of charge transfer in pumping the ion transitions of iodine in a mixture with He is, by analogy with Se and Hg, beyond doubt. Recall that I^+ was the first laser species for which charge transfer was postulated as a pump mechanism [46]. Arguments in favour of that process in steady-state PC discharges are those typically put forward for second-kind collisions: high He pressures, a high saturation current and the close disposition of I^+ levels to the energy of He^+.

Sources of iodine that have been used are pure iodine (the source region is cooled rather than heated) and CdI_2, where the operating temperature is 260 °C. However, in the second case, the presence of Cd vapour leads to a decrease in T_e and the effectiveness of I^+ laser oscillation is reduced [8].

In a similar manner to selenium, CW oscillation on tellurium ions was observed on many more lines compared with the pulsed regime [6, 47]. This may be attributed to the molecular composition (Te_2) of the vapour in the pulsed case. Laser oscillation occurs on 19 transitions with He buffer gas, and on 20 transitions with Ne (see Table 5.1). The most powerful lines lie in the yellow-green spectrum, and the laser power is a few milliwatts. For Te II, the optimum He pressure and the optimum current are typical for processes that involve long-lived particles. The dependence of the transition characteristics on the type of buffer gas indicates the importance of collisions with gas ions or metastables. The upper levels of all the laser transitions with Ne buffer gas are located to within $\pm 4kT_e$ of the energy of Ne^+, and are populated by charge transfer. Two of the transitions have an excitation energy 1.3 eV higher than Ne^+, which can be accounted for by charge transfer from Ne^+ ions to Te $5p^4\ {}^1D_2$ metastable atoms [47]. Only one of the nine upper laser levels of transitions with He buffer gas are energetically close to He^+, while the rest are below it. A possible excitation mechanism is either charge transfer from He_2^+, or charge transfer from He^+ followed by cascading radiative transitions. Estimation of the concentrations of He_2^+ and He^+ makes the second process more likely (see section 2.2). However, the observed optimum He pressures and Te vapour pressures indicate a more complex and as yet undefined excitation mechanism [48].

Laser oscillation is obtained on two ion transitions of Sb and on one ion transition of Bi. It does not depend on the type of buffer gas, and is observed at low pressures of He, Ne or their mixtures at high current density (to 15 $A\,cm^{-2}$) [17]. At such high current densities and relatively low gas pressures, electrophoresis (see section 2.2) of the buffer gas is quite strong, which means a bypass channel is required to connect the ends of the discharge tube (Figure 5.1d) so as to lower the pressure difference between the two ends.

The major excitation mechanism is the same as in the pulsed regime, i.e. electron-impact excitation from the atomic ground state, because the excitation energies of the transitions at 612.9 and 699.5 nm of Sb II and at 571.9 nm of Bi II are 19 eV, 18.5 eV and 17 eV, respectively. For the ion transitions of P and As in the CW regime, electron-impact excitation was proposed [6], which is the major process in the pulsed regime. The proposal is supported by the following facts: first, the upper laser levels lie too far below the ionization potential of helium (more than 1 eV for P II transitions and 2.5–3 eV for As II transitions), and lie above the ionization potential of Ne; second, the laser characteristics are practically insensitive to the ratio of the pressures of He and Ne in a mixture of the buffer gases.

5.3 OPTIMUM CONDITIONS FOR LASER OSCILLATION

The dependences of laser output power and gain coefficient on the current and the pressure of the gas mixture for all MVILs exhibit a form which is very close to bell-like. However, the optimum pressures and currents are specific for each medium: the lowest *pd* product corresponds to electron-impact pumping, the largest *pd* corresponds to charge transfer pumping, and intermediate values of *pd* correspond to Penning ionization pumping, which is characterized itself by the lowest optimum discharge current. We will now outline the causes of saturation of gain, laser power and efficiency, and their dependences on the discharge conditions and tube geometry for the three excitation mechanisms—Penning ionization, charge transfer and electron impact. We begin with the gain coefficient g_0, which determines the power. The gain coefficient is related to the pumping rate W_2 and the relative inversion density Δ:

$$g_0 = 3.7 \times 10^{-2}\,(g_1/g_2)\,\lambda^2(\Delta\nu_D)^{-1}W_2\tau_2A_{21}\Delta\,; \tag{5.4}$$

$$\Delta = 1 - \frac{g_2}{g_1}\frac{\tau_1 W_1}{\tau_2 W_2}\,; \tag{5.5}$$

where $\Delta\nu_D$ is the Doppler width of the emission line.

We will now analyse equation (5.4) for a gaseous active medium with a discharge current i, buffer gas pressure p_B, metal vapour pressure p_M and tube inside diameter d. If Δ is a constant, then there are two cases to consider. The first case corresponds to $\Delta \approx 1$, i.e. $\tau_1 W_1 \ll \tau_2 W_2$, when W_1 is of the same order as W_2, and $\tau_1 \ll \tau_2$. The second case occurs when τ_1 and τ_2 are of the same order and W_1/W_2 is a constant, i.e. the lower laser level is populated by the decay of the upper laser level due to stimulated emission. If Δ is a constant over the range of variation of the parameter x, we find that

$$g_0(x) \sim I_{sp}(x) T_g^{-1/2}(x) , \tag{5.6}$$

since the intensity of spontaneous emission on the laser transition is $I_{sp} = A_{21} W_2 \tau_2$ and the Doppler width $\Delta\nu_D \sim T_g^{-1/2}$. Hence for (5.4) we find

$$g_0(x) \sim T_g^{-1/2}(x) W_2(x) . \tag{5.7}$$

5.3.1 Dependence on Current

We shall now find the limits of the above-mentioned range, and combine them with the optimum conditions for laser oscillation. Consider lines pumped by Penning collisions (for example, 441.6 nm Cd II), which involve transitions with Beutler terms, and where the lower laser level is populated by electron impact from the ion ground state, directly by Penning ionization (cross-section $\xi_1 = 0.12$) and by cascade transitions from the upper level ($\xi_2 = 0.18$). Then, the ratio of the pump rates to the lower and upper levels, W_1 / W_2, can be written as

$$\frac{W_1}{W_2} \approx \frac{\Theta n_e^2}{n N_{He*}} \frac{k_{i1}(^2S \longrightarrow {}^2P)}{\xi_2 K_p} + \frac{\xi_1 + \tau_2 A_{21} \xi_2}{\xi_2} , \tag{5.8}$$

where $\Theta = n_{Cd+} / n_e$; at optimum cadmium vapour pressure, $\theta \approx 1/2$ (see Figure 2.4). Thus, at low (albeit optimum for laser oscillation) electron density n_e, the inversion $\Delta = 1$ for suitable ratios of the radiative lifetimes, even though the lifetime of the $^2D^*$ upper laser level is reduced in the discharge: $\tau_1/\tau_2 \approx 10^{-1}$–$10^{-2}$. If n_e increases (for example, when the vapour density or the discharge current increases), Δ decreases. Note that resonance trapping has an influence on τ_1 for $n_e > 4 \times 10^{12}$ cm^{-3}.

It is found experimentally that relation (5.6) is valid for currents up to $i/d \approx 0.6$ A cm^{-1} ($n_e < 2 \times 10^{12}$ cm^{-3}) for the 441.6 nm Cd II transition with an optimum (or lower) cadmium oven temperature. The saturation with current at $i/d \approx 0.4$ A cm^{-1} is explained by the saturation of the density of excited helium atoms. When the oven temperature is greater than that optimum for laser oscillation, relation (5.6) does not hold, since Δ decreases due to the rise

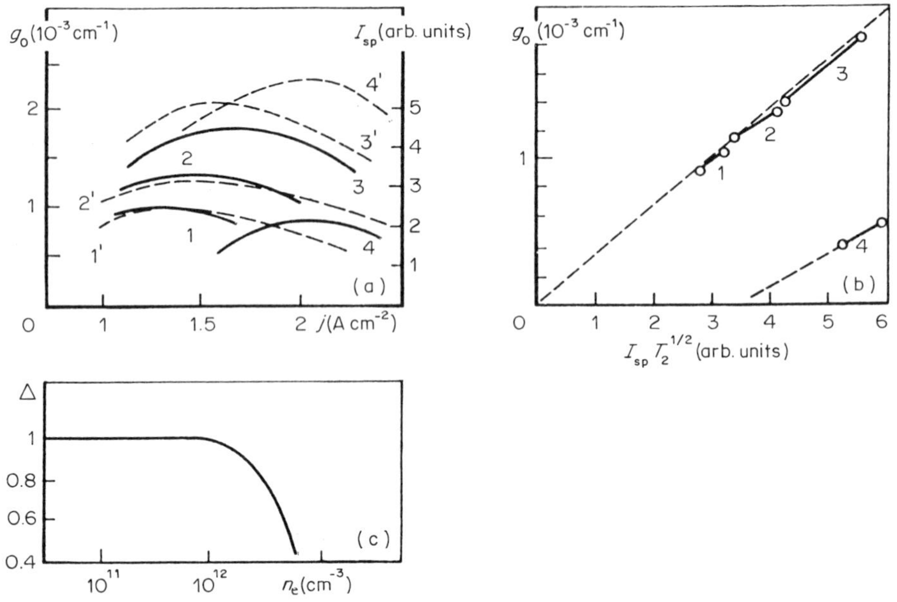

Figure 5.3 (a) The dependence on discharge current of the unsaturated gain coefficient, g_0 (solid curve), and of the spontaneous emission intensity, I_{sp}, (broken curve) of the Cd II 441.6 nm transition. (b) The relation between g_0 and I_{sp}. (c) The calculated dependence of relative inversion, Δ, on electron density, n_e. (a)–(c) pd = 0.12 kPa; cadmium vapour pressures: 0.2 Pa (1); 0.3 Pa (2, 2′); 0.4 Pa (3, 3′); 0.75 Pa (4).

in population of the lower laser level (Figure 5.3). Variation of the helium pressure ±50% away from the optimum pressure does not cause any changes in Δ, i.e. Δ is a constant.

For transitions excited by charge transfer (e.g. the 522.7 nm Se II line), the lifetimes of the upper and lower levels do not differ greatly. Thus, for that line, $\tau_1 = 2$ ns and $\tau_2 = 18.5$ ns [43] because this transition involves only one electron, unlike the dielectronic 441.6 nm Cd II transition. The lower level of the 522.7 nm Se II line is populated by electron impact from the ion ground state, and by cascade transitions from upper levels by charge transfer collisions. The lower laser level is shared by a group of lines: 522.7, 506.6 and 464 nm. Their partial cross-section for charge transfer is designated by $\xi_2^{(k)}$, and τ_2 and A_{21} by $\tau_2^{(k)}$ and $A_{21}^{(k)}$, respectively. For each one of these lines, we find that

$$\frac{W_1}{W_2} = \frac{n_e}{n}\,\frac{k_{i1}}{\xi_2 K_c} + \frac{\sum_k \xi_2^{(k)}\,\tau_2^{(k)}\,A_{21}^{(k)}}{\xi_2}. \tag{5.9}$$

The ratio W_1/W_2 is proportional to the first power of n_e (not to the second power as in the case of the 441.6 nm Cd II line). Aside from a poor ratio of lifetimes, the low excitation rate of the lower level via electron impact from the ion ground state together with the small degree of resonance trapping cause W_1/W_2 and Δ to be constant for Se II lines over a large range of n_e (up to 8×10^{12} cm^{-3}). The proportionality of g_0 and $I_{sp}T_g^{-1/2}$ (I_{sp} = spontaneous emission intensity) for the charge transfer lines of Se II holds over a wider range of discharge currents (i/d up to approx. 1.2–1.5 A cm^{-1}) than for the 441.6 nm Cd II line. In the same regime, the dependences of χ_0 and $I_{sp}T_g^{-1/2}$ on helium pressure are also similar, i.e. $\Delta(p)$ is a constant. For Sb II and Bi II laser lines, and other lines excited by electron impact, the laser power does not saturate up to i/d = 3 A cm^{-1}. This means that Δ is a constant and relation (5.6) is valid.

Thus, in the case of Penning excited lines, the saturation of gain (and output power) with current is not determined by a reduction in the inversion due to a lower level effect, whereas for lines excited by charge transfer or electron impact an increase in the population of the lower level *is* responsible. Nevertheless, if g_0 and the laser power depend linearly on current, the relative inversion Δ is a constant, and we can use relation (5.7).

5.3.2 Dependence on Metal Vapour Pressure

All types of MVLs have similar bell-shaped dependences of output power on metal vapour pressure. The higher the gain coefficient and the lower the cavity losses, the wider the optimum range of oven temperatures. Consider constant τ_2 and A_{21} when Δ is fixed. We use the relations (1.5) and (1.10) for the excitation rate of the upper laser level W_2 and assume $K \sim T_g^{-1/2}$ to find, for the three types of pumping, from (5.7) that:

$$g_0(n_{Cd}) \sim N_{He^*}(n)n_{Cd}\,; \tag{5.10a}$$

$$g_0(n_{Se}) \sim N_{He^+}(n)n_{Se}\,; \tag{5.10b}$$

$$g_0(n_{Sb}) \sim T_g^{-1/2}n_e(n)k_{02}(n)n_{Sb}\,. \tag{5.10c}$$

It was shown in Chapter 2 that a reduction in electron energy accompanies an increase in vapour density. This causes the deactivation rate to increase and the rate of formation of ions and excited helium atoms to decrease, leading to the bell-like dependence of $g_0(n)$ (relations (5.10)). For all mixtures with optimum vapour density, N_{He^*}, N_{He^+} and $I_B/p \sim n_e k_{02}$ (where I_B is the intensity of buffer gas lines) are about half their values in a 'pure' helium discharge (designated by (0)):

$$N_{He*}^{opt} \approx N_{He*}^{(0)}/2;\ N_{He+}^{opt} \approx N_{He+}^{(0)}/2;\ (I_B/p)^{opt} = (I_B/p)^{(0)}/2\,. \tag{5.11}$$

If we extrapolate N_{He*}, N_{He+} and I_{He} with linear dependences on n, e.g. $N_{He*} = N_{He*}^{(0)} - An$, and take into account that, under optimum conditions, $\partial g_0/\partial n = 0$ for $n = n_{opt}$, then we find that $N_{He*}^{opt} = N_{He*}^{(0)}/2$. Using (5.11), we can calculate the optimum density of the lasant [49], the accuracy of which is not affected by the small difference of the real factor from the value of 0.5.

5.3.3 Dependence on Buffer-Gas Pressure

The bell-like dependence of g_0 on the buffer-gas pressure can be described using relation (5.7). Using W_2 for each of the three pump mechanisms in turn, and taking τ_2 to be constant with respect to gas pressure p, we find from (5.11) that

$$g_0(p) \sim N_{He*}^{(0)}(p)n^{opt}(p)\,; \tag{5.12a}$$

$$g_0(p) \sim N_{He+}^{(0)}(p)n^{opt}(p)\,; \tag{5.12b}$$

$$g_0(p) \sim T_g^{-1/2}\,\frac{I_{He}^{(0)}}{p}(p)n^{opt}(p)\,. \tag{5.12c}$$

The values of $N_{He*}^{(0)}(p)$, $N_{He+}^{(0)}(p)$, $I_{He}^{(0)}(p)/p$ and $n^{opt}(p)$ are shown in Figure 5.4. The optimum vapour density decreases with an increase in p for two reasons [49]. First, the fall in the excitation and ionization frequencies of atoms by electron impact with the increase in metal vapour density begins earlier, the higher the buffer-gas pressure is. Second, the deactivation (depopulation) frequency of excited atoms and ions by second-kind collisions reduces with buffer-gas pressure due to the increasing importance of diffusion to the wall and electron impact, which set a lower limit to the possible decay rate of metal atoms, i.e. the optimum density of metal atoms is lowered.

In those terms, for gas pressures higher than that optimum for laser oscillation, the gain and power reduction is caused by: charge transfer reactions, which lower the optimum metal vapour density in the discharge; Penning collisions, which reduce the density of excited helium as well; and by electron impact-induced transitions, which are mainly superelastic.

In the discussion above, we dealt only with the main processes and the minimum number of levels to give a qualitative model. The real situation can be far more complicated due to the inclusion of additional processes such as quenching transitions etc.

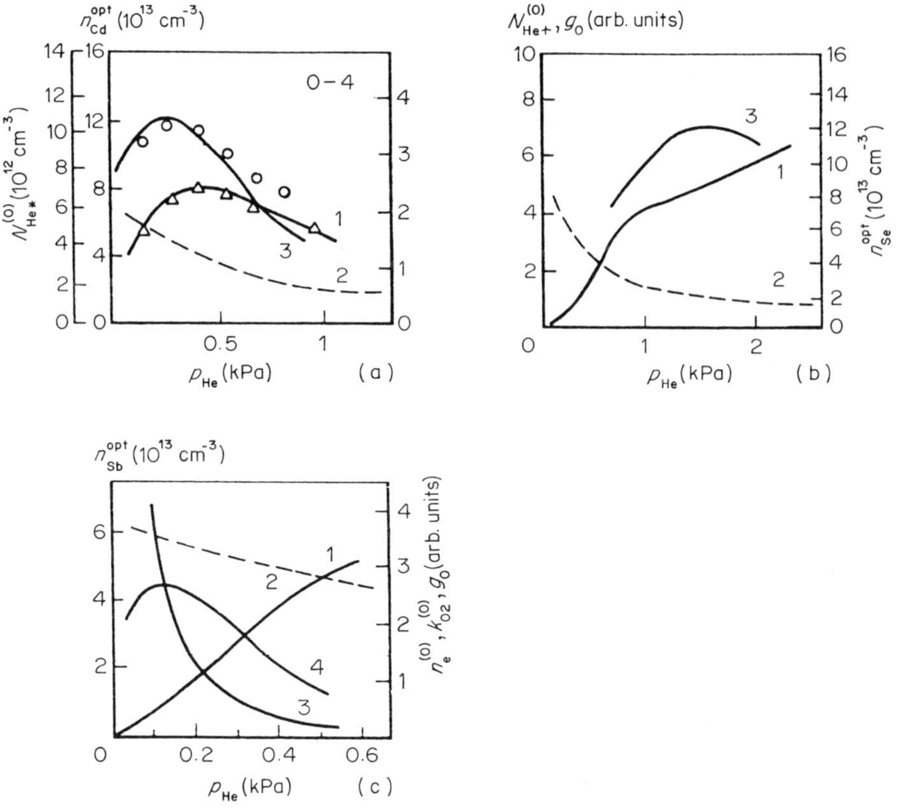

Figure 5.4 The dependence of plasma parameters and the gain coefficient, g_0, on helium pressure for three pumping mechanisms: (a) Penning ionization (441.6 nm, Cd II); 2.5 mm bore capillary tube: 1, $N_{He*}^{(0)}$; 2, n_{Cd}^{opt}; 3, pump rate of the 441.6 nm upper level, W_2; 4, g_0. (b) Charge transfer (522.7 nm, Se II): 1, $N_{He+}^{(0)}$; 2, n_{Se}^{opt}; 3, g_0. (c) Electron impact (612.9 nm, Sb II): 1, n_e; 2, n_{Sb}^{opt}; 3, $k_{02}^{(0)} \sim I_{He}^{(0)} / p$; 4, g_0.

5.3.4 Dependence on Tube Diameter

Although the optimum discharge parameters given in Table 5.2 differ, they do obey some similarity laws. Some of the parameters (such as pd = constant and nd = constant) can be derived from T_e = constant, whereas other parameters such as i/d = constant and $g_0 d$ = constant are specific to the excitation mechanism of the particular ion laser levels. It is found that the higher gain of small-bore tubes is mainly the result of the higher optimum density of metal atoms, n^{opt}. This fact is caused by the increase in the 'critical' values of n^{opt}, for which the ionization rate decreases and the decay rates of the active buffer-gas species increase (due to diffusion and electron de-excitation). Excited atoms and ions of helium decay at the rate of energy transfer to the lasant atoms with a frequency Kn, and through other channels (diffusion, electron-collisional transitions etc.) with a frequency ν. Then, the conditions that pertain to relation (5.11) will be valid if ν/Kn is a constant. Hence, in large-bore tubes, the smaller ν is, the smaller Kn is (the smaller is n^{opt}); when the tube diameter is increased, ν rises, which permits Kn to increase as well by allowing a larger n. According to equation (5.10), a larger n results in a higher g_0. The best case occurs when depopulation occurs via electron impact because, on the one hand, this process increases ν, and on the other, it causes the saturation of the excitation frequency to be shifted to higher n_e. Such a situation is partially realized in the Cd II laser on the 441.6 nm line, where $g_0 \sim d^{-4/3}$ [49].

5.3.5 Laser Action at Currents Above Optimum for Oscillation

In previous sections, we considered currents corresponding to n_e, when the gain g and laser output power are close to maximum (see Figure 5.3). However, the point where the inversion Δ intersects zero and crosses into negative values at above-optimum currents is no less informative regarding identification of the dominant pump mechanism. Let $(n_e)_{cr}$ denote the electron density at that crossing point, and show for the He–Cd laser as an example that its position depends mainly on the laser pump mechanisms involved.

Indeed, for $n_e = (n_e)_{cr}$, $\Delta = 0$, and from (5.5) it follows that

$$g_2\tau_1 W_1 = g_1\tau_2 W_2 , \tag{5.13}$$

which allows us to determine $(n_e)_{cr}$.

Taking into account that at $n_e \geq 10^{12}$ cm^{-3} the pump rate of the lower laser level by electron-impact from the cadmium ion ground state 5 $^2S_{1/2}$ (stepwise excitation) prevails over direct electron impact excitation and cascade transitions from higher lying energy levels, then W_1 can be written as

$$W_1 \approx n_e n_{Cd+} k_{il} = \theta n_e^2 k_{il} , \tag{5.14}$$

where for optimum cadmium atom concentration, $\theta \approx 1/2$, as stated earlier.

We shall now write down the expressions for W_2, corresponding to the four energetically possible and most probable pump mechanisms of the upper laser level $^2D^*_{5/2}$:

For Penning ionization,

$$(W_2)_p = N_{He^*} n \xi_2 K_p . \tag{5.15a}$$

For direct electron impact from the cadmium atom ground state $5\,^1S_0$:

$$(W_2)_{de} = n_e n k_{02}\,. \tag{5.15b}$$

The dependence on n_e of the pump rate into the Beutler ion levels by charge transfer, followed by cascade transitions down to the $^2D^*_{5/2}$ state is similar to (5.15b):

$$(W_2)_c = \theta_{He} n_e n \xi_2 K_c\,, \tag{5.15c}$$

where $\theta_{He} = N_{He+} / n_e$.

Finally, for the pump rate due to stepwise electron-impact excitation via the ion ground state $5\,^2S_{1/2}$:

$$(W_2)_{se} = n_e N_{Cd_0^+} k_{i2} = \theta n_e^2 k_{i2}\,. \tag{5.15d}$$

Substituting (5.14) and sequentially each formula (5.15a–d) into (5.13), we obtain expressions for determining $(n_e)_{cr}$ for each pump mechanism:

Pump:	Penning ionization	Direct electron impact	Charge transfer	Stepwise electron impact
$(n_e)_{cr}$ (cm^{-3}):	4×10^{12}–6×10^{13}	1.3×10^{14}–4×10^{15}	5×10^{13}–1.5×10^{15}	$>10^{16}$

The scatter of calculated $(n_e)_{cr}$ values for each pump mechanism is due to the low accuracy of the known cross-sections and, hence, of the rate constants k for the excitation processes. Nonetheless, it can be seen that $(n_e)_{cr}$ intervals for different processes do not overlap, allowing the pump mechanisms to be identified unambiguously for specific conditions.

The real value of $(n_e)_{cr}$ for the 441.6 nm line was determined experimentally at the moment of transition from light amplification to absorption with the aid of a discharge tube consisting of two sections (see Figure 7.3a), similar to that used in [49]. One of the sections defined an initial level of gain in the resonator and compensated its losses; the second served to measure the unsaturated gain versus current. The point of interest was determined by the moment when the level of gain and output power returned to their initial levels, corresponding in a 2.5 mm bore tube to a critical current density of about 6.3 A cm^{-2} and $(n_e)_{cr} \approx 6 \times 10^{12}$ cm^{-3}. Hence, further verification has been obtained that the Penning pump mechanism really is the main pump mechanism for the He–Cd 441.6 nm laser line.

When the cadmium vapour density n is raised beyond its optimum level, $(n_e)_{cr}$ goes down, due to an increase in the concentrations of both electrons and cadmium ions (parameter θ)—see Figure 2.4. For example, for $n \approx 10^{15}$ cm^{-3}, $(n_e)_{cr}$ decreases by a factor of 1.5.

With $n_e > (n_e)_{cr}$, the relative population inversion becomes negative, i.e. absorption appears for further growth in n_e. The cause of absorption, as can be seen, is an increase in the population of the lower laser level, $N_1 \geq g_1/g_2N_2$.

The absorption coefficient can be measured by placing a tube either outside [50] or inside [51] the laser cavity of an amplifying tube (see, for example, Figure 7.3a). Thus, at the moment of threshold laser oscillation, the level of absorption becomes equal to the initial level of amplification.

In the first technique, there is no restriction on the levels of absorption coefficient. In the second method, the accuracy is enhanced substantially, but the absorption coefficient cannot exceed approx. $g_0\ l_1/l_2$ (where g_0 and l_1 are the unsaturated gain and length of the amplifying section and l_2 is the length of the absorbing section). According to [50], the absorption values do not exceed several % cm^{-1}.

In [50], one more reason for absorption of the 441.6 nm line is postulated for $n \geq 10^{15}$ $cm^{-3,}$ absorption by excited Cd_2* molecules. However, when different isotopes of cadmium (^{112}Cd and ^{114}Cd) were employed [51], the absorption and amplification lines did not overlap and no absorption was registered. This is proof that the absorption is due to overpopulation of the lower laser level.

5.4 LASER POWER AND EFFICIENCY

Since the gain coefficient characterizes the specific laser output power, it is apparent that the maximum laser output power is generated when the gain is at a maximum, provided that the dimensions of the medium are maintained constant. When varying the tube diameter *d*, the volume of the active medium alters in proportion to d^2, and the dependences of the laser power and gain coefficient on diameter are not the same. The smaller the tube diameter, the larger the cavity losses. This is the reason that the ratio of laser power to the product of length and diameter, P_{las}/ld [52], is valid for all but very short tubes, when $g_0 >> \beta_0$.

Since E_z and the input power do not depend on the buffer-gas pressure as strongly as the laser power (see Figures 2.3 and 5.5), the maxima in laser power and efficiency as functions of buffer gas pressure should occur close to each other. The maximum efficiency is obtained at a higher vapour pressure than for the peak gain and laser power because E_z and the input power fall with an increase in *n* (Figure 5.5). This effect, for a MVIL at higher buffer-gas pressures, is less pronounced due to a weak dependence on E_z (see Figure 2.3). With an increase in current, the laser power and the gain coefficient increase linearly and then saturate. The efficiency at maximum laser power is about half that in the regime of a linear increase in output power.

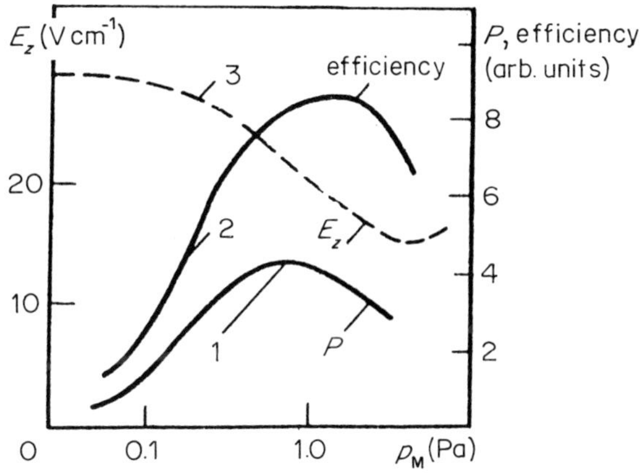

Figure 5.5 Typical dependences of efficiency and laser power on vapour pressure.

In tubes of the same length, but different inner diameter,

$$\eta(d) = P_{las}(d)/[E_z(d)i^{opt}(d)l] \,. \tag{5.16}$$

Applying the similarity laws $E_z d$ = constant and i^{opt}/d = constant, we find that, first, the optimum input power per unit length is not a function of diameter, i.e.P_{in}/l = a constant function of tube bore d and, second, the peak efficiency and peak laser power are produced in tubes of the same diameter.

In section 2.6, we calculated the effectiveness of the conversion of discharge energy into the energy of metal ions in the upper laser level, $\eta^*(=\eta_d\eta_3)$, which depends on the properties of the PC discharge in a mixture of a noble gas and the metal vapour, and, to a much lesser extent, on the specificity of the metal itself (excluding ξ). The overall efficiency of the laser, η, is the product of η^* and the efficiency of conversion of energy of the metal ions in the upper laser level to laser emission:

$$\eta = \eta^* \eta_4 \eta_q \,, \tag{5.17}$$

where $\eta_q = h\nu/\{\varepsilon\}$ is the quantum efficiency, i.e. the ratio of the energy of the laser quantum to the energy of the active buffer-gas particles (ions or metastable atoms). The coefficient η_4 is the ratio of the decay rate of the upper laser level due to stimulated emission to the pump rate of the same level. That coefficient can be measured in perturbation experiments (see section 5.2). It is at a maximum for the Beutler transitions in Cd II and Zn II for which $1/\tau_1 >> A_{21}$, and decreases due to electron impact and energy transfer collisions. As can be seen from Table 5.3[1], for visible gas lasers, the

Table 5.3 Efficiency and effectiveness of energy transfer stages in CW MVLs excited in PC discharges.

Laser	λ (nm)	Efficiency, η^a (10^{-4})	η_d	η_3	η_4	$\eta_q{}^b$
He-Cd	325.0	0.5	0.03	0.06	0.1	0.19
	441.6	3.5	0.03	0.08	0.3	0.14
He-Zn	589.4	0.3	0.03	0.06	0.2	0.106
	747.8	1	0.03	0.08	0.2	0.084
He-Se	497.6–530.5 (6 lines)	1	0.015	0.25	0.25	0.1
He-Ne	632.8	1.5–2	0.03	0.5	0.1	0.095

[a] Experimental values.
[b] With respect to the energy of the donor particle (He_m for Penning ionization; He^+ for charge transfer).

highest efficiencies are those of the Cd II laser (441.6 nm) and the He–Ne laser (632.8 nm). These transitions are populated by excited helium levels, but the optimum conditions are essentially different with respect to current (see Table 5.2). The larger optimum current in the cadmium vapour laser is determined by the higher saturation density of metastable helium atoms due to their role in Penning ionization of the cadmium atoms.

Thus, the relatively high energy losses in the discharge lead to a relatively low overall efficiency of CW MVILs with PC-discharge excitation. However, the absolute value of this efficiency is not of great importance for low-power lasers, because they are used predominantly in science and technology where they are chosen for their ease of design and high quality of laser radiation. Cadmium and selenium ion lasers operate under discharge conditions fairly similar to those of He–Ne lasers and supplement the blue, green and red wavelengths they produce. Since He–Cd and He-Zn lasers operate under very similar conditions, they are very promising for the production of multi-colour laser output by mixing Cd and Zn vapours in the same tube (for more on this see Chapter 7).

5.5 LASER RADIATION NOISE

One of the problems with cataphoretic MVILs is the high level of noise in the output laser emission. Ordinarily, the noise consists of a broadband low-frequency component, whose amplitude decreases with increasing frequency as $1/f$, and a number of discrete components at frequencies of the order of tens and hundreds of kilohertz due to discharge current modulation which is greatest for active media with negative dynamic resistance. Unfortunately, the He–Cd laser has one of the highest noise levels. Analysis of experiments shows that the presence of the low ionization potential metal atoms is the ultimate reason for the cataphoretic laser's noise, although to date there has not been a complete explanation of the phenomenon. Thus, in [52], the noise was attributed to an instability in the atomic helium density, resulting in a fluctuation in the electron density generated by changes in the rate of Penning ionization. In [5, 16, 53], the noise was explained by a fluctuation in the metal vapour density in the condenser zone and in the capillary tube. The majority of authors suggest that a spatial non-uniformity is produced in the active medium due to ionization-diffusional waves (striations) [52, 54–56]. The situation is complicated by the occurrence of oscillations with frequencies of 100–500 kHz which yield difference (0–100 kHz) and sum (>200 kHz) frequencies. However, the chief contribution comes from the low-frequency fluctuations. When the metal vapour pressure is increased, the intensities of all the fluctuations in the Cd II laser increase, but in the Se II laser they decrease. As a rule, in the Cd II laser, there is no correlation between fluctuations in current and fluctuations in spontaneous and stimulated emission; in the Se II laser, where the total noise level is around half that in the Cd II laser, such a correlation *is* observed over a wide range of frequencies, but current oscillations with frequencies above 200 kHz are not active.

The experimental observations just mentioned can be explained in the following way. The introduction of a low-potential admixture to the discharge lowers T_e, which results in a reduction of noise in the Se II laser. In the Cd II laser, however, the reduction in T_e is compensated for by the destabilization of the Penning process due to fluctuations in vapour density and/or metastable atom density. This leads to an oscillating degree of ionization (the contribution of Penning ionization to the total ionization rate can reach 20–30%), and enhances the ionization waves [56], the optimum conditions for which are very close to those for operation of the He–Cd laser [20, 52]. Further evidence for the role of He metastable atoms in the noise generation mechanism is that the noise level for the ^{3}He isotope is 25% higher than that for ^{4}He [52].

The most widely applied countermeasure against noise in cataphoretic lasers, and in particular, lasers pumped by Penning ionization, is discharge thermal stabilization. For example, the tube can be placed in a thermally isolating container with discharge current stabilization. This causes a decrease in the fluctuations in density of metal atoms originating from condensation and their subsequent evaporation from the tube wall, ultimately leading to longitudinal uniformity of the metal atom density. Since the axial distribution of the vapour density also depends on vapour removal at the condenser, it also must be thermally stabilized. This last point can explain why minimum noise appears at a fixed condenser temperature

[5, 16]. The lowest noise level achieved in Cd vapour lasers (down to 0.14%) [53] was the result of a completely thermally stabilized tube, with discharge current stabilization over a broad frequency range, and with vibration and other sources of mechanical instabilities minimized. Discharge tubes of improved construction—segmented, with a perforated capillary, a capillary of variable cross-section, or with a ceramic capillary [21]—can produce laser oscillation with low levels of noise.

The noise level can be reduced by synchronization of the striation waves by means of discharge-current modulation. This results in only one wave, of frequency equal to the modulation frequency, and which precludes beating (interaction) of striations in the low-frequency regime (up to 100 kHz) [55, 57]. An increase in laser power by a factor of around two is also observed. The synchronization frequency F_s follows a similarity law: $F_s d = 41$ kHz cm [55]. Suppression of striation waves can be achieved by using hot cathodes instead of cold cathodes. More details on the mechanisms of noise generation in MVILs are given in [56]. The most effective suppression of noise on ion laser transitions, coupled with the largest number of laser lines from each metal, is obtained in lasers with transverse excitation (see Chapter 6).

5.6 TECHNOLOGY AND APPLICATIONS OF CATAPHORETIC LASERS

The most popular cataphoretic MVILs are Cd and Se vapour lasers. According to data from laser industry journals, over the 5 years from 1975 to 1979 the volume of sales of He–Cd lasers increased by 5 to 10 times. The number of different models sold was 20 in 1975 and reached 104 in 1995 [58]. The technical specifications of some Russian lasers are given in Table 5.4. The lasers are fabricated from glass, use cold cathodes and, as a rule, employ isotopically enriched cadmium. Most of the discharge tubes are of a symmetric construction, which permits the direction of vapour transport to be reversed. Ordinarily, the oven is heated by the discharge and by an external (or internal) heater. Because of the small aperture of the cavity (small diameter and long capillary) and the high heat loading per unit length, the laser mirror positions must be very stable. This requires Invar-rod or silica-tube cavity supports to be used, which are thermally insulated from the active medium. The power supply of a cataphoretic laser is a simple high-voltage rectifier with current stabilization. Stabilization of the vapour admitted into the discharge is achieved by control of the oven voltage in accordance with changes in discharge voltage, which is proportional to E_z (see section 2.2).

Table 5.4 Parameters of commercial single-mode cataphoretic He–Cd lasers [13] and of experimental He–Se lasers [69].

Parameter	LPM-11	LG-63	LG-70	LG-61	LG-62	LGN-504	He-Se
λ (nm)	441.6	441.6	441.6	441.6	441.6	325.0	497.6 499.3 506.8 517.6 522.7 530.5
Laser power (mW)	20	8	40	50	100	15	5 (total)
Polarization[a]	L	N	N	L	N	L	L
Noise level (%) (20 Hz–1 MHz)	20	–	–	10	10	–	5
Divergence (angular minute)	–	3	1.4	2	2	2	3

[a] L and N stand for linear and natural polarization, respectively.

5.6.1 Recording on Photosensitive Materials

One of the most widespread and promising applications of the cataphoretic Cd (441.6 nm) laser is high-speed recording of information on photomaterials with high spatial resolution [59–63]. This laser is a competitor of the blue-green argon-ion laser and the red helium–neon laser. The Cd laser oscillates at a shorter wavelength in the single-mode regime, which produces the smallest focus of the three types of laser, and consequently, the highest density of information storage of the three. The higher sensitivity of photomaterials in the violet speeds up the recording process also. Thus, the exposure of a photoresist for the same time requires the following laser powers [59, 60]: 10 mW of 441.6 nm Cd-ion laser light (input power 200 W); 40 and 100 mW of 458 and 488 Ar-ion laser light (input power 3–5 kW). The noise level of the He–Cd laser is not higher than that of the Ar ion laser. A warm-up time is not only a characteristic of a He–Cd laser, but also of Ar ion lasers, since their cathodes require time to heat up. Application of the He–Cd laser to information storage is also cost-effective. According to the Liconix company, the cost of 1 mW h, for a Cd ion laser, is about $0.25, while for an Ar ion laser it is $0.45 to $0.50 (1975 figures [60]). For output powers up to 10 mW, He–Ne lasers are the most competitive, but their wavelengths are longer than those of green-blue lasers, which leads to lower resolution, lower recording speed and a requirement for expensive photomaterials for the red part of the spectrum. However, for output powers of 30–50 mW, the costs of Cd and Ar ion lasers become equal, and the use of He–Ne lasers is also impractical for economic reasons.

The He–Cd laser is used for producing holograms on photoresist and holographic video-discs with high resolution. Together with the Ar ion laser and the He–Ne laser, it is employed in recording colour images on photomaterials, and in making multi-colour holographic microimages. In this area, He–Se cataphoretic lasers are more suitable than Ar ion lasers due to their multi-wavelength oscillation in the 0.49–0.53 μm spectral region.

5.6.2 Other Applications [58, 64]

The cataphoretic He–Cd laser has been claimed to be 7 times more effective than other lasers employed for Raman, Brillouin and Rayleigh scattering. In Raman recombination scattering [64], the He–Cd laser is superior due to its higher photon energy as well as its wavelengths (441.6 and 325.0 nm), which lie in the area of greatest reflectivity of diffraction gratings and the highest sensitivity of photodetectors. The UV line of the He–Cd laser is used to study the fluorescence spectra of many liquids.

Laser emission at 325 nm can be used for isotope separation. Also, many devices have been proposed for monitoring levels of petrol in water and of NO_2 in the air by laser-induced fluorescence, and for blood analysis, etc.

REFERENCES

[1] W. T. Silfvast, *Appl. Phys. Lett.,* **13**(5), 169–71 (1968).
[2] E. K. Karabut, V. F. Kaydan, V. F. Kravchenko *et al., Elektron. Tekh., Ser. 1 Elektronika SVCh,* No. 9, 103–8 (1969) (in Russian).
[3] J. P. Goldsborough, *Appl. Phys. Lett.,* **15**(6), 159–61 (1969).
[4] T. P. Sosnowski, *J. Appl. Phys.,* **40**(13), 5138–44 (1969).
[5] J. R. Fendley Jr., I. Gorog, K. G. Hernqvist and C. Sun, *RCA Rev.,* **30**(3), 422–8 (1969).
[6] M. F. Sem, V. F. Keidan and V. C. Mikhalevskii, *Elektron. Tekh., Ser. 1 Elekronika SVCh,* No. 1, 69–76 (1971) (in Russian).
[7] D. J. Brink and V. Hasson, *J. Phys. D.,* **8**(2), L35–8 (1975).
[8] G. J. Collins, H. Kano, S. Hattori, K. Tokutome, M. Ishikawa and N. Kamide, *IEEE J. Quantum Electron.,* **8**(7), 679–80 (1972).
[9] I. G. Ivanov and M. F. Sem, *Elektron. Tekh., Ser. 4 Elektrovakuumiye i Gazorazryadniye Pribori,* No. 10, 42–6 (1974) (in Russian).
[10] T. P. Sosnowski and M. B. Klein, *IEEE J. Quantum Electron.,* **7**(8), 425–6 (1971).
[11] V. S. Aleinikov and V. V. Zubov, *Report on the First All-Soviet Seminar on Physical Processes in Metal Vapour Lasers* (Rostov-on-Don: Rostov State University) (1971) (in Russian).
[12] K. G. Hernqvist and D. C. Pultorak, *Rev. Sci. Instrum.,* **43**(2), 290–2 (1972); **41**(5), 696–7 (1970).
[13] M. K. Dyatlov, V. G. Kas'yan, N. G. Kashnikov *et al., Elektron. Prom.,* No. 5–6, 60–2 (1982).

[14] M. Dowley and W. Shull, Patent No. 3,883,818 USA, Kl. HO1s .
[15] G. G. Kiseleva, Yu N. Kulikov and V. A. Stepanov, Patent No. 392879 USSR, M.Kl.[3], HO1s *Gas laser, Otkritiya Izobr.* No. 25 (1975).
[16] K. G. Hernqvist, *Appl. Phys. Lett.*, **16**(11), 464–7 (1970); *IEEE J. Quantum Electron.*, **8**(9), 740–3 (1972).
[17] I. G. Ivanov and M. F. Sem, *Elektron. Tekh., Ser. 4 Elektrovakuumiye i Gazorazryadniye Pribori,* No. 9, 102–3 (1972) (in Russian).
[18] V. F. Keidan, V. S. Mikhalevskii, M. F. Sém and A. P. Shelepo, *Sov. J. Quantum Electron.*, **3**(1), 40–2 (1973).
[19] K. G. Hernqvist, *RCA Rev.*, **34**(3), 401–7 (1973).
[20] W. T. Silfvast and L. H., Szeto *Appl. Phys. Lett.*, **19**(10), 445–7 (1971).
[21] E. L. Latush, G. N. Tolmachev and V. Y. Khasilev, *Instrum. Exp. Tech. (USSR),* **19**(4), pt 2, 1189 (1976).
[22] A. A. Radzig and B. M. Smirnov, *Reference Data on Atoms, Molecules and Ions,* Berlin: Springer-Verlag (1985); *Data of Atoms and Atomic Ions,* Moscow:Energoatomizdat (1986) (in Russian).
[23] M. B. Klein and D. Maydan, *Appl. Phys. Lett.*, **16**(12), 509–11 (1970).
[24] A. Dienes and T. P. Sosnowski, *Appl. Phys. Lett.*, **16**(12), 512–14 (1970).
[25] M. K. Dyatlov, E. P. Ostapchenko and V. A. Stepanov, *Opt. Spectrosc.,* **29,** 539–40 (1970).
[26] T. Goto, H. Kano, S. Yoshino, J. K. Mizeraczyk and S. Hattori, *J. Phys. E,* **10**(3), 292–5 (1977).
[27] L. Csillag, M. Jánossy, K. Kántor, K. Rózsa and T. Salamon, *J. Phys. D.,* **3**(1), 64–8 (1970).
[28] V. V. Zhukov, I. G. Ivanov and M. F. Sem, *Zh. Prikl. Spektrosk.,* **26**(3), 544–7 (1977) (in Russian).
[29] I. G. Ivanov, E. L. Latush, V. F. Papakin and M. F. Sém, *Sov. Phys. J.,* **15**(8) 1149–53 (1972).
[30] S. V. Kasyanenko, G. M. Malyshev and Yu A. Tolmachev, *Opt. Spectrosc.,* **57**(5), 578–9 (1984).
[31] C. E. Moore, *Atomic Energy Levels,* NBS. No. 467 Washington DC: National Bureau of Standards, Vols I. (1949), II. (1952) and III. (1958).
[32] T. Goto and T. Sakurai, *J. Phys. D.,* **15**(11), 2413–21 (1982).
[33] V. V. Elagin and A. E. Fotiadi, *Opt. Spectrosc.,* **65**(1), 93–6 (1988).
[34] L. Lis, *Acta Phys. Pol. A,* **44**(2), 173–6 (1973).
[35] G. N. Tolmachev, E. L. Latush and V. S. Mikhalevskii, *Sov. Phys.–Tech. Phys.,* **19**(8), 1120–1 (1975).
[36] V. S. Aleinikov, G. N. Tolmachev, E. L. Latush and V. S. Mikhalevskii, *Sov. Phys.–Tech. Phys.,* **20**(9), 1245–6 (1976).
[37] R. C. Jensen, G. J. Collins and W. R. Bennett Jr., *Phys. Rev. Lett.,* **23**(7), 363–7 (1969).
[38] W. T. Silfvast, *Appl. Phys. Lett.,* **15**(1), 23–5 (1969).
[39] A. L. Bloom and J. P. Goldsborough, *IEEE J. Quantum Electron.,* **6**(3), 164 (1970).
[40] W. T. Silfvast and M. B. Klein, *Appl. Phys. Lett.,* **17**(9), 400–3 (1970).
[41] M. B. Klein and W. T. Silfvast, *Appl. Phys. Lett.,* **18**(11), 482–5 (1971).
[42] J. Berkowitz and W. A. Chupka, *J. Chem. Phys.,* **48**(12), 5743–4 (1968).
[43] I. Ogura, *Jpn. J. Appl. Phys.,* **13**(1), 164–9 (1974).
[44] A. Ferrario, *Opt. Commun.,* **7**(4), 376–8 (1973).
[45] E. L. Latush, M. F. Sem and G. D. Chebotarev, *Sov. Phys. J.,* **27**(5), 429–35 (1984).

[46] G. R. Fowles and R. C. Jensen, *Proc IEEE,* **52**(7), 851–2 (1964).
[47] W. T. Silfvast and M. B. Klein, *Appl. Phys. Lett.,* **20**(12), 501–4 (1972).
[48] S. N. Atamas', I. G. Ivanov, E. L. Latush and M. F. Sem, in *Radiation and Collisional Data of Atoms and Molecules. Tellurium,* ed. A. P. Ubelis, pp. 141–55, Riga: Latvian State University (1989) (in Russian).
[49] I. G. Ivanov, V. G. Il'yushko and M. F. Sém, *Sov. J. Quantum Electron.,* **4**(5), 589–93 (1974).
[50] J. Mizeraczyk, G. Jakob, E. Schmidt and J. Mentel, *J. Appl. Phys.,* **73**(11), 7180–3 (1993).
[51] I. V. Aleksandrov, N. V. Burdastyh and I. G. Ivanov, *Laser. Phys.,* No. 8, 18 (1994) (in Russian).
[52] T. F. Johnston and W. P. Kolb, *IEEE J. Quantum Electron.,* **12**(8), 482–93 (1976).
[53] D. C. Brown and N. Ginsburg, *Appl. Phys. Lett.,* **24**(6), 287–9 (1974).
[54] M. K. Dyatlov, V. G. Kas'yan and V. G. Levin, *J. Appl. Spectrosc.,* **30**(6), 719–24 (1979).
[55] V. A. Vas'kov, S. A. Gonchukov, E. V. Kurbatov and E. D. Protsenko, *Sov. J. Quantum Electron.,* **12**(2), 221–3 (1982); **18**(12), 1515–19 (1988) .
[56] V. V. Tuchin, *Dynamic Processes in Gas Discharge Lasers,* Moscow: Energoatomizdat (1990) (in Russian).
[57] M. K. Dyatlov, E. P. Ostapchenko, W. A. Stepanov *et al.*, Patent No. 3,848,716 USA, 5/11/1971 Kl. HOIs .
[58] E. A. Savicheva and S. P. Timofeeva, *Foreign Helium–Cadmium Lasers, Obz. Elektron. Tekh.,* Ser. 10, No. 2(403), (1976) (in Russian); *Laser Focus World,* The Buyers Guide 1995, Tulsa, USA: Pennwell, pp. 50–108.
[59] W. F. Hug, *High Speed Read/Write Techniques for Advanced Printing and Data Handling, Proc. SPIE,* **390,** 121–7 (1983) .
[60] M. W. Dowley, *High Speed Read/Write Techniques for Advanced Printing and Data Handling, Proc. SPIE,* **390,** 116–20 (1983).
[61] V. E. Iosirov, *Tekh Kino Televi.,* No. 4, 25–31 (1983) (in Russian).
[62] *Solid State Technol.,* **31**(2), 37–9 (1988).
[63] N. I. Mygai, Yu S. Oganov and S. V. Prozorov, *Lazerinform,* No. 48, 7–8 (1994) (in Russian).
[64] L. M. Rabkin, V. P. Dmitriev, V. S. Aleinikov, V. V. Zhubov and V. V. Bibikova, *Instrum. Exp. Tech. (USSR),* **18**(1), pt. 2, 299–300 (1975).
[65] V. M. Klimkin, V. E. Prokop'ev and S. S. Monasterev, *Efficient Gas Discharge Lasers,* Tomsk: Russian Academy of Sciences, pp. 6–12 (1978) (in Russian).
[66] K. G. Hernqvist, *Laser Focus,* **9**(9), 39–40 (1973).
[67] V. I. Pugnin, S. A. Rudelev and V. F. Stepanov, *J. Appl. Spectrosc.,* **18**(5), 667–8 (1973).
[68] I. G. Ivanov and M. F. Sem, *Elektron. Tekh., Ser. 4 Elektrovakuumiye i Gazorazryadniye Pribori,* No. 2, 12–16 (1974) (in Russian).
[69] V. B. Eprimyan, I. G. Ivanov, S. V. Prozorov *et al., Prib. Tekh. Eksp.,* No. 2, 259 (1981) (in Russian)

NOTES

1. The coefficients $\eta_d \ldots \eta_4$ given in Table 5.3, and later in Table 6.7, have been approximated to give an idea of the energy loss channels in MVILs.

6

Metal Vapour Ion Lasers Excited by Fast Electrons

6.1 METAL VAPOUR ION LASERS WITH HOLLOW-CATHODE DISCHARGES

The excitation of MVILs with hollow-cathode (HC) and transverse RF (TRF) discharges occurs in the negative glow (NG) regime in which fast electrons are accelerated in an electric field that is perpendicular to the optical axis of the laser (see section 2.4). These lasers are therefore transversely excited. As was discussed in Chapter 2, since the mechanisms of creation of fast electrons and their energy distribution in HC and TRF discharges are similar, the characteristics of these two types of laser are very similar.

Laser oscillation was first obtained in a HC discharge on atomic transitions of neon [1]. Later, it was found that transverse discharges are particularly effective for exciting ion transitions, especially those of metals. The earliest investigations were on Hg, Cd and Zn [2–12]; to date, over 23 metals have been studied in this regime, and the number of ion laser transitions has reached 259 (Table 6.1).

In HC and TRF discharge lasers, the optimum metal vapour pressure ($\approx$10–40 Pa) is two orders of magnitude higher than that of positive-column (PC) discharge lasers. For lasants with operating temperatures (i.e. the temperature corresponding to the mentioned vapour pressure) below 1000 °C (see Appendix A.2), the simplest technique for vapour production is evaporation [3–23]. In the cases of Cu, Ag, Au, Cr, etc., the operating temperature is very high; by choosing the cathode to be made of the lasant material, cathode sputtering can be exploited to produce the vapour (see section 2.4) [2, 24–43]. The operating temperature can be reduced by using volatile chemical compounds of the lasant metal that are dissociated in the discharge (see [44] and also [12, 16, 21, 33]). Laser oscillation can also be obtained in HC discharges on ion transitions of Ar, Kr and Xe [45], but is characterized by considerably lower specific output powers, unsaturated

Table 6.1 Ion laser lines in HC and TRF-excited MVLs.

Element	Active medium	Wavelength range (nm)	Number of lines	References
		Hollow-cathode-discharge excitation		
Hg II	He–Hg	615.0–1555.4	3	[4, 5, 6]
Cd II	He–Cd(Cd halide)[a]	325.0–1991.0	16	[2–4, 7–9, 44]
Zn II	He–Zn(Zn halide)[a]	491.2–2243.5	14	[2–4, 8–12]
I II	He–I	448.8–837.7	25	[9, 13, 14]
Se II	He–Se	460.4–649.0	13	[9]
As II	He–As(AsH_2)	538.5–710.2	8	[6, 16]
Ga II	Ne–Ga	633.4–719.8	2	[17, 18]
Tl II	He–Tl	473.7–1175.0	8	[91]
	Ne–Tl(TlCl)	582.9–707.0	4	[19–21, 91]
Be II	He–Be	467.7	1	[93]
	Ne–Be	1209.6	1	[93]
Ca II	Kr–Ca	1183.9; 1195.0	2	[22, 94]
Sr II	Kr–Sr	1087.6–1245.0	4	[22, 23, 94]
Cu II	Ne–Cu	248.5–274.1	12	[24, 27, 28–33
	He–Cu(Cu halide)[b]	450.6–2000.6	34	49, 116]
Ag II	Ne–Ag	227.7–502.7	6	[25, 29, 34,
	He–Ag[b]	224.3–2079.4	26	35, 48, 116]
Au II	He–Au[b]	226.4–886.8	24	[26, 29, 31, 48]
Al II	Ne(He)–Al	358.7–1063.9	15	[36, 38]
Ni II	Ne–Ni	796.2–797.5	2	[39]
Sn II	He–Sn[b]	579.9–1074.0	5	[40]
Pb II	He–Pb[b]	537.2–1322.1	5	[41, 42]
Cr II	He–Cr[b]	844.7–871.1	6	[43]
Ba II	Xe–Ba	614.2; 649.63	2	[22]
Ge II	He–Ge	513.17–589.34	3	[117]
Te II	He–Te	484.8–570.8	3	[117]
	Ne–Te	544.98–937.85	12	
Bi II	He–Bi	571.9–680.9	3	[118]
		Transverse-RF-discharge excitation		
Hg II	He–Hg	615.0; 794.5	2	[102]
Cd II	He–Cd	325.0–806.7	8	[102, 105]
Zn II	He–Zn	491.2–758.8	7	[102, 106]
Tl II	Ne–Tl	595.0; 695.0	2	[102]
	He–Tl	512.2	1	
Se II	He–Se	506.8–530.5	4	[102]
Cu II	He–Cu	467.3–780.8	3	[103]

[a] Laser oscillation is also observed on Cd II 325.0 and 441.6 nm, and on Zn II 589.4 and 747.8 nm lines in the absence of a buffer gas due to electron–impact excitation.
[b] Heavy noble-gas additives (Kr, Ar *etc.*) are utilized.

gains, etc. compared with metal ion lasers. Except for transitions in Cd and Zn, lasing will only occur in mixtures with noble buffer gases. Their role is the same as in PC discharges, and also to promote cathode sputtering and the consequent production of metal vapour. In Cd and Zn vapour, lasing can occur both with a buffer gas and without it (in the 'pure' metal vapour). It has been established that, for the majority of laser transitions in the NG in mixtures with buffer gases, and for all regimes (CW, quasi-CW and with microsecond pulses), the main excitation processes are charge transfer and Penning ionization. In HC (and TRF) discharges with evaporation of the metal vapour lasant, the regime of microsecond pulses is easily realized. In this regime, specific powers of 0.5 $W\,cm^{-3}$ and average powers of hundreds of milliwatts can be attained at pulse recurrence frequencies of tens of kilohertz [46, 47]. In principle, CW oscillation is feasible for all transitions populated by charge transfer and Penning ionization, with the exception of transitions that require very high pump rates.

6.1.1 CW MVILs with Evaporation of the Metal Lasant

The production of metal vapour by evaporation allows various laser tube designs to be employed, in which the pressures of the various mixture components can be controlled and optimized independently.

The first experiments were performed with pulsed discharges, although the goal was the development of CW MVILs. The first HC discharge MVIL used Hg as the lasant [5], and comprised a hollow cylinder as the cathode, and two anodes (Figure 6.1a). The evaporators were located between the cathode and each anode, and heated by the discharge. The laser exhibited good stability, but was not axially uniform, particularly when long (5 to 10 times the tube bore) cathodes were used (see below). In another tube configuration, a single transverse anode and a hollow cathode have been employed (Figure 6.1b) [3, 7, 20, 46]. In [3] an anode rod was placed along the length of a cathode slot. In [7, 46], the anode was a cylinder with the cathode placed coaxially inside it and the NG was formed within the cathode cavity. The width of the cathode slot should not be very small so that the discharge can penetrate the cathode cavity; otherwise parasitic discharges appear around the slot edges. However, the slot width should not be too large, otherwise, for slot widths approaching the cavity diameter, the effectiveness of the hollow cathode is greatly reduced. For small-diameter tubes, the effectiveness can be maintained by making the cavity a deep trough [48, 49].

The active length and volume of a tube with an axial discharge into a hollow cathode (of the type shown in Figure 6.1a) can be scaled up using one of three techniques. The first is to use a construction with a 'perforated' cathode and a coaxial anode (Figure 6.1c) [50–53]. The diameters of the holes

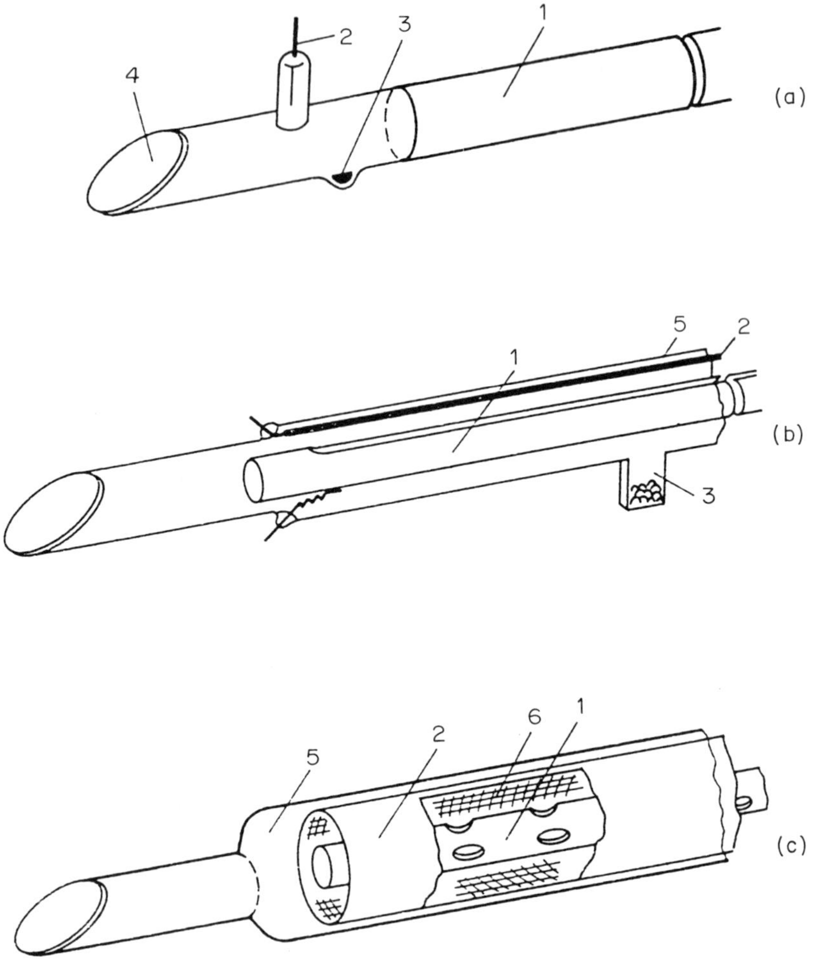

Figure 6.1 (a)–(g) Discharge tubes for HC discharge lasers with metal evaporation: 1, cathode; 2, anode; 3, metal reservoir; 4, output coupler at Brewster's angle; 5, glass sleeve; 6, wick; 7, ceramic tube; 8, gas outlet; 9, auxiliary anode; 10, cathode segments; 11, gas inlet.

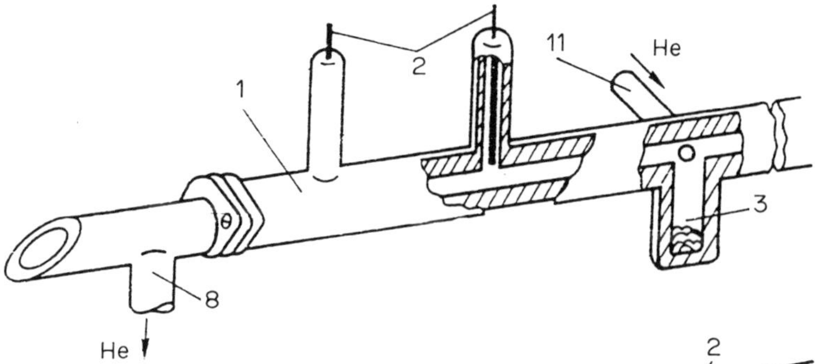

(d)

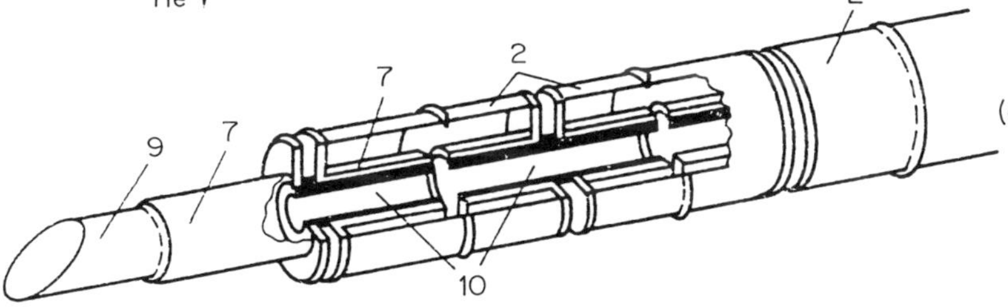

(e)

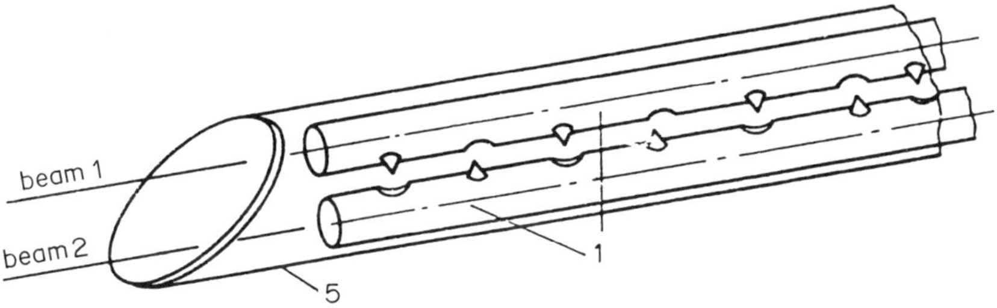

(f)

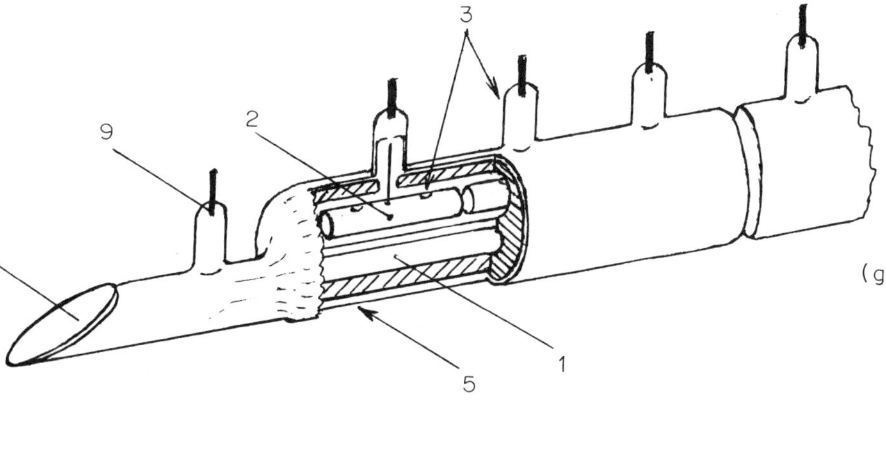

(g)

in the cathode are optimized in the same way as for the slot width. The second is a 'flute' type construction, where the cathode can be the outer vacuum jacket, and thus will prevent the discharge from attatching to the outside surface of the cathode (Figure 6.1d) [8, 9, 54–56]. Note that the first HC discharge laser using a He–Ne mixture employed a flute construction [1]. The third uses an array of segments to produce the discharge (Figure 6.1e) [57, 58]. To prevent the discharge from extending to the outer surfaces of the cathodes in the first and third types of construction, and thereby from reducing efficiency and from increasing the probability of arcing and sputtering, their surfaces are coated with pyrex or a ceramic, or the separation between the cathode and anode can be made as small as the width of the cathode dark space. It is important that the anodes in the second and third tube constructions be connected to the power supply by separate ballast resistors. It has been found that the transition of the discharge from a glow to a low-voltage arc in transverse slot arrangements occurs less often if a metal grid is placed between the anode and cathode. The potential of the metal grid is allowed to float, and it equalizes the plasma potential longitudinally, precluding local plasma overheating [59].

To prevent discharges to the outer surface of a cathode, the HC discharge can be arranged inside a longitudinal cathode cavity machined from a steel bar, having some anode rods placed along the cavity axis in a special 'anode cavity' adjacent to the cathode one and drilled from the same bar (see Figure 6.1g) [60, 61].

The lasant metal is placed in one or more evaporators, or inside the anode (sometimes the cathode) in the form of wire or pellets, where the lasant is heated by the discharge. If the holes of the perforated cathode do not enable the diffusion of vapour that is necessary, then some more holes must be drilled, but of smaller diameter [51].

Confinement of metal vapour, which also prevents laser window contamination, within the cathode cavity can be realized by means of so-called 'confinement sections' in a longitudinal self-heated discharge, with anodes placed between the tube windows and the main HC [55, 57]. The cataphoresis in PC discharges of these sections 'pushes' the metal vapour back into the HC and, as a result, the vapour density falls exponentially towards the windows (zone I in Figure 2.7). As was demonstrated in section 5.3, at high metal vapour pressures, absorption appears on the 441.6 nm line in the PC of a He–Cd discharge. However, the absorption coefficient does not exceed several % cm^{-1}, which has an insignificant effect on overall gain and power output from a HC discharge laser [62, 63].

In a tube with coaxial anodes, the metal (cadmium) can circulate by use of the 'heat-pipe' effect, which reduces the rate of metal loss [50]. Cadmium condenses on the ends of the anode which is at a temperature higher than 320°C (the melting point of cadmium). Then the metal returns to the middle of the anode by the wicking action of a multilayer stainless steel mesh which covers the inside of the anode (Figure 6.1c). There is no cathode sputtering in Cd ion lasers that have small bore (1.2 mm) cathodes coated in Cd, nor loss of vapour due to parasitic discharges away from the desired NG region [55]. For current densities up to 40 mA cm^{-2}, the lifetime of such a tube is more than 1000 h. Improved thermal stability of the discharge can be obtained from the use of a cathode made of bulk metal [55, 60, 64, 65].

In the plasma of a HC discharge, the pressure of the low-potential gas component has a strong influence on cathode emission. Thus, vapour density fluctuations occurring longitudinally lead to discharge current fluctuations of reverse sign. These fluctuations in flute-type laser tubes can be minimized by flowing the gas mixture through the cathode cavity [9]. The gas flow also removes contaminants from the cathode. The flow rate of a mixture of metal vapour and buffer gas in [9] was 50–350 cm s^{-1}. In such flowing lasers with Cd, Zn and I vapours, the highest laser power amongst CW HC discharge lasers of around 0.1 W has been obtained [14, 66]. The output power increases with increasing flow rate, and then saturates at around 100 cm s^{-1} [66]. Thereafter, the laser power falls due to the longitudinal gradient in the laser gas mixture (100's of Pa, see section 2.2), and the saturation of vapour in the cathode cavity. Because consumption of the laser mixture is high in such cases (up to 15 cm^3 MPa min^{-1}), these flow rates are only practical in scientific research.

Thus, the many versions of HC discharge tubes can essentially be divided into two main groups, which are distinguished by the relative positions of the positive and negative electrodes. If the anodes (or anode sections) are situated in parallel to the cathode cavity axis, and electrons and ions move normal to the cavity axis, then we have 'transverse discharge' (TD) tubes. In 'longitudinal discharge' (LD) tubes, however, the main cathode (or each cathode segment if the tube has a few cathode segments) is situated between the anodes along the axial direction, i.e. the anodes and the cathode segments are placed alternately along the tube axis, and the electrons emitted by the cathode surface then undergo axial motion towards the anodes. The experimental studies demonstrate that both types of HC tube are almost equal with respect to discharge stability (i.e. switching from a glow to arc discharge) and consequently with respect to stability of the laser output. But the question as to which kind of tube provides higher laser powers and efficiencies remains no less important, and the issue is now actively discussed [67–70].

In order to understand the specific features of TD and LD tubes as applicable to HC discharge MVILs, we should consider the axial distribution of plasma parameters in the HC LD tube. Such a consideration reveals that inside each cathode segment, axial inhomogeneity of discharge current density from the cathode surface occurs and an axial electric field E_z is present along with the radial field in the HC. The E_z field is necessary for axial movement of electrons from the central parts of the cathode segment to the anode. The value of the cathode potential fall U_{CDS} (CDS = cathode dark space) in these parts is therefore smaller by the value of the potential fall in the NG. A smaller U_{CDS} value leads to smaller surface density currents in the central parts of the cathode segment compared with those close to anode.

Note however that, despite the existence of an axial electric field, the NG is not similar to the PC of a longitudinal discharge, for the EEDF, and ionization and excitation processes of atomic and ion levels are, as with E_z, due to electrons emitted by the cathode and accelerated across its CDS. Measurements show that the ratios of atomic and ion line intensities in LD and TD tubes do not differ significantly. Even so, the absolute values of spectral intensities decrease slightly faster than does the current density the further inside the cathode segment excitation and emission occur. Thus, the differences in non-uniformity of current and electron densities in the NG determine a major difference in excitation and ionization rates of buffer-gas atoms in TD and LD HC tubes, and consequently different pump rates of the laser transitions in each case.

Note also that in solid (or 'deaf') cathode segments, without additional holes, the axial NG electric field E_z induces cataphoresis of the metal vapour in the centre of the segment and causes the axial distribution of metal vapour to be opposite to that of the current density.

Thus, comparing TD and LD HC tubes, it is quite clear that in LD tubes, current and metal vapour densities are 'optimum' only along a short length of the cathode, whereas in a TD tube, close to optimum densities can be provided throughout the length of the tube. Only in one case, at currents much lower than optimum, where laser output power exhibits a supralinear dependence on current, has LD tube construction been shown to be preferable to TD construction [67]. In general, the output power of TD HC laser tubes is 1.5–1.8 times higher than that of LD HC laser tubes of similar active dimensions; the laser efficiency is also higher in the former (by 1.3 to 1.5 times) [68, 70]. In addition, due to the absence of any axial electric field in a TD laser, the lineshape of ion laser lines is not distorted, which is very important for metrologic applications of gas lasers.

A special place in the group of HC MVILs is held by lasers with helical cathodes; the properties of such discharges were discussed briefly in section 2.4. Such cathodes have been successfully employed in He–Cd, He–Zn, He–TlCl, He–Kr and some other lasers [69, 71–75]. 'Double-helical' cathodes for alternating current excitation have also been designed, as well as 'triple-helical' cathodes for three-phase excitation [72, 73]. In [74], an interesting solution to the problem of metal vapour generation and seeding within a helical cathode was found. A helical cathode with small holes acted as the reservoir of lasing metal (cadmium), and a continuous flow of vapour from the cathode surface into the discharge volume was thereby established. The metal vapour density inside the cathode volume was regulated by varying the cathode temperature, determined by electrical input power to the discharge.

Since the discharge voltage in HC discharge lasers does not alter significantly for different experimental arrangements, it is convenient to use the discharge current per unit cathode area as a characteristic of the pump power. In CW lasers, the discharge current (and the input power) is limited by overheating of the medium, instabilities, arcing, the power supply, etc. Thus, many investigators have used a quasi-CW regime with 'long' (0.1–10 ms) pulses at low pulse recurrence frequency (≤100 Hz). The current pulse amplitude can be higher than in the CW regime, and the excitation conditions in the pulse correspond to those of CW operation. In these experiments, even at the highest current densities (≈300 mA cm^2) [14], only a few of the laser transitions are saturated, whereas for the others, the laser power increases with current.

Most HC discharge lasers have been investigated following long examinations of pump mechanisms in PC discharge (pulsed and CW) lasers, and it has been accepted that the pump mechanisms are generally the same. We shall survey some of the similar characteristics of laser transitions of the most effective metals (Table 6.2). For Penning-excited lines, the optimum vapour pressure is less than that for charge-transfer-excited lines. This is the result of a sharper fall in the rate of formation of helium metastable atoms compared with helium ions with an increase in metal vapour pressure, since the electron energy losses to metal atoms in the range 20–30 eV (corresponding to the maximum of excitation of helium metastables) are higher than the energy lost by electrons with energies above 50 eV, which are responsible for ionization. For that reason, the excitation rate of helium metastables (and consequently of the upper levels pumped by Penning ionization) is nearly the same in HC discharges as in PC discharges, so yielding very similar energetic characteristics for these transitions (see Tables 5.2 and 6.2 for the Beutler transitions of Cd II and Zn II).

In the NG region, excitation is characterized by a weak dependence of buffer-gas pressure on cathode diameter. As mentioned above (section 2.4), under 'lasing' conditions, buffer gas ions decay by charge transfer to metal atoms. Thus, in contrast to PC discharges, where different gas pressures lead to different optimum metal vapour pressures (they reduce with increasing gas pressure), in HC discharges the pump rate is only determined by the rate of ionization of the buffer gas (relation (2.58)), which is a maximum when the radial profile of the ionization rate is almost 'flat' [76]. For that reason, in HC discharges, the dependence of gain on cathode diameter is not very pronounced. However, in practice, gain increases with a reduction in cathode diameter, which can be explained by the fact that, for small diameters, higher levels of pumping can be produced by increasing the electron density without the appearance of arcing.

Table 6.2 Output characteristics of HC discharge lasers with metal evaporation in the presence of helium buffer gas.

Line	λ (nm)	Helium pressure (kPa)	CW regime		Quasi-CW regime		References
			P (mW) (P_{sp} (mW cm^{-3}))	*j* (mA.cm^{-2})	*P* (mW) (P_{sp} (mW cm^{-3}))	*j* (mA cm^{-2})	
Hg II	615.0	1–1.7	20 (5.7)	7.5	80 (23)	200	[119]
			4 (33)	250	–	–	[60]
	794.5	1.33	–	–	5 (1.5)	200	
Cd II	441.6	1.6–2	5 (62)	40 [a]	–	–	[55]
			33(18.6)	34	–	–	[77]
			92.5(12)	26	–	–	[56]
			102 (14)	47 [a]	–	–	[66]
			115(20)	26.5	–	–	[70]
	533.7, 537.8	2.5–4	12 (6.8)	34			[77]
			35 (6.2)	26.5			[70]
			60.4(7.9)	26			[56]
			97(13)	47			[66]
					10(1.45) [b]	–	[74]
	635.5, 636.0	1	8 (4.5)	34	–	–	[77]
			13(2.3)	26.5	–	–	[70]
			41.3(5.4)	26	–	–	[56]
	723.7–887.8	1	25 (4.2)	–	–	–	[57, 66]
Zn II	491.2	3.3	15 (4.3)	50	145 (42) (491.2 and 492.4)	120	[120]
	492.4	3.3	25 (7.2)	50			
	589.4	1.6	10 (4.3)	50	15 (4.5)	80 [a]	
	747.9	1.6	35 (10)	50	60 (17)	80 [a]	
	758.8	1.6	12 (3.5)	50	80		
I II	576.1	1.33	7 (2)	60	100 (20)	275	[13, 14]
	467.5–887.7	1.33	33 (9.5)	60	1000 (200)	275	

[a] Discharge current density from cathode surface is optimum (i.e. specific laser output power is a maximum).
[b] Helical cathode [74], $i = 2.5$ A.

The highest optimum helium pressure corresponds to the green 533.7 and 537.8 nm doublet of Cd II, which is populated in the NG by charge transfer, followed by collisional relaxation to the 4 $^2F^o$ levels, similar to the case in the afterglow of a positive column. Therefore, the pump rate of the green doublet is proportional to $W_{\mathrm{He1}}\tilde{n}_e$ (where $\tilde{n}_e$ is the density of slow electrons). As was mentioned above, the radial profile of the slow electron density resembles the profile of the helium ionization rate (*i.e.* the charge transfer rate) at higher (about 1.5 times) helium pressures, and the peak of the ionization rate is shifted to near the wall while the magnitude of the ionization rate throughout the discharge volume is reduced slightly. Therefore, $W_{\mathrm{He1}}n_{\mathrm{e}}$ over the volume of the NG peaks at a higher helium pressure than $W_{\mathrm{He+}}$. In that case, the decrease in ionization rate at high pressures is over-compensated for by the increase in the electron de-excitation rate due to the higher density of slow electrons, and finally the optimum conditions move into a higher (by 1.5 times) pressure regime.

As can be seen in Table 6.2, the highest specific output powers ($\geq$10 mW cm^{-3}) from HC discharges are obtained on the 441.6, 533.7 and 537.8 nm lines of Cd II, the 491.2, 492.4 and 747.8 nm lines of Zn II, the 615.0 nm line of Hg II and the yellow-green lines of I II. Compared with PC discharges, the specific output power on the Beutler $D^* \longrightarrow P$ transitions of Cd II and Zn II that are populated by Penning collisions are practically the same, while for transitions pumped by charge transfer, HC discharge lasers have much higher specific output powers (e.g. on the 615.0 nm Hg II line, 33 mW cm^{-3} is generated in HC discharges and only 4 mW cm^{-3} in PC discharges). Taking into account the absence of saturation with current for most of the lines, a considerable increase in power (to 100's of mW) can be expected on charge-transfer-excited laser lines by making the pumping more intense.

Despite the good operating characteristics of iodine laser transitions, it is not very suitable as a lasant in HC discharges due to its chemical incompatibility with the metal components of the discharge tube (tungsten is found to be the most resistant metal [14]). For the same reason, and probably due to the occurrence of homonuclear molecules, HC-discharge Se and Te lasers have worse characteristics than their PC versions.

A remarkable feature of CW HC (and TRF) discharge lasers is that they are almost totally free from high-frequency (fractions of a kilohertz to a few MHz) noise in the laser output radiation, which occurs in PC discharges due to plasma fluctuations, discharge striations, oscillations of the double layer, etc. In addition, the low level of noise in a HC discharge is promoted by (a) the reduced role, compared with in PC discharges, of Penning ionization in contribution to the overall ionization rate of cadmium (Penning ionization being a destabilizing factor); (b) the growth of voltage across the discharge with an increase either in cadmium vapour density or in current (due to the positive dynamic discharge impedance of a HC discharge (see section 2.4)); and (c) the absence or insignificance of electric field in the NG, especially in TD HC tubes. Thus the overall noise level in HC (and TRF) discharge lasers is an order of magnitude lower (less than 2–3% in amplitude and 1% of the root mean square value) than that in PC discharge lasers, before any special measure is taken to stabilize the laser power [10, 77–80]. The use of a constant-voltage supply allows even higher discharge stability and a lower noise level, as does the connexion of the Cd heater either directly in series with the HC discharge unit or through a controller, so that the heater current is proportional to the discharge current [56, 81]. This provides a constant metal vapour pressure, together with current and voltage stability of the discharge (U_{CDS}). Control of the Cd heater can be done easily by linking it to the level of the discharge voltage (see Figure 2.17a).

In HC discharge lasers constructed using bulk metal, the high thermal conductivity gives better long-term stability of the vapour pressure and laser output power, i.e. reduced low-frequency noise is attained in comparison with glass cataphoretic laser tubes. The high optimum helium pressure, especially for the green $4\,^2F \longrightarrow 5\,^2D$ doublet of Cd II, gives rise to strong Lorentz broadening of the gain line, and to single-frequency operation of the laser. Note that such a self-selection of the oscillation modes in this doublet occurs with almost 100% efficiency, i.e. all other modes are effectively absent. In [82], three wavelengths (441.6 nm, 533.7 nm and 537.8 nm) of a HC discharge He–Cd laser were simultaneously stabilized by locking to the gain maximum for any one of these wavelengths. The variance for the beat signal between two such lasers varied from 4.7 to 6.9 MHz over a 100 s period. This represents a frequency stability of a few parts in 10^8. If the stability could be improved to a few parts in 10^7 simultaneously on different laser lines, then this laser might become a suitable coherent replacement for Cd- and Kr-discharge lamps used for precision measurements, e.g. for characterizing product standards, etc.

As can be seen in Table 6.2, the discharge conditions (buffer-gas pressure and current) can be chosen to be appropriate for simultaneous oscillation on laser lines across a wide spectrum, and has allowed the development of ‘white’ light lasers. Alternatively, more than one cathode can be mounted in parallel inside a single discharge tube; for instance, in the Cd II laser with two parallel perforated cathodes in a single casing, (Figure 6.1e) operating with a

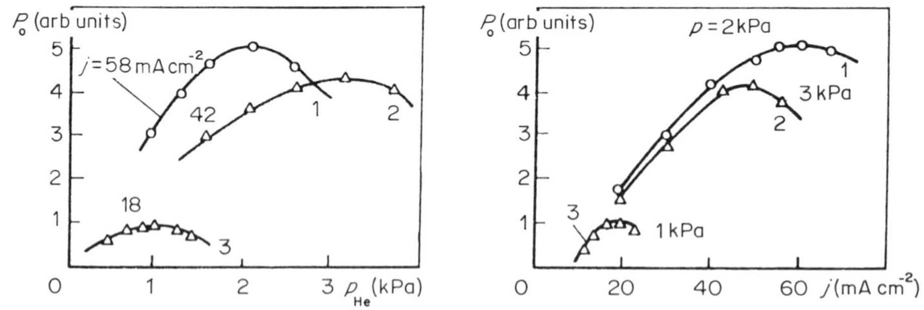

Figure 6.2 Typical dependences on discharge current and gas pressure of the intensities of the blue (1), green (2) and red (3) output lines of a He–Cd HC discharge laser.

50 Hz AC power supply, two alternating pulsed beams are produced, on the blue (441.6 nm) and green (533.7 and 537.8 nm) lines [83]. Simultaneous oscillation on a number of lines of a single lasant, and from a mixture of various lasants, has also been demonstrated. Such HC discharge lasers are the subject of Chapter 7.

In recent years, progress has been made in the development of HC discharge He–Cd MVILs with long-life (over 1000 h), stable operation ([55, 64, 65], and as described in advertising for these lasers by the Japanese companies NDK and Koito). However, further progress in this field depends on determining (a) how to control the cathode temperature and metal vapour density in the discharge precisely; and (b) how to minimize cathode erosion and the loss of Cd vapour and He during the laser lifetime.

6.1.2 MVILs with Cathode Sputtering

Cathode sputtering was first used in MVLs to produce Cd and Zn vapour [2]; later, the vapours of Cu, Ag, Au etc. were produced by cathode sputtering, and many UV laser lines have been obtained as a result.

The first laser incorporated a water-cooled (to avoid overheating and deformations) cylindrical hollow cathode with an insert of the metal to be sputtered (cadmium or zinc), and an anode rod placed inside the cathode cavity (Figure 6.3a) [2]. To increase the specific input power (and the metal density), the cathode cavity was made within a massive water-cooled copper body with circular cross-sectional diameter of 1 to 6 mm [27, 30] (Figure 6.3b), or in the form of a narrow (2 mm) rectangular slot of a depth (6 mm) that was larger than its width (Figure 6.3c) [48, 49]. This has led to the production of stable discharges with spatial homogeneity up to cathode lengths of 60 cm, and a current density in the 'long' pulse regime of up to 2 A cm^{-2}, without any localization or expulsion of the plasma. The cathode was either entirely made of the lasant metal (for Cu, Al, Ni) or had a long slotted insert of the metal of interest (Ag, Au, etc.) in a water-cooled copper block; alternatively, the lasant can be deposited electrolytically on copper (see [48] and also [39, 43]. For a copper vapour laser with a power input of a few kilowatts, sectioned discharge tubes have been used. They comprise a water-cooled cylindrical copper cathode or a rectangular active zone formed by two parallel cathode plates that are also water-cooled (Figure 6.3d) [48, 84]. Each tube section is connected to the rectifier through separate ballast resistors.

To reduce cathode corrosion, which deforms the slot and causes laser instabilities, a laser tube has been proposed that is excited by three-phase current from the mains power supply (Figure 6.3e). In that tube, the discharge is initiated between three parallel copper-rod electrodes [48]. The

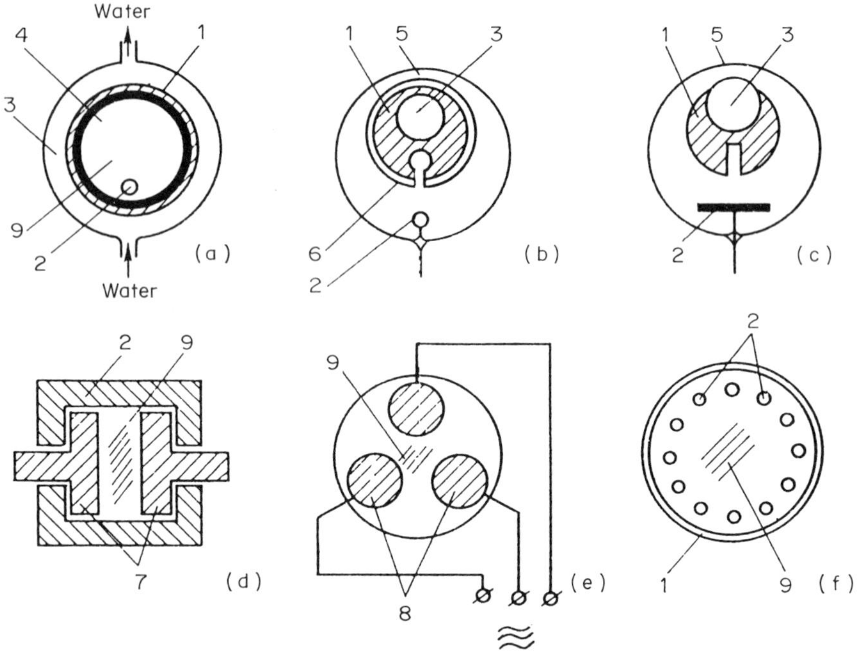

Figure 6.3 (a)–(f)Cross-sections of discharge tubes of HC discharge and cathode sputtering lasers: 1, cathode; 2, anode; 3, water-cooled sleeve; 4, cathode insert; 5, glass or quartz laser envelope; 6, quartz cathode envelope; 7, cathode plates; 8, electrodes; 9, NG zone.

electrode polarities change alternately, and thus rod deterioration is slowed. The power supply is very simple.

The operating characteristics of the best HC discharge lasers and cathode sputtering lasers for the most intense ion transitions of Cu, Ag and Au are given in Table 6.3, together with the cathode surface current density. Although problems appear with current and input power optimization (see section 2.4), it is found that the highest UV power in the CW regime is attained, as in the case of the quasi-CW regime, for Cu II lines: 270.3 nm (34 mW CW at a specific output power of 5 mW cm^{-3}; 0.5 W quasi-CW at 56 mW cm^{-3}); and on a group of UV lines in the range 248.5–270.3 nm (0.2 W CW at a specific output power of 14 mW cm^{-3} [85]; 1.25 W quasi-CW at 140 mW cm^{-3}). The shortest wavelength, CW UV line, for any class of laser, is the 224.3 nm line of Ag II. The 248.5 nm line of Cu II has the largest unsaturated gain coefficient in the quasi-CW regime at a current of 2 A cm^{-2}: 8.3% m^{-1} (the threshold current density is 200 mA cm^{-2}). In the visible and

Table 6.3 Output characteristics of HC discharge lasers with cathode sputtering.

Line	λ (nm)	Buffer-gas pressure, p_{He} (kPa)	CW regime P (mW) (P_{sp} (mW cm^{-3}))	CW regime j (A cm^{-2})	Quasi-CW regime P (W) (P_{sp} (mW cm^{-3}))	Quasi-CW regime j (A cm^{-2})	References
Cu II	248.5		—	—	0.5 (56)	1.1	[30, 85,
	259.1	1.6 (Ne)	21 (3)	0.25		1.1	86–88, 90]
	260.0		5 (0.7)	0.25	0.25 (28)		121, 125,
	270.3	1.6 (Ne)	34 (5)	0.25	0.5 (56)	1.1	
	248.5–270.3	1.6 (Ne) –	200 (14)	0.41	0.8 (267)	1.14	
		1.6 (Ne)			0.9 (62.5)	0.6	
					0.58 (17.1)[a]	—	
	780.8	2 (He) +	1000 (17.4)	0.1	5.2 (250)	0.7	
		0.03 (Ar)	400 (106)	0.3	—	—	
		0.8 (He) +			0.026 (10)[b]	0.146	
		0.04 (Ar)					
		2.0 (He) +			0.058 (61)[c]	0.63	
		0.08 (Ar)					
		2.4 (He)			1.3 (38.3)[d]	—	
Ag II	224.3	2.7 (He) +	1.46 (0.5)	0.14	0.05 (17)	1.1	[30, 122]
		0.03 (Ar)			0.065 (7.5)	1.3	
	318.1		—	—	1.3 (140)	1.3	
	408.5	1.6 (Ne)	—	—	0.83 (90)	1.3	
	478.8		—	—	2.3 (250)	1.3	
	800.5–840.3	1.6 (He) +	100 (15)	—	2.0 (95)	0.7	
		+ 0.01 (Xe)					
Au II	282.3	3 (He) +	19.2 (1)	0.11	—	—	[29, 31,
	284.7	0.03 (Ar)					48, 89]
	289.3	—	—	—	0.6 (20)	1.1	
	291.8	—	—	—			
	282.3	2 (He) +	—	—	0.024 (9.5)[e]	0.25	
		0.02 (Ar)					

[a] Helical cathode, i = 165 A [90].
[b] Hollow-anode cathode tube [86].
[c] High-voltage segmented HC tube [87, 88].
[d] Helical cathode, i = 180 A [90].
[e] High-voltage segmented HC tube [89].

near-IR spectral regions, the highest power is produced on the 780.8 nm Cu II line: 1 W CW, and over 5 W quasi-CW. Its threshold current density is the lowest at 20 mA cm^{-2}, and the peak CW specific power corresponds to a current density of 0.3 A cm^{-2}.

However, the optimum vapour density is usually not achievable using cathode sputtering. This was demonstrated in [48], where a 13 cm long cathode was heated to 1350–1500 °C, and the optimum vapour density of Cu *was* reached. In that case, the laser efficiency of the Cu II 780.8 nm line increased by an order of magnitude compared with that of the sputtering regime. This means that the maximum output powers of MVILs that utilize Cu, Ag and other metals can only be approached in a heated regime. Several techniques have been applied to increase the density of metal atoms in the discharge, but only one has proved to be effective: the addition of small amounts of heavy noble gases (Ar, Kr, Xe), which increases the mass of bombarding ions. At the same time, a fall in the rate of ionization of helium occurs, but is compensated for by the higher metal vapour density that is generated.

For IR oscillation in Cu II and Al II, and for UV oscillation in Cu II and Au II, three other designs of discharge tube have been employed. The first type is the high-voltage, hollow anode-cathode, one of which is shown in Figure 6.3(f); the second makes use of segmented HC tubes of various configurations [36, 86–89]; the third uses helical hollow cathodes [71, 72, 90]. These constructions cause the threshold current for oscillation to decrease, and the voltage to increase; thus the efficiency does not change. It is not clear why the threshold current reduces—whether the ionization rate of the buffer gas increases (which seems very unlikely) or the metal atom density is higher due to the higher energy of ions bombarding the cathode. There have been experiments to employ halides in HC discharges to provide Cu atoms [33], in the same way as in self-terminating copper lasers. However, the low-potential species added to the discharge together with electrode corrosion lead to a reduction in ionization of the buffer gas and in the laser power.

6.1.3 MVILs Excited by Microsecond Current Pulses

For HC discharge lasers operating in the CW regime, the optimum pump power is reached for very few transitions. The major reasons for this are spatial discharge inhomogeneities and arcing, and the problems that arise with the levels of heat removal necessary. On other lines, the laser power cannot approach the level obtained in quasi-CW operation (see Tables 6.2 and 6.3). In discharges excited by short pulses (durations 0.1–10 μs), discharge inhomogeneities do not develop, and it is possible to reach optimum pumping. The average input power and laser power can be regulated by variation of the pulse recurrence frequency (PRF).

Although microsecond current excitation was employed in the first HC discharge lasers of Hg, Cd and Zn [3, 5], this regime has only been used extensively in investigations with the vapours of Tl [19, 70, 91, 92], Be [93], Ca and Sr [94] and in other relatively recent studies [46, 47, 95].

Just recently, the excitation and plasma characteristics of the most effective

HC discharge lasers (using vapours of Hg, Zn and Tl) have been examined with microsecond pulses [46, 47, 92, 95]. The purpose was to optimize the excitation (amplitude, duration and recurrence frequency of the current pulses), and the diameter of the cathode cavity in tubes constructed with transverse side discharges (Figure 6.1a) and slotted cathodes (Figure 6.1b). It was found that the largest peak laser powers are generated with current pulses of τ = 1–2 μs duration (Figure 6.4), and the average power increases monotonically with an increase in τ. The maximum efficiencies were also obtained when τ = 1–2 μs. The optimum current density, for most transitions, was 3 to 5 times larger than in the CW regime. This resulted in a similar increase in specific peak laser output power (Table 6.4). For the 615.0 nm Hg II line, the peak power increased by a factor of 15, and its optimum current density was greatest (0.9 A cm^{-2}).

Table 6.4 The most intense laser lines in HC discharges, excited by 1 μs current pulses.

Ion	λ (nm)	Buffer gas	j (mA cm^{-2})	g_0 (dB m^{-1})
Hg II	615.0	He	900	30
	794.5	He	900	35
Cd II	441.6	He	200	5[a]
	533.7	He	150	4.5
	537.8	He	150	4.5
	636.0	He	95	2.5
	728.4	He	150	6
	806.7	He	150	7.5
Zn II	491.2	He	430	7
	492.4	He	430	7.5
	589.4	He	540	3.5
	610.2	He	220	2
	747.9	He	540	5[a]
	758.8	He	280	5
Tl II	595.0	Ne	340	15
	695.0	Ne	340	9
	707.0	Ne	340	3.8
Cu II	780.8	He	1000	33.5
Sr II	1201.7	Kr	–	24
Ga II	633.4	Ne	600	13
	719.8	Ne	600	10.7

[a]When excited in the absence of a buffer gas, g_0 = 15–20 dB m^{-1}.

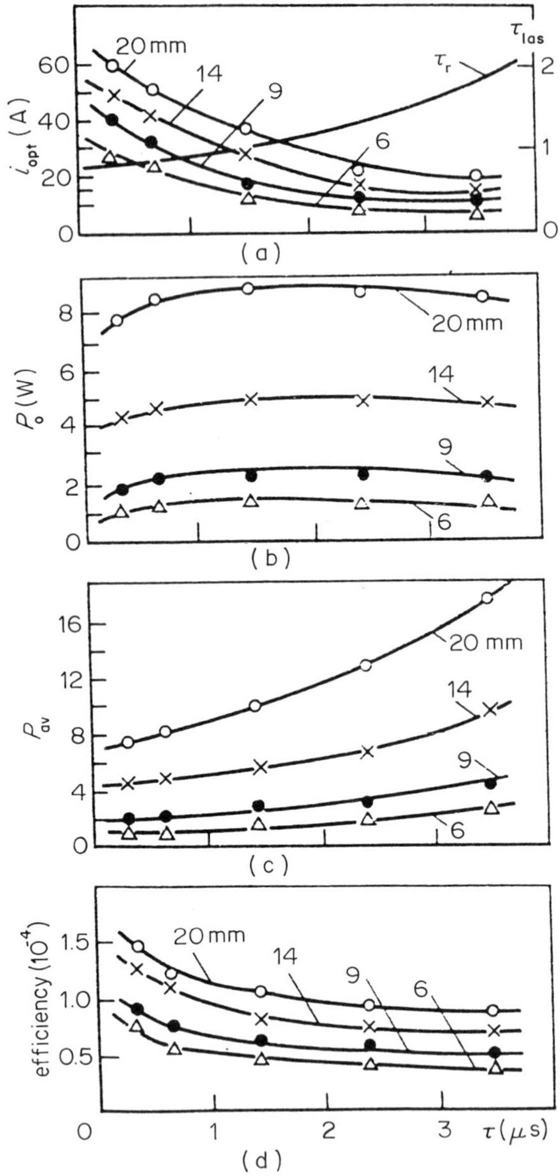

Figure 6.4 Dependences on current pulse length, τ, of: (a) optimum pumping current, i_{opt}, and laser pulse duration, τ_{las}, (b) peak laser power, (c) average laser output power, (d) efficiency, of the Hg II 615 nm line; p_{He} = 1.1 kPa for various cathode-diameter tubes.

According to section 2.5, the HC discharge regime is characterized by a large number of fast electrons, the density of which peaks 0.5–1 μs after the beginning of the current pulse, and by a monotonically increasing number of slow electrons, $\tilde{n}_e$, which are relatively few during the first microsecond. Slow electrons are responsible for electron impact de-excitation, which causes most lines to saturate and cease oscillating at critical electron densities, $\tilde{n}_e^{cr}$, which are specific to the various transitions (see Figure 2.26, for instance). For longer current pulses, the optimum current amplitude should be smaller in order that $\tilde{n}_e \leq \tilde{n}_e^{cr}$, as observed experimentally (Figure 6.4a). This is strong evidence for the role of electron de-excitation on all the transitions. At the beginning of a microsecond current pulse, when $\tilde{n}_e$ is small, electron de-excitation is important at higher currents (several times), and thus the excitation rate due to fast electrons in those pulses for which $\tilde{n}_e < \tilde{n}_e^{cr}$ will rise by the same factor. The optimum current is lowest for charge-transfer transitions of the Cd II ion, and highest for the 615.0 nm charge-transfer line of Hg II. Nevertheless, saturation of laser power at 615.0 nm in Hg II still occurs due to electron de-excitation. As investigations [95] show, even for extended current pulses of optimum amplitude, after 2 to 3 μs the inversion ceases because slow electrons depopulate the upper laser level, and increase the population of the close-lying 6 ^2D levels (Figure 6.5). For the line 794.5 nm, whose levels lie closest to the 6 ^2D levels, quenching is of higher importance (see Figure 2.26), and therefore the gain and laser power in HC discharges are lower than in PC discharges. Note that as a result of the EEDF form peculiar to HC discharge lasers, quenching takes place not only in the discharge afterglow (as in a PC discharge) but also during the current pulse.

For Penning excited transitions in mixtures with He, the saturation and subsequent fall in laser power at higher currents are the result of a decrease in He metastable densities.

In pulsed HC discharges, the laser characteristics also depend on other factors. The potential of the CDS which determines the energy of the primary electrons is several times higher in the first several microseconds than the steady-state potential (see Figure 2.24). Thus, the rate of ionization of the buffer gas is very high, while the production rate of metastable atoms remains the same. As was shown in section 2.4, cathodes with high CDS potentials enhance the pump rate. Thus, an increase in laser power results from the use of high potential cathodes, and the highest laser powers are obtained with graphite cathodes, which have the largest cathode falls. A Hg II laser with a graphite cathode produces a specific laser power 30% higher than when a tantalum cathode is used [96]. The high electric field in the NG (up to 100 V.cm^{-1}) causes additional heating of electrons at the beginning of the current pulse, enhances the pump rate and shifts the threshold for electron de-excitation to higher currents (i.e. $\tilde{n}_e^{cr}$ is raised). The high electric

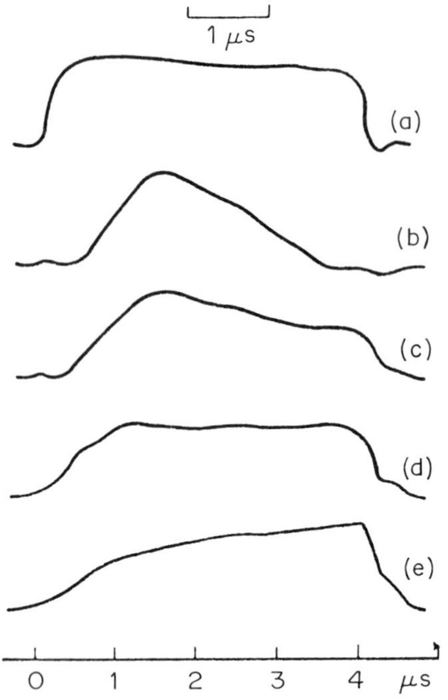

Figure 6.5 Oscillograms of (a) the current pulse, (b) oscillation in a HC discharge of the Hg II 615 nm line, (c) population of the 7 $^2P_{3/2}$ level (spontaneous emission intensity at 615 nm), (d) rate of charge transfer (284.7 nm line intensity) 7 $^2S \rightarrow 6\ ^2P$ and (e) the population of the 6 ^2D level (222.5 nm Hg II line intensity).

field gives rise to cataphoresis, which depletes metal atoms in the axial zone and deforms the radial profile of the pump rate.

When the PRF is varied from 0 to 100 kHz [47], it is found that the peak laser power is a maximum at low frequencies. The average laser power increases with frequency linearly at first, then saturates (at the maximum power), and then decreases (Figure 6.6). The fall in laser power with frequency is the result of a monotonic fall in peak laser power with frequency[1], which occurs more rapidly the longer the current pulse. This is caused by electron de-excitation, which is stronger with the increase in the density of slow electrons in longer current pulses, as well as by some decrease in the rate of buffer-gas ionization with the fall in the CDS potential (see Figure 2.27). Another negative factor is the depletion of the metal density on the axis of the discharge due to radial cataphoresis in the NG field. The optimum frequency tends to fall with increasing cathode diameter (for example, for the 615.0 nm line, with a cathode diameter of 9 mm, the optimum

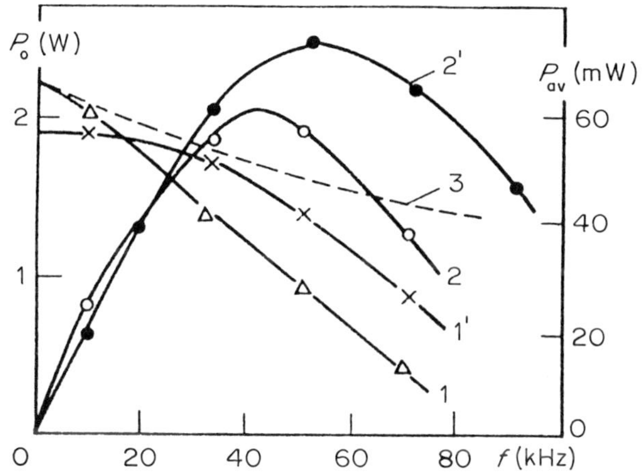

Figure 6.6 Dependences of the pulse (1, 1′) and average (2, 2′) laser output power on pulse recurrence frequency, f, for $\tau = 1$ μs (1, 2) and 0.3 μs (1′, 2′); cathode bore 9 mm, (3) experiment with double pulse excitation [4].

PRF is 41 kHz, whereas for a 20 mm cathode, the optimum PRF is 34 kHz).

As can be seen in Tables 6.4 and 6.5, the highest gains and specific laser output powers are obtained on the Hg II 615 nm, Cd II 441.6 nm and Tl II 595.5 nm lines. The gains and powers can be enhanced by using single isotopes; for Cd and Hg even isotopes are used because of isotopic splitting [5, 19, 97]. The 615 nm Hg II line consists of three well-defined and approximately equal-intensity line components, corresponding to the ^{198}Hg, ^{200}Hg and ^{202}Hg isotopes, and one weak component due to ^{204}Hg [5]. Thus the substitution of natural mercury by the single, even isotope results in an increase in gain on the 615.0 nm line by a factor of 3.3. Unlike mercury, which naturally consists mainly of even isotopes, naturally occurring Tl contains the two odd isotopes ^{203}Tl and ^{205}Tl (30% and 70% occurrence, respectively). This leads not only to a weak isotopic splitting, but also to a much stronger hyperfine splitting of each of the energy levels into two sublevels (because the nuclear momentum quantum number $I = 1/2$) [98]. Therefore, each ion laser transition of Tl normally contains only two lines (generation on the third, weak component of hyperfine structure does not occur due to competition from the other transitions). Thus, the strongest laser line of Tl II, 595 nm, has two components with intensities in the ratio of 10:8 and a 0.05 nm spacing between them [19].

The high gain (up to 0.3 dB cm^{-1}), together with the slow decrease in peak power and the monotonic rise in average power with increasing cathode diameter allows the successful scaling of MVILs to large cathode diameters (up to 40 mm) with short lengths (around 10–20 cm), i.e. with low-length to

Table 6.5 Powers and efficiencies of MVILs excited with pulsed HC discharges.

λ (nm)		$d_{cathode}$ (cm) ($l_{cathode}$) (cm)	i_{pk} (A) (τ(μs))	PRF (kHz)	P_{pk} (W) (P_{pk}^{sp} (mW.cm^{-3}))	P_{av} (mW) (P_{av}^{sp} (mW.cm^{-3}))	Efficiency (%)
Cd II	441.6	0.9 (40)	20 (1)	13	3.5 (140)[a]	46 (1.8)	0.02
	533.7	0.9 (40)	15 (1)	13	0.25 (10)	6.5 (0.26)	0.002
	537.8	0.9 (40)	15 (1)	13	0.25 (10)	6.5 (0.26)	0.002
Hg II	615.0	0.6 (40)	60 (1)	4	5.6 (500)	27 (2.4)	0.01
		1.8 (20)	100 (0.5)	35	9.2 (170)	320 (6)	0.01
		1.8 (57.5)	250 (0.5)	15	17 (320)	700 (5.25)	0.023
Tl II	595.0	1.45 (40)	100 (0.5)	18	9.8 (140)[b]	250 (3.6)	0.025

[a] Up to 250 mW cm^{-3} with a 0.6 cm cathode at low PRFs.
[b] Up to 500 mW cm^{-3} with the same conditions [46].

tube-bore (*l/d*) ratios of 3 to 4. The characteristics of the most efficient HC discharge lasers with mixtures of metal/noble gas, excited by microsecond current pulses, are given in Table 6.5. They can generate high specific peak and specific average powers (up to 0.5 W cm^{-3} and 0.6 mW cm^{-3} kHz^{-1} for Hg II and Tl II, respectively), which allow volume scaling to produce average output powers of a few watts.

The $D^* \longrightarrow P$ Beutler laser transitions of Cd II (441.6 nm and 325 nm) and of Zn II (589.4 nm and 747.9 nm) were historically the first HC discharges to use no buffer gas, and for which gains much higher than those with mixtures of helium or other gases were obtained [3]. Laser oscillation only occurs during the current pulse over a broad range of vapour pressures (1.5 to 270 Pa) due to the ionization of cadmium and zinc atoms by fast electrons to the $5s^2\ {}^2D^*$ state, as a result of the large cross-section (1.5×10^{-16} cm^2 [99]) at high electron energies. However, complications with experiments with pure metal vapours prevent complete data on the laser energy characteristics from being obtained.

6.2 METAL VAPOUR ION LASERS WITH TRANSVERSE-RF EXCITATION

Although capacitive RF discharges were initially employed to heat the lasants in the first elemental ion lasers [100, 101], the first MVILs pumped by capacitive TRF discharges were developed in 1976 [102]. The laser constructions were very simple: two parallel electrodes were placed along the length of the outside surface of a silica tube (Figure 6.7a), and were excited by an RF generator

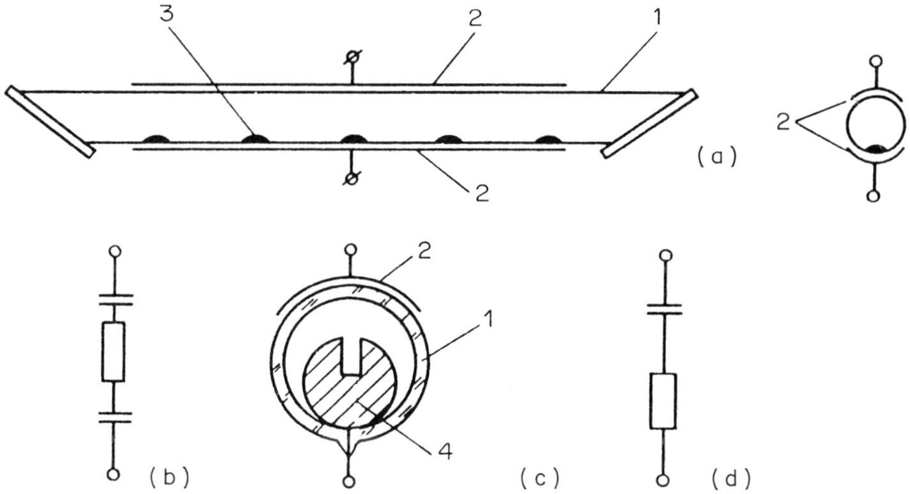

Figure 6.7 Transverse RF discharge laser tubes (a), (c) and their equivalent circuit representations (b), (d): (a) tube with external electrodes; (b) tube with internal and external electrodes; 1, quartz tube; 2, external electrode; 3, metal pieces; 4, internal electrode.

operating in the range 1–30 MHz. Metal pellets were placed into side-arm heaters, or directly on the surface inside the silica tube. The metal vapours entered the tube by virtue of tube heating and ion bombardment of the pellets. In [103], it was reported that, as in HC discharge lasers, TRF discharge lasers can use sputtering of one of the electrodes to admit the metal vapours to the discharge (the inner one). The sputtering electrode had a slot-like cavity and was fabricated from the metal lasant (Figure 6.7c).

To date, 27 laser transitions have been obtained in six metals in the CW and quasi-CW regimes of TRF discharges (see Table 6.1). Copper vapours can be obtained by the RF sputtering of copper electrodes; vapours of other metals are mainly produced by evaporation. The characteristics of lasers in the TRF regime are similar to those in the HC regime, although TRF discharges are far more stable with respect to arcing. As can be seen in the electrical circuit equivalents of the TRF discharge (Figures 6.7b and d), this stability is caused by the capacitance distributed along the tube (the length of the electrode), i.e. by the reactive ballast impedance.

The most intense laser lines are the 441.6 nm and 533.7/537.8 nm lines of Cd II and the 491.2/492.4 nm doublet of Zn II. The high effectiveness of charge transfer and Penning ionization in HC and TRF discharges is due to efficient ionization and excitation of the buffer gas, despite the presence of high densities of metal vapours ($>10^{15}$ cm^{-3}).

Most investigators have studied cadmium because its blue and green lines are some of the most powerful, and the optimum temperature of this metal (around 320°C) is very suitable for TRF discharge operation [80, 104–108]. In [108] in the burst-mode regime, with 1 ms trains of RF excitation, the optimum laser conditions for ion lines of Cd were examined in tubes with diameters from 1.7 to 5.7 mm (Table 6.6). It was found that the laser power peaks at $p_{He}d$ = 5.2, 10.6 and 8 kPa mm, respectively, for the 441.6, 533.7 and 537.8 nm transitions. The optimum input power per unit length for 441.6 nm does not depend on the tube diameter and is around 20 W cm^{-1}. For 533.7 and 537.8 nm, the input power per unit length is proportional to the tube diameter as, respectively, 125*d* and 95*d* (where *d* is in centimetres). The maximum specific output power for 441.6 nm is 43 mW cm^{-3}, and has been obtained in the tube of the smallest diameter, whereas for 537.8 nm it is 4.5 mW cm^{-3} for a 3.8 mm diameter tube. The laser power depends weakly on frequency over the range 1 to 30 MHz, where at higher frequencies the same input power is applied at lower electrode voltages.

In the CW regime, in order to prevent overheating of the cadmium, and for better thermal stability, one of the electrodes is placed inside the tube, as in tubes that utilize sputtering. This electrode has a longitudinal slot and holes for the cadmium. The characteristics of such a laser tube are given in Table 6.6 [109].

For the same reasons as in HC discharge lasers, in CW TRF discharge lasers single-mode oscillation on the green lines of Cd II and multi-colour

Table 6.6 The most intense TRF discharge He–Cd laser lines.

		Quasi-CW regime [108]		CW regime	
λ (nm)	$p_{He}d$ (kPa mm)	P_{in}^{sp} (W cm^{-3})	P (mW) (P_{sp} (mW.cm^{-3}))	P_{in}^{sp} (W cm^{-3})	P (mW) (P_{sp} (mW.cm^{-3}))
441.6	5.2 [108]	52	57 (19)	—	10 (1.7)
		d = 3.8 mm[a]		[80]	
	6.7 [80]	—	—	—	—
	8.0 [109]	142	38 (43)	—	—
		d = 1.7 mm			
533.7/537.8	8; 10.6 [108]	300	19 (14.5)	510 [109]	45 (13) [109]
	14.7 [109]	d = 3.8 mm		10 (1.7) [80]	
	20 [80]				

[a] Here *d* is the width of the inner electrode slot [109]; in all other cases *d* is the inside diameter of a quartz tube with external electrodes [80, 108].

operation on the blue, green and red lines of Cd II are feasible. The noise level on all transitions is of the lowest achievable magnitude [79, 80], approximately the same as for CW HC discharge lasers.

As for HC lasers, the excitation of TRF lasers with microsecond pulse trains results in higher specific output powers (up to 110 mW cm^{-3} for the blue-green transitions of Cd II) than in the CW regime, with specific input powers of 290 W cm^{-3} [107]. The optimum PRF of the RF trains was found to be 10's of kHz, and in such a regime with the appropriate thermal conditions, specific average output powers of up to 5 mW cm^{-3} can be obtained on each line. Thus, TRF discharge lasers are as efficient as HC discharge lasers, but TRF devices are simpler in construction and arcing occurs more rarely than in HC lasers.

Based on what we have presented in sections 6.1 and 6.2, we can say that the search for new transitions in HC and TRF discharge lasers, and their optimization, has come to an end. The investigations have resulted in an increase in the number of laser lines produced by gas lasers in the visible spectrum. Three-colour lasing in the CW regime and CW generation at wavelengths down to 224.3 nm near the air–UV cut-off wavelength are achievements of HC excited MVILs. The CW specific laser power per unit length reaches 10's of mW cm^{-1}, and some of the laser lines can deliver a few watts of CW output. Microsecond pulse excitation in TRF lasers, as in HC lasers, allows high peak powers (0.6 W cm^{-3}) to be attained at average output powers of 100's of mW, as well as avoiding the many technical problems of HC devices.

6.3 THE EFFICIENCY OF TRANSVERSELY EXCITED METAL VAPOUR ION LASERS

The general conditions for laser oscillation of high efficiency with CW HC and TRF discharge lasers are the same as for PC discharge lasers (see section 5.4), taking into account the rise in plasma impedance with the increase in vapour pressure and the positive dynamic resistance of the discharge. The overall efficiency can be found from the relation (5.17).

As was shown in section 2.6, the output power of charge-transfer transitions depends on buffer-gas ionization rates, which are higher in HC and TRF discharges than in PC discharges. This is due to the larger discharge efficiency η_d and charge-transfer efficiency η_3 (see relation (2.69)). In Table 6.7, we present the values of η_d, η_3, η_4 and η_q as well as the overall efficiency, η, for the maximum output power conditions of Cd II and Hg II lasers (the optimum vapour pressure is produced by evaporation), which are the most favourable amongst HC and TRF discharge lasers. The great number of slow electrons in the NG of a HC discharge can lead to different (compared with the PC discharge) coefficients η_3 and η_4 for various transitions. The additional transitions from higher levels (as for the green doublet of Cd II) may increase η_3, whereas depopulating transitions from the upper laser level and transitions terminating on the lower laser level can cause η_4 to decrease [110]. For most of the charge-transfer lines, such as 615.0 nm in Hg II, the overall efficiency at maximum output power is less than when the input power is lower, and saturation of the laser power with current does not occur (Figure 6.8). The

Table 6.7 Efficiency and effectiveness of HC and TRF discharge lasers with mercury and cadmium vapours.

	λ (nm)	Pumping	Discharge	η^a (10^{-4})	η_d	η_3	η_4	η_q^b
Hg II	615.0	Charge transfer	Stationary HC	0.5 [60], [123]	0.058	0.22	0.1	0.083
			Pulsed HC [c]	2 [47]	0.12	0.22	0.15	0.083
Cd II	533.7/ 537.8	Charge transfer with cascade transitions	Stationary TRF	1 [108]	0.13	0.3	0.3	0.093
Cd II	441.6	Penning ionization	Stationary HC	1 [123]	0.036	0.17	0.35	0.14
			Stationary TRF	1 [108]	0.072	0.17	0.35	0.14

[a] Experimental values.
[b] Quantum efficiency relative to the energy of the donor particle (He^+ for charge transfer; He_m for Penning ionization).
[c] For low PRFs when the peak output power is a maximum.

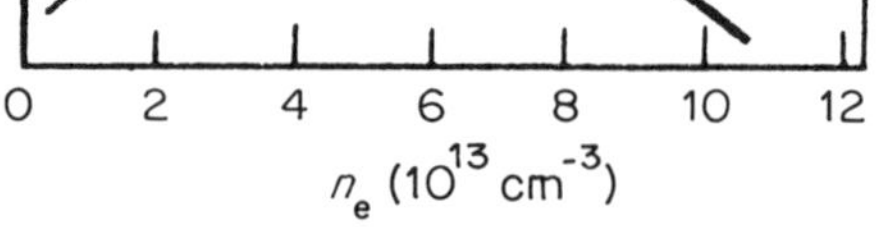

Figure 6.8 Typical dependence of efficiency on n_e for the: 1, Hg II 615 nm line in a HC discharge and 2, Cd II 533.7/537.8 nm lines in a TRF discharge when strong downwards collisional relaxation causes additional population of the 4 ^2F upper laser level. 3, laser output power.

values of η_3 and η_4 given in Table 6.7 are approximations only. This means that the efficiency of each transition should be evaluated individually.

As discussed in section 2.6, the pulsed regime has a higher discharge efficiency, as well as better optimization of the output power with discharge current. However, the analysis of laser efficiency in the pulsed regime is more difficult than in the steady-state regime. Thus, with an increase in τ, the efficiency initially falls (see Figure 6.4), since the peak laser power (which follows the ionization rate, Figure 6.4c) and the duration of the laser pulse change little, whereas the input energy increases. To establish a large inversion (and high η_4), the current should be reduced, so decreasing the input power, and η_1. Although η_2 increases, it cannot compensate for the overall degradation which yields a lower total efficiency. The decrease in the field of the NG also results in a decrease in η_1 (see Figure 2.24), and thus the microsecond pulse regime is preferable. Overall efficiency increases with cathode diameter d, since the optimum specific input power falls as $1/d$, corresponding to the condition $\tilde{n}_e$ = constant; this arises from the reduction in the rate of loss of electrons to the wall with increasing d. Therefore, the specific laser output power reduces more slowly (Figure 6.4). In this case, again η_4 = constant, and the increase in η arises from the increase in η_2, which is the result of a reduction in the average energy of the fast electrons.

As can easily be shown, when the PRF is changed, the overall efficiency $\eta \sim P_{pk}/P_{in}$ (where P_{pk} is the peak laser power, and P_{in} is the input power to the discharge). Hence, the maximum efficiency is at low PRFs, when P_{pk} is at a maximum. At the optimum frequency, where the average laser power is a maximum, the overall efficiency falls by a factor of about 2 due to the decrease in the density of fast electrons (i.e. η_1 is lower) and the increase in the density of slow electrons (i.e. η_4 falls also).

6.4 ELECTRON-BEAM-EXCITED LASERS

Metal vapour ion lasers have also been excited by intense collimated beams of electrons (e-beams) [111–114]. e-Beam currents of around 1 A at electron energies of several keV are typically employed. Two types of laser configuration have been proposed: with an e-beam permeating the gas along the optical axis of the laser (longitudinal excitation), and with an e-beam perpendicular to the optical axis (transverse excitation) (Figure 6.9). Longitudinal configurations employ one or two e-beam guns that can cover electron drift volumes of up to 100 cm^3. Metal vapour is introduced to the electron drift zone (externally heated) by entrainment in a buffer gas that is first made to flow through the metal source evaporator; the gas flows in from one or two directions with a speed of around 400 litre min^{-1} (Figure 6.9).

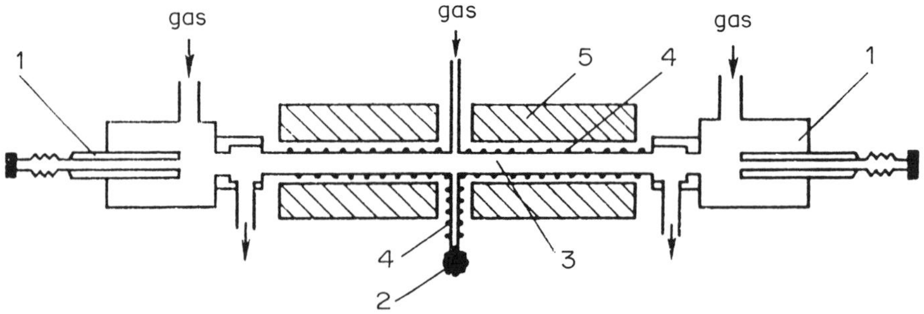

Figure 6.9 Longitudinal e-beam excited lasers [113]: 1, e-beam gun; 2, metal evaporator; 3, e-beam drift zone; 4, heater; 5, solenoid.

The glow discharge electron gun incorporates a hollow cathode (initially graphite, and recently aluminium) operated at a potential of several kilovolts. In the cavity of a hollow cathode, an ordinary HC discharge (a 'plasma' electrode) produces electrons of a few hundred volts energy for the outer, highly obstructed discharge, where most of voltage falls. To make it obstructed, the outside surface of the cathode is covered in ceramic. The electrons from the cathode cavity are accelerated in the outer CDS up to several kilovolts, in the form of an e-beam entering the active region. However, the investigators report that the e-beam regime occurs over a very narrow range of gas pressures and currents due to discharge instabilities that lead to the ordinary low-voltage current regime. The e-beam is collimated in the drift zone by a longitudinal magnetic field of 0.45 T, but the field also appears in the zone of the outer CDS (magnitude 0.025 T), and hence alters the voltage–current characteristics of the gun. This leads to a decrease in the e-beam energy and a transformation to the low-voltage discharge regime.

Under such conditions, laser oscillation is achieved in seven metal ions on 30 transitions (Table 6.8). The evaporator temperature corresponds to a vapour pressure of 10–150 Pa, i.e. the same as in HC and TRF discharge lasers. As in HC and TRF lasers, fast electrons are used to produce excitation, but the diffusive expansion of the zones of electron generation and excitation in e-beam lasers is much higher. For practically all the lines (except 491.2, 492.4 and 758.8 nm of Zn II), saturation is not reached for even 2–3 kW input power. The highest average output power obtained was 1.2 W on the blue-green doublet of Zn II from an active volume of 57 cm^3, produced by two contra-directed e-beams. The effectiveness of the longitudinal beam was 60% [113] (measured calorimetrically). However, the overall efficiency was low (around 0.034%), apparently due to the higher energy losses compared with HC and TRF discharges in other stages of the

Table 6.8 Spectral and energy characteristics of MVILs pumped by longitudinal e-beams (helium is the buffer gas) [111–113].

Wavelength range (nm)		Number of lines	Most intense lines: λ (nm)	P (mW) (P_{sp}, mW.cm^{-3})
Cd II	441.6–806.7	6	533.7; 537.8	80 (2.8)
Zn II	491.2–758.8	4	491.2; 492.4	1200 (21.1)
Hg II	615.0	1	615.0	250; 6[a] (1.3)[a]
As II	651.17	1	—	—
I II	576.07; 612.75	2	—	—
Se II	497.57–783.88	15	497.57; 530.53	20 (0.2)
Ag II	840.39	1	840.39	140 (2.4)

[a] Transverse e-beam excitation.

conversion of e-beam power to laser power. Since the length of electron drift in longitudinal guns is proportional to electron energy in the beam, it is possible to increase the e-beam energy. However, the optimum electron energy (especially for metastable-level excitation) is only tens or hundreds of electron volts. Transverse e-beam guns with contra-directed beams have electron energies closer to the optimum of 1 keV; the drift length is about 1 cm and is formed by two long, concave electrodes [113]. The increase in the efficiency of ionization compared with longitudinal e-beam devices can be attributed to oscillations of electrons between the CDSs of the oppositely placed electron guns, in the same way as in a HC discharge at low pressures. However, the overall efficiency of the transverse type is low, about 4×10^{-6}.

Thus, the e-beam experiments have doubled the efficiency of conversion of discharge energy to e-beam energy, compared with HC discharges. Laser powers of up to 1.2 W have been achieved from large active volumes. However, the overall efficiencies compared with HC and TRF discharge lasers are not higher. This can be explained by the intense electron de-excitation due to thermal electrons of low (about 0.1 eV) energy which exist in e-beam lasers [111, 113].

6.5 THE TECHNOLOGY AND APPLICATIONS OF TRANSVERSELY EXCITED METAL VAPOUR ION LASERS

Transversely excited MVILs are only now making the transition to practical devices. Amongst them are included the pulsed HC Hg II and TRF Cd II lasers, as well as CW HC and TRF discharge Cd II lasers.

The possibilities of active volume scaling, which is afforded by the high gain, and of an increase in average and peak powers with an enlargement in cathode diameter, make laser generation over a wide range of output powers possible: from 5 to 300 mW average power and up to 30 W peak power (Table 6.9). If the outer anode cylinder is the vacuum enclosure as well, then effective transfer of heat out of the system can be achieved. The power supply of such a laser is not special: a storage capacitor is discharged through a hydrogen thyratron (see Figure 3.2). As a trigger circuit, a blocking oscillator can be used with an output transformer (for thyratrons with low grid currents) or a cascade circuit incorporating another thyratron (which is smaller than the primary one).

Table 6.9 Parameters of pulsed lasers using mercury vapour with HC-discharge excitation [124], and using cadmium vapour with TRF-discharge excitation [79].

	Examples of He-Hg HC lasers			TRF He-Cd laser	
Laser wavelength (nm)	615	615.0[a]	615.0[a]	441.6	533.7 and 537.8
Peak pulsed laser power (W)	5	10	30	—	—
Average laser power (mW)	6–20	5–50	300	20	30
				(50 total)	
Laser line-width (MHz)	3000	3000	3000	1000	100
Laser beam diameter (mm)	6	15	20	8	
PRF (kHz)	1–3.5	0.5–5	20	20	
Laser pulse duration (μs)	0.5	0.5	0.5	3–10	
Cathode cavity size (mm^2) (length by diameter)	220 × 8	170 × 18	320 × 20	800 × 10	
Active element cooling	air convection				
Excitation circuit switch (Russian manufacture)	TGI 1-100/8	TGI 1-100/8	TGI 1-1000/25	6C33c	

[a] Oscillation possible on the 794.5 nm line with proper laser cavity mirrors ($\Delta\nu = 600$ MHz).

The discharge tubes for pulsed TRF discharge lasers are of simpler construction than for HC discharge lasers, and the main consideration is to minimize the capacitance between the electrode and plasma, because the upper limit to input power to the discharge at a frequency ω is given by [107] $P_{in} = U^2\varepsilon\omega/4t$ (where U is the amplitude of the RF voltage and ε and t are the dielectric constant and thickness, respectively, of the tube wall material). With an increase in pump frequency, the input power increases and the laser power tends to saturate. The RF generator is usually based on the common grid circuit (Figure 6.10), and matching of the generator to the load is performed by an adjustable Π filter loop.

One of the more interesting applications of large-bore Hg II HC lasers was demonstrated in [115]: the laser was used in an active optical system for the reproduction of graphic images of a sample placed within the laser optical cavity. The maximum resolution density in the optical system was 100 lines mm^{-1}, which corresponds to a total information volume of 10^6 bits carried by the laser beam.

The possible applications discussed in section 5.6 for cataphoretic lasers can also be satisfied successfully by CW HC and TRF metal vapour lasers, on account of the large range of lines generated and their output characteristics. These lasers are more suitable than He–Ne lasers due to their shorter wavelength lines, i.e. their higher resolution capability and speed of lithographic processing, and because the more expensive photosensitive materials are not necessary. Their noise levels are an order of magnitude lower than those of cataphoretic lasers, and even of He–Ne lasers. CW HC and TRF lasers are capable of higher specific laser output powers, and so can be more compact than He–Ne or He–Cd cataphoretic lasers of the same power. Hollow cathode lasers also operate at lower voltages (about 500 V),

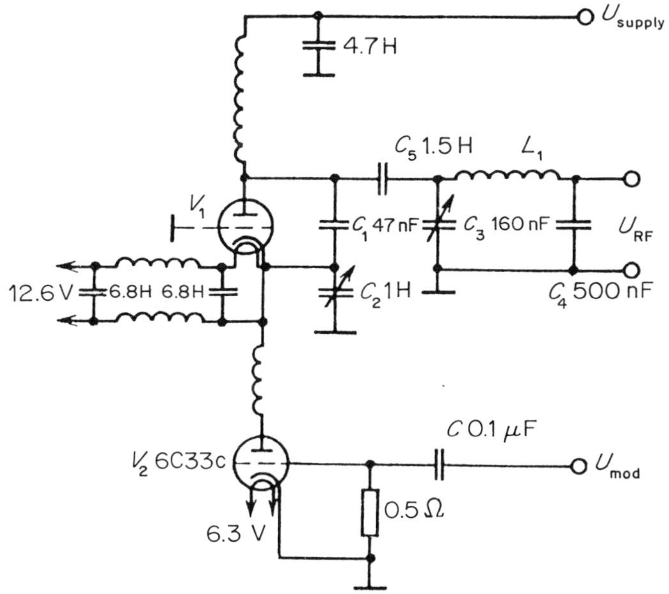

Figure 6.10 Basic scheme of a modulator for the excitation of TRF discharge metal vapour lasers [107].

which makes their power supplies simpler. The solid-state power supplies of TRF lasers can also be of lower voltage and size if they operate at frequencies of 100's of MHz.

Hollow cathode and TRF Cd lasers and Cd–Hg ion lasers can operate simultaneously on blue, green and red lines at low noise levels. This is very important for colour image processing, for various measurement devices (such as multiple beam interferometers) and for multi-colour holography, etc. The UV emission ($\lambda < 320$ nm) of CW lasers that utilize Cu, Ag or Au vapours can be used for the pumping of dyes (such as *p*-terphenyl, 2-(4-biphenyl)-5-phenyl-1,3,4-oxadiazole (PBD) and PPF), dimers and crystals emitting in the range 300–400 nm; for photo-lithography to increase speed and resolution; and in the quality control of optical materials using fluorescence [48].

REFERENCES

[1] V. P. Chebotaev and V. V. Pokasov, *Radiotekh. i Elektron.,* **10**(5), 958–60 (1965) (in Russian).

[2] É. K. Karabut, V. C. Michalevskii, V. F. Papakin and M. F. Sem, *Sov. Phys.–Tech. Phys.,* **14**(10), 1447–8 (1970).

[3] M. F. Sem and V. S. Mikhalevskii, *Zh. Prikl. Spektrosk.,* **6**(5), 668–9 (1967) (in Russian); *Elektron. Tekh., Ser. 1 Elektronika SVCh,* No. 9, 98–102 (1969) (in Russian).

[4] E. K. Karabut, V. F. Kaydan, V. F. Kravchenko *et al., Elektron. Tekh., Ser. 1 Elekronika SVCh,* No. 9, 103–8 (1969) (in Russian).

[5] R. L. Byer, W. E. Bell, E. Hodges and A. L. Bloom, *J. Opt. Soc. Am.,* **55**(12), 1598–602 (1965).

[6] W. K. Schuebel, *IEEE J. Quantum Electron.,* **7**(1), 39–40 (1971).

[7] W. K. Schuebel, *Appl. Phys. Lett.,* **16**(11), 470–2 (1970); *IEEE J. Quantum Electron.* **6**(9) 574–5 (1970).

[8] Y. Sugawara, V. Tokiwa and T. Iijima, *Jpn. J. Appl. Phys.,* **9**(12), 1537 (1970); **9**(5), 588–9 (1970).

[9] J. A. Piper and C. E. Webb, *J. Phys. D,* **6**(4), 400–7 (1973).

[10] R. C. Jensen, G. J. Collins and W. R. Bennett Jr., *Appl. Phys. Lett.,* **18**(2), 50–1 (1971).

[11] T. Iijima and Y. Sugawara, *J. Appl. Phys.,* **45**(11), 5091–2 (1974).

[12] J. A. Piper and M. Brandt, *J. Appl. Phys.,* **48**(11), 4486–94 (1977).

[13] J. A. Piper, *J. Phys. D,* **7**(2), 323–8 (1974).

[14] J. A. Piper and C. E. Webb, *IEEE J. Quantum Electron.,* **12**(1), 21–5 (1976).

[15] J. A. Piper and C. E. Webb, *J. Phys., B,* **6**(5), L116–20 (1973).

[16] J. A. Piper, *Opt. Commun.,* **31**(3), 374–6 (1979).

[17] Liu Jian-bang, *Appl. Phys. B,* **29**(4), 251–3 (1982).

[18] S. P. Zinchenko, I. G. Ivanov and M. F. Sem, *Summaries of Papers. Population Inversion and Oscillation on Atomic and Molecular Transitions,* Part 1, ed. O. V. Ravodina and I. I. Murav'ev, p. 146 Tomsk: Institute of Atmospheric Optics (Sept. 1986) (in Russian).

[19] V. V. Vainer, S. P. Zinchenko, I. G. Ivanovand M. F. Sém, *J. Appl. Spectrosc.*, **31**(5), 1432–34 (1979).
[20] M. G. Grozeva, N. V. Sabotinov and N. K. Vuchkov, *Opt. Commun.*, **29**(3), 339–40 (1979).
[21] M. G. Grozeva, N. V. Sabotinov, P. K. Telbizov and N. K. Vuchkov, *Opt. Commun.*, **31**(2), 211–13 (1979).
[22] K. U Baron and B. Stadler, *Opt. Commun.*, **18**(1), 160–2 (1976).
[23] E. L. Latush, R. Solanki and G. J., Collins *Phys. Lett. A.*, **73**(5,6), 387–8 (1979).
[24] J. R. McNeil, G. J. Collins, K. B. Persson and D. L. Franzen, *Appl. Phys. Lett.*, **28**(4), 207–9 (1976).
[25] J. R. McNeil, W. L. Johnson, G. J. Collins and K. B., Persson *Appl. Phys. Lett.*, **29**(3), 172–4 (1976).
[26] R. D. Reid, J. R. McNeil and G. J. Collins, *Appl. Phys. Lett.*, **29**(10), 666–8 (1976).
[27] L. Csillag, M. Jánossy, K. Rósza and T. Salamon, *Phys. Lett. A.*, **50**(1), 13–14 (1974).
[28] J. R. McNeil and G. J. Collins, *IEEE J. Quantum Electron.*, **12**(6), 371–2 (1976).
[29] K. Jain, *Appl. Phys. Lett.*, **34**(6), 398–9 (1979).
[30] R. Solanki, W. M. Fairbank Jr. and G. J. Collins, *IEEE J. Quantum Electron.*, **16**(12), 1292–4 (1980).
[31] K. Jain, *IEEE J. Quantum Electron.*, **16**(4), 387–8 (1980).
[32] R. Solanki, E. L. Latush, D. C. Gerstenberger, W. M. Fairbank Jr. and G. J. Collins, *Appl. Phys. Lett.*, **35**(4), 317–9 (1979).
[33] J. A. Piper and D. F. Neely, *Appl. Phys. Lett.*, **33**(7), 621–3 (1978).
[34] W. L. Johnson, J. R. McNeil and G. J. Collins, *Appl. Phys. Lett.*, **29**(2), 101–2 (1976).
[35] R. D. Reid, W. L. Johnson, J. R. McNeil and G. J. Collins, *IEEE J. Quantum Electron.*, **12**(12), 778–9 (1976).
[36] K. Rósza, M. Jánossy, L. Csillag and J. Bergou, *Opt. Commun.*, **23**(2), 162–4 (1977); *Phys. Lett. A,* **63**(3), 231–2 (1977).
[37] W. K. Schuebel, *Appl. Phys. Lett.*, **30**(10), 516–9 (1977).
[38] D. C. Gerstenberger, R. D. Reid and G. J. Collins, *Appl. Phys. Lett.*, **30**(9), 466–8 (1977).
[39] K. Jain, *Appl. Phys. Lett.*, **34**(12), 845–6 (1979).
[40] K. Gnädig and Lin Fucheng, *Opt. Commun.*, **34**(2), 218–20 (1980).
[41] K. Gnädig, *Opt. Commun.*, **35**(3), 427–9 (1980).
[42] Lin Fucheng, *Opt. Commun.*, **36**(5), 387–90 (1981).
[43] K. Jain, *Appl. Phys. Lett.*, **37**(4), 362–4 (1980).
[44] J. A. Piper, *Opt. Commun.*, **19**(2), 189–92 (1976).
[45] *Laser Handbook*, ed. A. M. Prokhorov, Moscow: Sovyetskoya Radio (1978) (in Russian).
[46] V. V. Vainer, S. P. Zinchenko, I. G. Ivanov and M. F. Sem, *Sov. J. Quantum Electron.*, **10**(5), 581–6 (1980).
[47] S. P. Zinchenko, I. G. Ivanov and M. F. Sem, *Sov. J. Quantum Electron.*, **10**(8), 1055–7 (1980).
[48] D. C. Gerstenberger, R. Solanki and G. J. Collins, *IEEE J. Quantum Electron.*, **16,** 820–34 (1980).
[49] J. R. McNeil, G. J. Collins, K. B. Persson and D. L. Franzen, *Appl. Phys. Lett.*, **27**(11), 595–8 (1975).
[50] S. C. Wang and A. F. Siegman, *Appl. Phys.*, **2**(3), 143–50 (1973).

[51] K. Fujii, *et al. Proc. II Int. Conf. on Lasers '79,* 454–7 McLean, VA: STS. Press (1980); *IEEE J. Quantum Electron.,* **16**(6), 590–22 (1980).
[52] W. Sasaki, I. Ebina and T. Ohta, *Metal Vapor, Deep Blue and Ultraviolet Lasers,* ed. J. J. Kim, R. Kimball, P. J. Wisoff, *Proc. SPIE,* **1041,** 117–22 (1989).
[53] W. Sasaki, K. Itani and T. Ohta, *Metal Vapor, Deep Blue and Ultraviolet Lasers,* ed. J. J. Kim, R. Kimball, P. J. Wisoff, *Proc. SPIE,* **1041,** 123–30 (1989).
[54] K. Fujii, T. Takahashi and Y. Asami, *IEEE J. Quantum Electron.,* **11**(3), 111–4 (1975).
[55] K. G. Hernqvist, *IEEE J. Quantum Electron.,* **14**(2), 129–32 (1978).
[56] A. Fuke, K. Mashuda and Y. Tokita, *Trans. Inst. Electron. and Commun. Eng. Jpn Sect. E,* **69**(4), 365–6 (1986); *Jpn. J. Appl. Phys. Part 1,* **26**(1), 96–101 (1987).
[57] S. Fukuda and M. Miya, *Jpn. J. Appl. Phys.,* **13**(4), 667–74 (1974).
[58] H. H. Telle, I. D. Hopkin, P. Ramalingam, H. K. Fun and C. Grey Morgan, *J. Phys. D,* **21**(10S), 167–70 (1988).
[59] K. Fujii and K. Watanabe, *Proc. VII Int. Conf. on Gas Discharges and Applications,* 401–2, London: Imperial College of Science and Technology (1982).
[60] V. V. Vainer, I. G. Ivanov, V. S. Mikhalevskii and M. F. Sem, *Sov. J. Quantum Electron.,* **13**(4), 417–18 (1983).
[61] V. V. Vainer, I. G. Ivanov, Patent No. 1582941 Russ. Fed., Kl. HO1s 3/22, 18.07.88.
[62] J. Mizeraczyk, G. Jakob, E. Schmidt and J. Mentel, *J. Appl. Phys.,* **73**(11), 7180–3 (1993).
[63] I. V. Aleksandrov, N. V. Burdastyh and I. G. Ivanov, *Laser. Phys.,* No. 8, 18 (1994) (in Russian).
[64] H. Tsuda and J. A. Piper, *J. Phys. E,* **22**(7), 462–5 (1989).
[65] J. Mizeraczyk, J. Mentel, E. Schmidt, N. Reich and C. Carlsson, *Meas. Sci. Technol.,* **5**(8), 936–41 (1994).
[66] J. R. Grace and A. I. McIntosh, *J. Phys. D,* **12**(12), 2043–51 (1979).
[67] J. Mizeraczyk, M. Neiger and J. Steffen, *IEEE J. Quantum Electron.,* **20**(11), 1233–5 (1984).
[68] I. G. Ivanov, V. V. Vainer , G. A. Kalinchenko, in *Proc. III Int. Conf. on Lasers and their Applications, Plovdiv, Bulgaria* 6–7 (1988).
[69] C. Boulmer-Leborgne, B. Dabreuil and J. C. Pellicer, *J. Phys. D,* **22,** 512–18 (1989).
[70] V. V. Vainer, I. G. Ivanov, G. A. Kalinchenko and M. F. Sem, *Metal Vapor Lasers and Their Applications: CIS Selected Papers,* ed. G. G. Petrash *Proc. SPIE,* **2110,** 128–49 (1993).
[71] M. Grozeva and N. V. Sabotinov, *Opt. Commun.,* **41**(1), 57–8 (1982); **51**(6), 417–19 (1984).
[72] M. Grozeva and N. V. Sabotinov, *Symposium Optika '84, Budapest, Hungary, 24–27 Apr. 1984, Proc. SPIE,* **473,** 174–6 (1984).
[73] M. Stefanova and P. Pramatarov, *Phys. Lett. A,* **139**(8), 391–4 (1989).
[74] I. Angelov, M. Grozeva and N. V. Sabotinov, *Meas. Sci. Technol.,* **4**(6), 682–4 (1993).
[75] M. Grozeva, N. V. Sabotinov and M. Jánossy, *Opt. Quantum Electron.,* **18**(6), 455–9 (1986).
[76] G. A. Kalinchenko, V. V. Vainer and I. G. Ivanov, *Summaries of Papers. Population Inversion and Oscillation on Atomic and Molecular Transitions,* Part 1, ed. O. V. Ravodina and I. I. Murav'ev, p. 247, Tomsk: Institute of

Atmospheric Optics (Sept. 1986) (in Russian); *Sov. J. Plasma Phys.*, **16**(4), 460–4 (1990) (in Russian).
[77] S. C. Wang, *Laser Focus,* **19**(3), 14 (1983).
[78] V. V. Vainer, I. G. Ivanov and V. S. Mikhalevskii, *Summaries of Papers. Population Inversion and Oscillation on Atomic and Molecular Transitions,* Part 1, ed. O. V. Ravodina and I. I. Murav'ev, p. 147 Tomsk: Institute of Atmospheric Optics (Sept. 1986) (in Russian).
[79] *Helium–Cadmium Laser with Transverse RF Excitation,* Prospectus of VDNX (Exhibition of National Economic Achievements, Moscow), Rostov-on-Don: Rostov State University (1977).
[80] M. K. Dyatlov, V. G. Kas'yan and V. G. Levin, *Sov. Tech. Phys. Lett.,* **3**(7), 264 (1977).
[81] V. V. Vainer, S. P. Zinchenko, I. G. Ivanov and M. F. Sem, Patent No. 1031397 Russ. Fed., Kl. HO1s 3/22, 30.04.81 .
[82] B. K. Ward, J. A. Piper and P. E. Ciddor, *Opt. Lett.,* **13**(8), 625–7 (1988).
[83] N. V. Subotinov, J. Konieczka, J. Mizeraczyk and N. K. Vuchkov, *Sov. J. Quantum Electron.,* **8**(7), 918–19 (1978).
[84] B. Auschwitz, H. J. Eichler and W. Wittwer, *IEEE J. Quantum Electron.,* **17**(4), 546–53 (1981).
[85] H. J. Eichler, H. Koch, R. Molt and J. L., Qiu *Appl. Phys. B,* **26**(1), 49–56 (1981).
[86] Z. Zhang, N. D. Perry and R. C. Tobin, *J. Appl. Phys.,* **71**(1), 64–9 (1992).
[87] K. A. Peard, Z. Donkó, K. Rózsa, L. Szalai and R. C. Tobin, *IEEE J. Quantum Electron.,* **30**(9), 2157–65 (1994).
[88] K. A. Peard, K. Rozsa and R. C. Tobin, *J. Phys. D,* **27**(2), 219–27 (1994).
[89] K. A. Peard, R. C. Tobin, K. Rózsa and Z. Donkó, *IEEE J. Quantum Electron.,* **30**(5), 1181–6 (1994).
[90] H. J. Eichler , J. Hamisch, R. Macdonald, M. Sollinger and C. Yuging, *J. Appl. Phys.,* **61**(5), 2069–71 (1987).
[91] I. G. Ivanov and M. F. Sem, *Elektron. Tekh., Ser. 4 Elektrovakuumiye i Gazorazryadniye Pribori,* No. 2, 12–16 (1974) (in Russian).
[92] S. P. Zinchenko, I. G. Ivanov and M. F. Sem, *Metal Vapor Lasers and Their Applications: CIS Selected Papers,* ed. G. G. Petrash, *Proc. SPIE,* **2110,** 150–64 (1993).
[93] V. V. Zhukov, V. G. Il'yushko, E. L. Latush and M. F. Sem, *Sov. J. Quantum Electron.,* **5**(7), 757–60 (1975).
[94] V. V. Zhukov, E. L. Latush and M. F. Sem, *Zh. Prikl. Spektrosk.,* **32**(4), 738–40 (1980) (in Russian).
[95] S. P. Zinchenko, I. G. Ivanov, E. L. Latush and M. F. Sem, *Opt. Spectrosc.,* **58**(2), 180–3 (1985).
[96] V. V. Vainer, I. G. Ivanov and M. F. Sem, *Proc. XV Int. Conf. on Phenomena in Ionized Gases,* Part 2 pp. 869–70, Minsk: Institute of Physics of the BSSR Academy of Sciences (1981).
[97] V. F. Papakin and M. F. Sém, *Sov. Phys. J.,* **13**(2), 230–1 (1972).
[98] A. A. Radzig and B. M. Smirnov, *Reference Data on Atoms, Molecules and Ions,* Berlin: Springer-Verlag (1985); *Data of Atoms and Atomic Ions,* Moscow:Energoatomizdat (1986) (in Russian).
[99] V. S. Aleinikov and V. V. Ushakov, *Opt. Spectrosc.,* **29,** 111–12 (1970); R. F. Pottie, *J. Chem. Phys.,* **44**(3), 916–22 (1966).
[100] R. C. Jensen and G. R. Fowles, *Proc. IEEE,* **52**(11), 1350 (1964).
[101] G. R. Fowles and R. C. Jensen, *Proc. IEEE,* **52**(7), 851–2 (1964).

[102] E. L. Latush, V. S. Mikhalevskii, M. F. Sem, G. N. Tolmachev and V. Ya Khasilev, *JETP Lett.*, **24**(1), 69–71 (1976).
[103] V. S. Mikhalevskii, M. F. Sem, G. N. Tolmachev and V. Ya Khasilev, *J. Appl. Spectrosc.*, **32**(4), 321–3 (1980).
[104] A. N. Korol'kov, S. A. Rudelev and V. A. Stepanov, *Elektron. Tekh.* Ser. 4, **9,** 12–16 (1977) (in Russian).
[105] A. A. Aleksandrov, V. V. Elagin and A. E. Fotiadi, *Pis'ma Zh. Tekh. Fiz.*, **6**(3), 160–1 (1980) (in Russian).
[106] N. Sabotinov and P. Telbizov, *Bulg. J. Phys.*, **11**(5), 559–64 (1984) (in Russian); *Opt. Commun.*, **59**(4), 290–2 (1986).
[107] V. S. Mikhalevskii, G. N. Tolmachev and V. Y. Hasilev, *Prib. Tekh. Eks.*, No. 5 207–9 (1980).
[108] V. S. Mikalevskii, G. N. Tolmachev and V. Ya Khasilev, *Sov. J. Quantum Electron.*, **10**(7), 884–7 (1980).
[109] G. P. Strokan and G. N. Tolmachev, *Optoelectronics, Instrumentation and Data Processing,* No. 1, 56–58 (1984).
[110] E. L. Latush, M. F. Sem and G. D. Chebotarev, *Sov. Phys. J.*, **27**(5), 429–35 (1984).
[111] Z. Yu, J. J. Rocca and G. J. Collins, *J. Appl. Phys.*, **54**(1), 131–6 (1983).
[112] J. J. Rocca, J. Meyer and G. J. Collins, *Phys. Lett.*, **87A**(5), 237–9 (1982); *Appl. Phys. Lett.*, **40**(4), 300–2 (1982).
[113] J. J. Rocca and G. J. Collins, *Optoelectronics, Instrumentation and Data Processing,* No. 1, 1–16 (1984).
[114] B. Wernsman, T. Prabhuram, K. Lewis, F. Gonzalez, M. Villagran and J. J. Rocca, *IEEE J. Quantum Electron.*, **24**(8), 1554–6 (1988).
[115] R. A. Myers, H. Wieder and R. V. Pole, *IEEE J. Quantum Electron.*, **2**(8), 270–5 (1966).
[116] R. Solanki, E. L. Latush, W. M. Fairbank Jr. and G. J. Collins, *Appl. Phys. Lett.*, **34**(9), 568–70 (1979).
[117] Liu Jian-bang, *Appl. Phys. B,* **32**(4), 211–15 (1983).
[118] Zhu Xu-hui and Liu Jian-bang, *Appl. Phys. B,* **29**(4), 291–2 (1982).
[119] J. A. Piper and C. E. Webb, *Opt. Commun.*, **13**(2), 122–5 (1975).
[120] J. A. Piper and P. Gill, *J. Phys. D,* **8**(1), 127–34 (1975).
[121] K. G. Hernqvist, *IEEE J. Quantum Electron.*, **13**(11), 929 (1977).
[122] K. Jain, *Appl. Phys. Lett.*, **36**(1), 10–11 (1980).
[123] I. G. Ivanov and V. V. Vainer, *Proc. Int. Conf. Lasers '83,* pp. 442–4 McLean: STS. Press (1984).
[124] S. P. Zinchenko, I. G. Ivanov and M. F. Sem, *Prib. Tekh. Eks.* No. 2, 225 (1982) (in Russian).
[125] K. Jain and S. A. Newton, *Appl. Phys. B,* **26**(1), 43–8 (1982).

Notes

1. For MVILs, the earlier-employed method of optimization of PRF using double-pulse excitation [4] produces higher power levels than the more reliable burst-mode technique [47].

7

Multi-Colour CW Metal Vapour Ion Lasers

7.1 MULTI-COLOUR LASER OSCILLATION IN METAL VAPOURS

The rapid development of multi-colour devices for information processing, colour displays and colour holography, as well as multiple wavelength interferometry [1, 2] requires compact, low-noise 'white' light lasers that operate in the milliwatt regime. The 'white' light is a result of an additive mixing of the three basic colours—blue, green and red—located as close as possible to the apexes of the 'colour triangle' (410, 520 and 650 nm), see Figure 7.1. From that point of view, ion lasers that make use of argon–krypton mixtures, with several watts of output power, are not suitable due to their high input power, and because the blue 488 nm line of Ar II cannot cover the violet sector of the colour triangle.

In general, MVILs operate on a larger number of CW lines in comparison with noble-gas ion lasers. This has stimulated research on simultaneous multi-colour oscillation in either the vapour of single elements or the vapours of mixtures of metals. One of the simplest designs for a multi-colour MVIL utilizes several discharge tubes that are configured longitudinally with a common optical axis [3, 4], but this design is inconvenient due to its large length.

As can be seen from Table 5.1, among PC-discharge-excited lasers, the ones that use Se and Zn (the latter has no violet lines) cover most of the visible spectrum. The Se ion laser oscillates in the visible in the range 460–649 nm, but six of the most intense lines are in the green, and the intensities of the blue and red lines are 10 times lower. By using a dispersive optical resonator [5], the power in the green lines can be made to be about the same as the powers of the red and blue lines. As a result, 'white' light output power at 460.4, 517.6 and 649 nm reduces to 3 mW from a 100 cm long capillary tube with 1 kW input power. This situation is not very satisfactory, since, with a non-selective resonator, the output power from the same tube would typically reach 70 mW, with an efficiency 20 times higher.

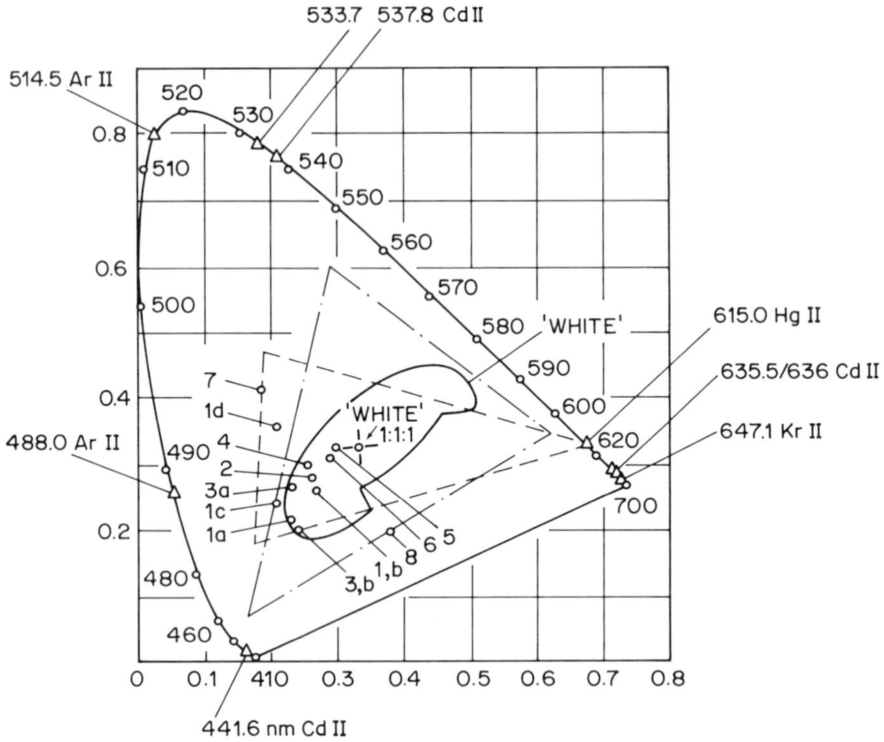

Figure 7.1 CIE chromatic diagram labelled with some Ar and Kr ion laser lines, and with the reproduction characteristics of 'white' light MVILs with HC discharge excitation. Points (1) to (6) denote corresponding positions in Table 7.1 for a HC He–Cd laser; (7) is the region of colour reproduction for a HC He–Cd–Hg laser (see Table 7.2); (8) is the colour reproduction region used in SECAM colour television.

More success in the development of 'white' light lasers has been achieved by using HC and TRF discharge excitation. As mentioned in Chapter 6, because of the more intense charge transfer and Penning ionization in those discharges, the Cd II laser can generate efficiently at 441.6 nm and on the green and red doublets, 533.7/537.8 nm and 635.5/636 nm. CW laser oscillation on the blue, green and red lines of the cadmium ion have been reported in numerous papers (see, for example, [6–8]), but it was not until the work described in [9], where wide-band mirrors were employed in the optical resonator and simultaneous laser oscillation occurred, that intensive studies began to investigate and develop He–Cd lasers with transverse-discharge excitation

[10–20]. However, the optimum excitation conditions, especially for current and helium pressure are very different for these lines (see Figure 6.2). Thus, the typical optimum of helium pressure for the green lines is about 3 times higher than that for the red lines, and the optimum current density for the blue, green and red lines is 58, 42 and 18 mA cm^{-2}, respectively [12]. This, however, provides opportunities for controlling the ratios of intensities for the different colours in the beam (see Figure 7.1). As can be seen in Table 7.1, the output power of the red line is always lower than that of the blue and green lines. Therefore, in order to produce evident 'white' light, the powers of the coloured transitions must be reduced to a ratio of blue, green and red powers of about 0.6:1:4 [2], which is more suited to the wavelength dependence of sensitivity of available detectors in the red spectral region. Nevertheless, the transverse-discharge-excited Cd II laser is very promising as a 'white' light laser because of its low level of noise, the narrow laser lines and the high coherency of the output beam (up to 400 cm) [21]. As an example, we can make reference to the use of the HC He–Cd 'white' light laser in a full-colour printer [22]. In [19] simultaneous lasing was observed on the two zinc ion transitions at 492.4 nm and 610.2 nm — these are complementary colours which together provide the visual impression of white light. Several points, corresponding to different laser power ratios for these two lines, are indicated on the CIE chromatic diagram in the cited paper. However, such a laser is suitable only for specific purposes.

Table 7.1 Powers of the blue, green and red lines of HC-discharge-excited He–Cd lasers.

	Discharge conditions	Units	Laser power				References
			Blue	Green	Red	Total	
1a	1.5 kPa (He), 250°C	arbitrary	32	10	6	48	[9]
1b	1.5 kPa (He), 260°C	arbitrary	21	13	8	42	[9]
1c	2 kPa (He), 260°C	arbitrary	48	17	3	68	[9]
1d	2 kPa (He), 280°C	arbitrary	23	22	5	50	[9]
2	1.3 kPa (He), 6.5 Pa (Cd) $j = 35$ mA cm^{-2}	mW	18	8	5	31	[10]
3a	1.3 kPa (He), 15 Pa (Cd)	mW	24	10	4	38	[21, 12]
3b	$j = 26.5$ mA cm^{-2}	mW	115	35	13	163	[14]
4	2.8 kPa (He), 360°C $j = 24$ mA cm^{-2}	mW	7.5	4	2	13.5	[13]
5	1.33 kPa (He) $j = 26.3$ mA cm^{-2}	mW	92.5	60.4	41.3	194.2	[20]
6	1.2 kPa (He) $i= 1.2$ A	mW	2.8	2	1.1	5.9	[17]

7.2 METAL VAPOUR ION LASERS WITH COMBINED ACTIVE MEDIA

The other way to create a laser emitting several laser lines simultaneously is to use a mixture of different lasants in the same laser tube to generate the basic colours. Using PC-discharge excitation, such media could include He–Cd for the blue, He–Se for the green and He–Ne or He–Zn mixtures for the red; with HC-discharge excitation, suitable media are the He–Cd mixture for the blue and green, and the He-Hg mixture for the red.

For PC-discharge excitation, mixtures of He, Ne and Cd vapour require quite different discharge conditions for generation on the 632.8 nm Ne I line ($p_{He}d$ = 0.04 kPa cm and i/d = 0.1 A cm^{-1}) and on the 441.6 nm Cd II lines ($p_{He}d$ = 0.12 kPa cm and i/d = 0.4 A cm^{-1}). However, it is found that with an increase in gas temperature, the optimum helium pressure for a He–Ne mixture increases, the optimum pressure of neon reduces and the optimum discharge current rises [23]. The output power decreases of course. For the 350 °C gas temperature typical of a He–Cd laser, a He pressure of 330 Pa, a ratio of helium to neon of 76:1 and an i/d value of 200 mA cm^{-1}, simultaneous oscillation on both laser lines occurs, each with an output power of 5 mW; the laser power ratio of these lines can be varied over a wide range, depending on the discharge conditions (Figure 7.2) [23]. However, the laser power is much less than that of the maximum output power attainable with the lasants contained in separate laser tubes (for 632.8 nm, 30–40 mW; for 441.6 nm,

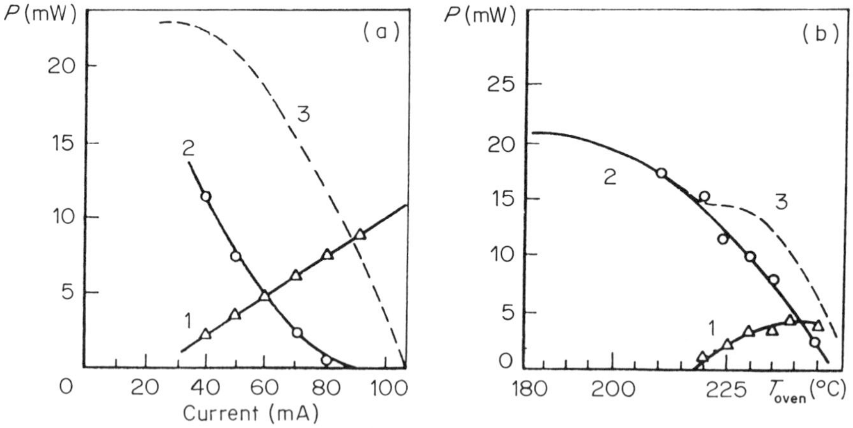

Figure 7.2 Dependences of laser output power from He–Ne–Cd mixtures on discharge current and on the temperature of the Cd reservoir [23], with p_{He} = 0.33 kPa, p_{He}/p_{Ne} = 76:1, and a 2.5 mm tube bore diameter: (a) 1 – Cd II 441.6 nm, 2 – Ne I 632.8 nm with T_{oven} = 235 °C, 3 – Ne I 632.8 nm in the absence of Cd vapour; (b) 1 – 441.6 nm, 2 – 632.8 nm, 3 – total power at i = 50 mA.

around 50 mW). The situation is similar for He–Ne–Zn mixtures [24], where simultaneous laser oscillation at 632.8 nm and 747.8 nm is produced only over a narrow range of discharge currents.

It can be seen from Table 5.2 that the values of *pd* for He–Cd and He–Se mixtures are rather close. On this basis, in [3] a two-sectioned discharge tube with a common condenser and cathode was proposed (Figure 7.3a). The difference in *pd* is achieved via the difference in the diameters of the cadmium and selenium sections: $d_{Cd} = (pd)_{Cd}/p$ and $d_{Se} = (pd)_{Se}/p$; the discharge currents through the two sections are separately controlled. For a capillary of given size, the full laser power generated on the 441.6 nm Cd II line can be extracted, but due to the smaller bore of the cadmium section, the Se laser beam is apertured, and full laser power cannot be extracted. In the same way, a laser using a He–Se–Ne mixture has been 'optimized' [25].

From the experiments described above, we shall now state the requirements for mixed active media.

One of the main reasons for the reduction in laser power in mixed MVILs is the occurrence of plasma-chemical reactions that deplete the metal vapours. The reactions can be less important if the chemical compounds formed dissociate back to the initial mixture components. Another restriction, especially for HC and TRF discharges, is the detrimental interaction of the electrodes with aggressive metal vapours.

Besides the chemical processes that can occur, the partition of the pumping between ion levels of several media causes the output powers of individual

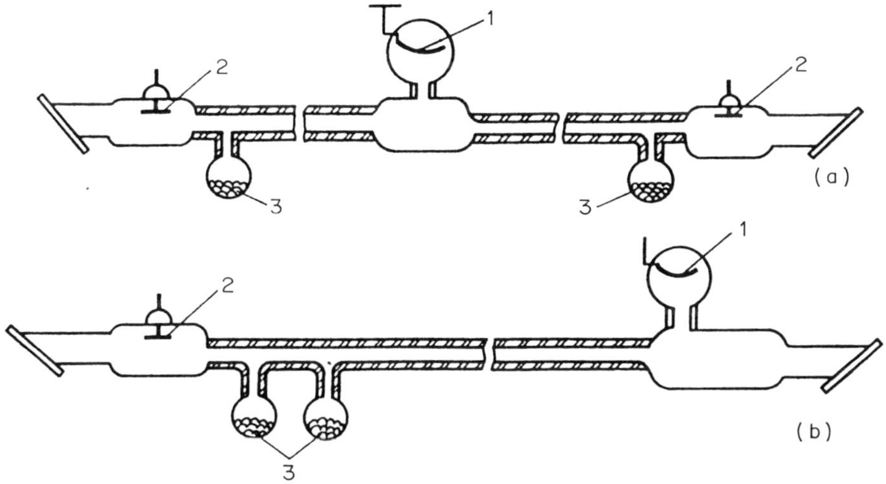

Figure 7.3 Cataphoretic discharge tubes for multiple-metal MVLs: (a) twin-segmented tube for media requiring different *pd* values [3]; (b) with twin evaporators [3, 28]; 1, cathode; 2, anode; 3, evaporators.

lines to be lower. The introduction of any low-potential species to the active region leads to a decrease in the electron temperature and thus to a fall in the densities of excited buffer-gas atoms and ions; a fall in the degree of pumping of high-lying levels results, together with a rise in the quenching rate of levels by collision with metal atoms (see equation (2.10)). The decrease in pumping is determined by the effective total vapour concentration, $n^* = \sum_i n_i a_i$, and the density of buffer-gas ions will be

$$N^+ = W^+(n^*) \,/\, \left(\nu_B^+ + \sum_i K_{ci} n_i \right). \tag{7.1}$$

Then, the partial pumping rate of the kth level of an ion of medium m (for instance, by charge transfer) is

$$W_m^{(k)} = \frac{W^+(n^*)}{\nu_B^+ + \sum_i K_{ci} n_m \xi_m^{(k)}} \tag{7.2}$$

By using (7.2), we can show that in combined active media, the pumping rate $W_m^{(k)}$ (and, consequently, the partial laser output power) is always less than that in individual laser media, and that the overall laser power is less than that of the most powerful lasant in the individual regime. At the same time, the power of individual lines can be balanced by changing the partial pressures. An additional decrease in individual line power is caused by the differences in the buffer-gas pressure and discharge current from their optimum values for each individual lasant. The differences in optimum buffer-gas pressures can be great, especially for lasants that operate by charge transfer (Se, I, Hg, Te, etc.) and electron impact (Sb, Bi). Thus, it is not expedient to use mixtures of charge-transfer-excited and electron-impact-excited lasants in combined laser tubes.

Lasers for combined active media ideally require the following criteria to be satisfied.

I. The lasants should be chemically non-aggressive. They can be evaporated simultaneously (even from the same oven side-arm) if an alloy is made according to F. Raoult's law [26], or they can be sputtered from a cathode made of the appropriate alloy. If chemical interactions take place in the discharge, the reactants formed must dissociate back to the initial reactants. In TRF discharges with internal electrodes, and in HC discharges, the metal vapours must not react with the electrode material.

II. The lasants must be of high laser efficiency, because the addition of other lasants in general reduces the pumping rate of individual lasants.

III. The lasants should have optimum discharge conditions as close as possible to each other (type and pressure of the buffer gas, discharge current, etc.).

The possibility of introducing different metals from separate ovens allows relatively high powers to be achieved on individual lines. This is an advantage of MVILs, compared with noble-gas ion lasers, where the gases are always present together in the laser tube.

7.2.1 PC-Discharge-Excited MVILs with Combined Active Media

According to the requirements outlined above, it is not recommended to utilize in the same tube (a) a He–Ne mixture with Cd [23], Zn [24] or Se [25] vapours; and (b) Cd and Se vapours [3]. In (a), there is a great difference in the optimum conditions for laser oscillation for the Ne atoms and metal ions, as well as the permanent presence of Ne in the tube, which makes effective operation of the metal vapour transitions impossible in the individual regime. In (b), chemical reactions between Cd and Se result in non-volatile compounds.

To preclude any chemical reactions in the discharge zone, it is necessary that the elements employed possess similar chemical properties, i.e. that they be of one subgroup of the periodic table of elements, or at least belong to adjacent subgroups. Such elements have similar dispositions of energy levels, ionization potentials and excitation potentials of the upper laser levels; this leads to similar (within each subgroup) mechanisms of pumping, and hence to similar optimum discharge conditions.

Various tube constructions can be used for mixed metal vapour lasers [27, 28]. The tube shown in Figure 5.1(h) is suitable for vapours of various metals regardless of their properties. It can be used for chemically active metals (such as Cd and Se), but only if their vapours are admitted separately. The vapours of each metal are admitted to the capillary tube through separate channels by oven heating and cataphoresis. Note that the current density through each channel must prevent the vapours from other ovens from entering the chosen channel.

If chemically inactive metals are employed, then the tube construction can be very simple (Figure 7.3b) [3, 28]. Here, only the first oven, which is near the anode, is blocked by the discharge against the vapours of other metals. For the other metals, there is no blocking. The vapour pressures of the lasants at the same temperature must be quite different (see section 5.2): the vapour pressure of the first oven must not be less than two orders of magnitude higher than that of the second oven (or the following ones). Thus, if the most volatile metal is located near the anode, it is possible to prevent condensation of its vapour in the ovens that follow [27, 28].

Hence, the first design of discharge tube (see Figure 5.1h) is more widely applicable than that of Figure 7.3(b). It does not allow contamination (or mixing) of the lasants in the ovens (which permits various volatile and chemically active metals to be used), places no restrictions on the oven design nor on the discharge current (the capillary tube current can be altered by changing the current in the other channels), and leaves the current for the operating oven at its optimum value.

In [27] a three-channel laser tube was reported (see Figure 5.1h) which used the three most effective lasants—cadmium, selenium and zinc vapours—for generation in the violet, blue-green, yellow and red spectral regions. The following regimes of operation were found to be practical.

I. Individual laser oscillation using Se II (in the range 497 to 531 nm), Cd II (441.6 nm) and Zn II (589 and 748 nm) could be obtained with powers of about 60% of the maximum possible from the tube (due to the non-optimum helium pressure, which could not be altered).
II. Simultaneous laser oscillation with Cd II (441.6 nm) and Zn II (748 and 589 nm) was possible using common excitation of cadmium and zinc vapours. The laser powers were about 35% of the maximum obtained in the regime of individual lasing. The time of switch-over from one regime to the other was determined by the cooling and warming up times of the ovens.

With a 'double' evaporator tube, operating simultaneously on Cd and Zn vapours (Figure 7.3b), the same laser lines could be generated, but with higher powers (around 40% more) because the helium pressure could be close to optimum (Figure 7.4) [28].

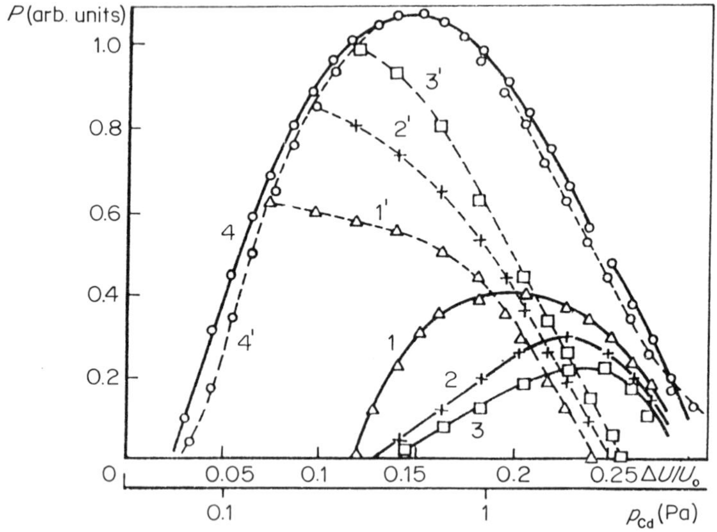

Figure 7.4 Laser power of the 441.6 nm Cd II (solid lines) and 747.8 nm Zn II (broken lines) transitions versus cadmium pressure, controlled by adjustment of the ratio $-\Delta E_z/E_z$: 1, 1′, p_{Zn} = constant; $-\Delta E_z/E_z$ = 7.5%; 2,2′, 10%; 3, 3′, 12.5%; 4, 4′, separate lasing regime with alternating admittance of metal vapours: $p_{He}d$ = 0.12 kPa cm [28].

7.2.2 Mixed Metal Vapours in HC Discharges

From the foregoing discussion of mixed metal lasers using PC discharge excitation, the 'white' light regime cannot be realized without a sharp fall in laser efficiency. Transversely excited lasers (HC and TRF discharges) that utilize Cd are more promising because the decrease in individual line laser powers due to a non-optimum He pressure is only around a factor of two (see Figure 6.2), and the only real issue is the increase in the power of the red transitions required.

In [29], a HC discharge laser utilizing CdI_2 was reported. Although the partial iodine vapour pressure is twice the cadmium vapour pressure, the power of the iodine line is not sufficiently high and the iodine attacks the cathode. More hopeful results have been obtained with the simultaneous excitation of cadmium ions (441.6 and 537 nm) and mercury ions (615 nm) in HC discharges [30]. The optimum helium pressure for mercury is close to those optimum for the blue and green cadmium lines (see Table 6.2). Thus, the optimum helium pressure for the mixed metal medium is 2.7 kPa. Despite the great difference in optimum currents, the range of common operation of the three lines is quite broad, which permits rapid alteration of line powers by discharge-current modulation.

In Figure 7.1, the area of the colour triangle that is covered by the output wavelengths of a HC discharge laser with cadmium and mercury vapours is indicated. The excitation conditions that ensure the power of the red line exceeds the powers of the blue and green lines, and so allow the three-colour radiation to simulate natural white light, are given in Table 7.2. This kind of laser has all the attributes of the HC-discharge cadmium ion laser: low noise, narrow linewidths and large coherence lengths.

Table 7.2 Output powers (arbitrary units) of blue, green and red lines of a HC-discharge He–Cd–Hg laser [14, 30].

		Laser power			
Composition of active medium	Current density mA cm^{-2}	Cd II 441.6 nm	Cd II 533.7/537.8 nm	Hg II 615.0 nm	Total
2.7 kPa (He) + 13 Pa (Cd) + 5.4 Pa (Hg)	53	1.1	1.3	0.9	3.3
	66	1.6	1.4	1.5	4.5
	80	1.9	1.0	2.1	5.0
2.7 kPa (He) + 13 Pa (Cd) + 13 Pa (Hg)	66	1.3	1.1	2.1	4.5

REFERENCES

[1] V. E. Iosirov, *Tekh. Kino Telev.*, No. 4, 25–31 (1983) (in Russian).
[2] R. A. Bartolini, *Proc. IEEE,* **70**(6), 589–97 (1982).
[3] N. V. S'botinov, P. K. Telbizov and S. D. Kalchev, *Sov. Phys.–Tech. Phys.*, **18**(12), 1646–7 (1974).
[4] K. Jain and S. A. Newton, *Appl. Phys., B* **26**(1), 43–8 (1982).
[5] H. P. Popp and E. Schmidt, *IEEE J. Quantum Electron.*, **15**(9), 840–2 (1979).
[6] É. K. Karabut, V. C. Michalevskii, V. F. Papakin and M. F. Sem, *Sov. Phys.–Tech. Phys.*, **14**(10), 1447–8 (1970).
[7] W. K. Schuebel, *Appl. Phys. Lett.*, **16**(11), 470–2 (1970); *IEEE J. Quantum Electron.*, **6**(9), 574–5 (1970).
[8] J. A. Piper and C. E. Webb, *J. Phys. D*, **6**(4), 400–7 (1973).
[9] K. Fujii, T. Takahashi and Y. Asami, *IEEE J. Quantum Electron.*, **11**(3), 111–14 (1975).
[10] S. C. Wang, *Laser Focus,* **19**(3), 14 (1983).
[11] M. K. Dyatlov, V. G. Kas'yan and V. G. Levin, *Sov. Tech. Phys. Lett,*. **3**(7), 264 (1977).
[12] I. G. Ivanov and V. V. Vainer, *Proc. Int. Conf. Lasers '83,* pp. 442–4, McLean, STS Press (1984).
[13] W. Kin-Hung and C. Grey Morgan, *J. Phys D,* **16**(1), L1–4 (1983).
[14] V. V. Vainer, I. G. Ivanov, G. A. Kalinchenko and M. F. Sem, *Metal Vapor Lasers and Their Applications: CIS Selected Papers,* ed. G. G. Petrash, *Proc. SPIE,* **2110**, 128–49 (1993).
[15] C. Boulmer-Leborgne, B. Dabreuil and J. C. Pellicer, *J. Phys. D,* **22**, 512–18 (1989).
[16] J. Mizeraczyk, J. Mentel, E. Schmidt, N. Reich and C. Carlsson, *Meas. Sci. Technol.*, **5**(8), 936–41 (1994).
[17] H. H. Telle, I. D. Hopkin, P. Ramalingam, H. K. Fun and C. Grey Morgan, *J. Phys. D,* **21**(10S), 167–70 (1988).
[18] W. Sasaki, I. Ebina and T. Ohta, *Metal Vapor, Deep Blue and Ultraviolet Lasers,* ed. J. J. Kim, R. Kimball and P. J. Wisoff, *Proc. SPIE,* **1041**, 117–22 (1989).
[19] W. Sasaki, K. Itani and T. Ohta, *Metal Vapor, Deep Blue and Ultraviolet Lasers,* ed. J. J. Kim, R. Kimball and P. J. Wisoff, *Proc, SPIE,* **1041**, 123–30 (1989).
[20] A. Fuke, K. Mashuda and Y. Tokita, *Trans. Inst. Electron. Commun. Eng. Jpn., Sect. E,* **69**(4), 365–6 (1986); *Jpn. J. Appl. Phys., Part 1,* **26**(1), 96–101 (1987).
[21] V. V. Vainer, I. G. Ivanov and V. S. Mikhalevskii, *Summaries of Papers. Population Inversion and Oscillation on Atomic and Molecular Transitions,* Part 1, ed. O. V. Ravodina and I. I. Murav'ev, p. 147, Tomsk: Institute of Atmospheric Optics (Sept. 1986) (in Russian).
[22] M. Takashima, S. Okada, H. Nishiyama, M. Matsuda and K. Fujii, *Rev. Sci. Instrum.*, **62**(5), 1238–45 (1991).
[23] S. A. Ahmed and A. J. Campillo, *Proc. IEEE,* **57**(11), 2084–5 (1969).
[24] G. J. Collins, *J. Appl. Phys.*, **42**(10), 3812–15 (1971).
[25] N. V. Sabotinov and P. K. Telbizov, *C. R. Acad. Bulg. Sci.*, **27**(6), 763–6 (1974).
[26] V. F. Kravchenko, *Candidate's dissertation Phys.-Math. Sci.,* Rostov-on-Don, Russia (1974).
[27] I. G. Ivanov, V. G. Il'yushko, V. S. Mikhalevskii, V. P. Mukhanov and M. F. Sem, *Instrum. Exp. Tech.*, **17**(4), pt 2, 1136–7 (1974).

[28] I. G. Ivanov, V. G. Il'yushko and M. F. Sem, *Sov. J. Quantum Electron.*, **4**(3), 411–12 (1974).
[29] J. A. Piper, *Opt. Commun.*, **19**(2), 189–92 (1976).
[30] V. V. Vainer, I. G. Ivanov, V. S. Mikhalevskii and M. F. Sem, *Sov. J. Quantum Electron.*, **13**(4), 417–18 (1983).

Appendices

A.1 EXCITATION POTENTIALS OF METASTABLE LEVELS, AND FIRST AND SECOND IONIZATION POTENTIALS OF ATOMS [1], IN UNITS OF ELECTRON-VOLTS.

Atom	Z	Metastable levels	Single ionization	Double ionization
H	1	10.2	13.5985	–
He	2	19.82; 20.62	24.5876	54.418
Li	3		5.3918	75.641
Be	4	2.726	9.3228	18.211
B	5	3.58	8.2981	25.155
C	6	1.264; 2.684; 4.183	11.26	24.383
N	7	2.384; 3.576	14.534	29.602
O	8	1.976; 4.19; 9.146	13.618	35.118
F	9	0.0501	17.423	34.971
Ne	10	16.62; 16.72	21.565	40.963
Na	11		5.13912	47.287
Mg	12	2.712; 2.717	7.6463	15.0354
Al	13	0.014; 3.608	5.98581	18.829
Si	14	0.781; 1.909; 4.132	8.1517	16.3460
P	15	1.41; 2.32	10.4868	19.770
S	16	0.049; 0.071; 1.145; 2.75; 6.525	10.36012	23.338
Cl	17	0.109; 8.962	12.968	23.814
Ar	18	11.55; 11.72	15.760	27.630
K	19		4.34070	31.63
Ca	20	1.879; 1.886; 1.899; 2.71	6.1132	11.872
Sc	21	0.021	6.5615	12.800
Ti	22	0.021; 0.048; 0.833	6.82	13.58
V	23	0.017; 0.04; 0.0686	6.74	14.66
Cr	24	0.941; 1.003	6.766	16.50
Mn	25	2.114; 2.143; 2.164; 2.178; 2.187	7.43408	15.64
Fe	26	0.86; 2.176; 2.28	7.9024	16.188
Co	27	0.1; 0.174; 0.224	7.86	17.084
Ni	28	0.165; 0.275; 1.676	7.637	18.169
Cu	29	1.389; 1.642	7.7264	20.294
Zn	30	4.006; 4.03; 4.078	9.3943	17.964

A.1 (*continued*)

Atom	Z	Metastable levels	Single ionization	Double ionization
Ga	31	0.102	5.99935	20.515
Ge	32	0.069; 0.175; 0.833; 2.029	7.8995	15.935
As	33	1.313; 1.353; 2.255; 2.312	9.789	18.59
Se	34	0.247; 0.314; 1.187; 2.783	9.752	21.16
Br	35	0.457	11.814	21.81
Kr	36	9.915; 10.56	14	24.360
Rb	37		4.177	27.290
Sr	38	1.798; 2.498	5.695	11.03
Y	39	0.0658	6.217	12.24
Zr	40	0.0707; 0.154	6.837	13.13
Nb	41		6.88	14.32
Mo	42	1.335; 1.467	7.099	16.16
Tc	43		7.28	15.26
Ru	44		7.366	16.76
Rh	45		7.46	18.08
Pd	46		8.336	19.43
Ag	47		7.5763	21.49
Cd	48	3.734; 3.8; 3.946	8.9939	16.908
In	49	0.274	5.7864	18.87
Sn	50	0.21; 0.425; 1.068; 2.128	7.344	14.632
Sb	51	1.055; 1.222; 2.033; 2.289	8.609	16.53
Te	52	0.584; 0.589; 1.309; 2.876	9.01	18.6
I	53	0.943	10.451	19.131
Xe	54	8.315; 9.447	12.1299	20.98
Cs	55		3.89394	23.15
Ba	56	1.12; 1.143; 1.19; 1.413	5.21171	10.004
La	57		5.577	11.1
Ce	58		5.539	10.8
Pr	59		5.47	10.6
Nd	60		5.525	10.7
Pm	61		5.58	10.9
Sm	62		5.644	11.1
Eu	63		5.6704	11.24
Gd	64		6.150	12.1
Tb	65		5.864	11.5
Dy	66		5.939	11.7
Ho	67		6.022	11.8
Er	68		6.108	11.9
Tm	69		6.184	12.1
Yb	70		6.254	12.18
Lu	71		5.426	13.9
Hf	72		6.8	14.9
Ta	73		7.89	–
W	74	0.207; 0.366; 0.412; 0.599; 0.771	7.98	–
Re	75		7.88	–

(*continued overleaf*)

A.1 (*continued*)

Atom	Z	Metastable levels	Single ionization	Double ionization
Os	76		8.73	–
Ir	77		9.05	–
Pt	78		8.96	18.56
Au	79	1.136; 2.658	9.2256	20.5
Hg	80	4.667; 4.887; 5.461	10.4376	18.76
Tl	81	0.966	6.1083	20.43
Pb	82	0.969; 1.32; 2.66; 3.653	7.4167	15.033
Bi	83	1.416;1.914; 2.686; 4.112	7.2856	16.7
Po	84		8.4168	–
At	85		9.0	–
Rn	86		10.75	–
Fr	87		4.0	–
Ra	88		5.2785	10.15
Ac	89		5.2	11.75
Th	90		6.1	11.9
Pa	91		6.0	–
U	92		6.194	11.9
Np	93		6.266	–
Pu	94		6.06	–
Am	95		6.0	–
Cm	96		6.02	–
Bk	97		6.23	–
Cf	98		6.30	–
Es	99		6.42	–
Fm	100		6.5	–
Md	101		6.6	–
No	102		6.6	–

A.2 THE RANGES OF TEMPERATURES FOR WHICH THE VAPOUR PRESSURES OF THE ELEMENTS LIE BETWEEN 0.13 AND 133 Pa [2]

The symbol ‘o’ marks the melting point of each element.

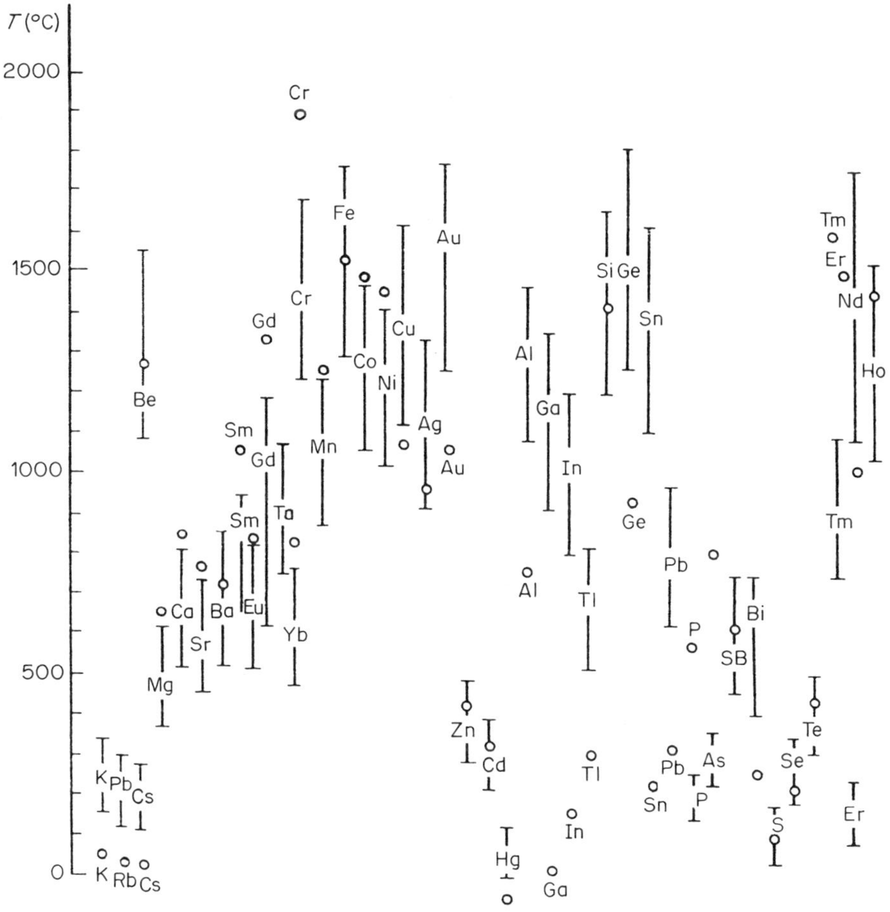

A.3 CROSS-SECTIONS AND RATE CONSTANTS FOR DE-EXCITATION, RECOMBINATION AND HEAVY-BODY REACTIONS IN MVIL DISCHARGES

Process	Cross-section (10^{-16} cm^2)	Rate constant[a] (cm^3 s^{-1}; cm^6 s^{-1})	References[b]
He (helium, Z = 2)			
$He^*(2\,^3S_1) + Ar \rightarrow He + Ar^+ + e$	7±4	–	[3]
$He^*(2\,^3S_1) + Kr \rightarrow He + Kr^+ + e$	8.8±0.9	–	[3]
$He^*(2\,^3S_1) + H \rightarrow He + H^+ + e$	22	–	[3]
$He^*(2\,^3S_1) + H_2 \rightarrow He + H_2^+ + e$	2.3±1.4	–	[3]
$He^*(2\,^3S_1) + Ne \rightarrow He + Ne^*(2s)$	3.3×10^{-11}	$\exp(-0.051/T)$	[3]
$He^*(2\,^1S_0) + Ne \rightarrow He + Ne^*(3s)$	2.1×10^{-10}	$\exp(-0.041/T)$	[3]
$He^*(2\,^3S) + He^*(2\,^3S) \rightarrow He^+ + He + e$	100±10 (300 K)	$(1.5\pm0.3) \times 10^{-9}$	[3]
$He^*(2\,^1S) + Ar \rightarrow He + Ar^+ + e$	25±10	–	[3]
$He^*(2\,^1S) + Kr \rightarrow He + Kr^+ + e$	34±20	–	[3]
$He^*(2\,^1S) + H \rightarrow He + H^+ + e$	33	–	[3]
$He^*(2\,^1S) + H_2 \rightarrow He + H_2^+ + e$	2.7±0.8	–	[3]
$He^*(2\,^1P) + Ar \rightarrow He + Ar^+ + e$	86 - 108	–	[3]
$He^*(2\,^1P) + Kr \rightarrow He + Kr^+ + e$	89 - 124	–	[3]
$He^+ + 2He \rightarrow He_2^+ + He$		$2.84 \times 10^{-33}(1/T)$	[4, 5]
$He_2^+ + Ne \rightarrow 2He + Ne^+$		6×10^{-10} (at 200 K)	[6]
$He^*(2\,^3S_1) + e \rightarrow He + e$		$(2.8\pm1) \times 10^{-9}$	[7, 3]
$He^*(2\,^1S_0) + e \rightarrow He^*(2\,^3S_1) + e$		3.8×10^{-7};	[3]
		$(7\pm1) \times 10^{-7}$	[3]
$He^+ + 2e \rightarrow He + e$		$8.8 \times 10^{-30} T_e^{-9/2}$ x $\exp(0.09/(T_e^5 + 0.015))$	[8][c]
$He_2^+ + e \rightarrow He + He^*$		10^{-10} - 10^{-8};	[9]
		$3.7 \times 10^{-11} T_e^{-3/2}$	[6]
$He^+ + e \rightarrow He + h\nu$	–	$3.25 \times 10^{-13} T_e^{-0.7}$	[10]
$He_2^* + e \rightarrow He + He + e$	–	$(3.8\pm0.8) \times 10^{-9}$	[3]
Ne (neon, Z = 10)			
$Ne^*(^3P_2) + Ar \rightarrow Ne + Ar^+ + e$	3.1±0.8	–	[3]
$Ne^*(^3P_2) + Kr \rightarrow Ne + Kr^+ + e$	1 - 15	–	[3]
$Ne^*(^3P_2) + H_2 \rightarrow Ne + H_2^+ + e$	2 - 35	–	[3]
$Ne^+ + 2Ne \rightarrow Ne_2^+ + Ne$	–	6.4×10^{-32} (at 300K)	[5]
$Ne^+ + 2He \rightarrow (HeNe)^+ + He$	–	2.0×10^{-32}	[5]
$Ne^+ + Ne + He \rightarrow Ne_2^+ + He$	–	3×10^{-31}	[5]
$Ne^+ + 2e \rightarrow Ne + e$	–	$8.8 \times 10^{-30} T_e^{-9/2}$ x $\exp(0.09/(Te^5 + 0.015))$	[8][c]
$Ne_2^+ + e \rightarrow Ne + Ne^*$	–	$3.53 \times 10^{-8} T_e^{-0.43}$	[9]
$Ne^+ + e \rightarrow Ne + h\nu$	–	$2.32 \times 10^{-13} T_e^{-0.7}$	[10]

A.3 (*continued*)

Process	Cross-section (10^{-16} cm^2)	Rate constant[a] (cm^3 s^{-1}; cm^6 s^{-1})	References[b]
Be (beryllium, Z = 4)			
$He^+ + Be \rightarrow He + Be^{+*}$	15±3	–	[11]
Na (sodium, Z = 11)			
$He^*(2\,^3S) + Na \rightarrow He + Na^+ + e$	14	–	[3]
$He^*(2\,^1S) + Na \rightarrow He + Na^+ + e$	17	–	[3]
Mg (magnesium, Z = 12)			
$He^+ + Mg \rightarrow He + Mg^{2+} + e$	16±5	–	[12]
$Ne^+ + Mg \rightarrow He + Mg^{+*}$	20	–	[13]
K (potassium, Z = 19)			
$He^*(2\,^3S) + K \rightarrow He + K^+ + e$	38±9	–	[14]
$Ne^*(3\,^3P_2) + K \rightarrow Ne + K^+ + e$	62±18	–	[14]
Ca (calcium, Z = 20)			
$He^+ + Ca \rightarrow He + Ca^{2+} + e$	18	–	[15]
$Ne^+ + Ca \rightarrow Ne + Ca^{2+} + e$	19	–	[15]
$Kr^+ + Ca \rightarrow Kr + Ca^{+*}$	19	–	[16]
$Xe^+ + Ca \rightarrow Xe + Ca^{+*}$	1.5; 2.2	–	[16, 17]
$He^*(2\,^3S_1) + Ca \rightarrow He + Ca^{+*} + e$	41	–	[17]
Zn (zinc, Z = 30)			
$He^+ + Zn \rightarrow He + Zn^{+*}$	33±5	–	[14]
	21; 5	–	[18, 19, 20]
Partial cross-sections for charge transfer to various levels of zinc ions			
17.5(5 ^2D):3.5(6 ^2P) (absolute units (10^{-16} cm^2))			[20]
44(6 $^2P_{3/2}$):5.9(6 $^2P_{1/2}$):290(5 $^2D_{5/2}$):160(5 $^2D_{3/2}$): 12(4 $^2F_{7/2}$):6.7(4 $^2F_{5/2}$):28(6 $^2S_{1/2}$) (relative arbitrary units)			[21]
$Ne^+ + Zn \rightarrow Ne + Zn^{+*}$	30±3	–	[14]
	23	–	[22]
$He^*(2\,^3S_1) + Zn \rightarrow He + Zn^+ + e$	44±6	–	[14, 20]
	29±1.5	–	[19]
$He^*(2\,^1S_0) + Zn \rightarrow He + Zn^+ + e$	53±18	–	[23]

(*continued overleaf*)

A.3 (*continued*)

Process	Cross-section (10^{-16} cm^2)	Rate constant[a] (cm^3 s^{-1}; cm^6 s^{-1})	References[b]
$He^*(2\,^1P_0) + Zn \rightarrow He + Zn^+ + e$	100±24	–	[23]
$Ne^*(3\,^3P_2) + Zn \rightarrow He + Zn^+ + e$	55±10	–	[14]
	41.8	–	[19]
Se (selenium, Z = 34)			
$He^+ + Se \rightarrow He + Se^{+*}$	40–50	–	[24]
Rb (rubidium, Z = 37)			
$He^+ + Rb \rightarrow He^*(2\,^3S_1) + Rb^+$	28±7	–	[14]
$He^*(2\,^1S_0) + Rb \rightarrow He + Rb^+ + e$	290±30	–	[25]
$He^*(2\,^1P) + Rb \rightarrow He + Rb^+ + e$	380±60	–	[25]
$Ne^*(3\,^3P_2) + Rb \rightarrow Ne + Rb^+ + e$	55±17	–	[14]
Sr (strontium, Z = 38)			
$He^+ + Sr \rightarrow He + Sr^{2+} + e$	27	–	[15]
$Ne^+ + Sr \rightarrow Ne + Sr^{2+} + e$	49	–	[15]
$Kr^+ + Sr \rightarrow Kr + Sr^{+*}$	25	–	[16]
$Xe^+ + Sr \rightarrow Xe + Sr^{+*}$	3; 2.3	–	[16, 17]
$He^*(2\,^3S_1) + Sr \rightarrow He + Sr^{+*} + e$	16	–	[17]
Ag (silver, Z = 47)			
$He^+ + Ag \rightarrow He + Ag^{+*}$	1.5±0.2	–	[26]
Cd (cadmium, Z = 48)			
$He^+ + Cd \rightarrow He + Cd^{+*}$	42±4	–	[14]
	0.3; 37	–	[18, 20]
	34	–	[27]

Relative cross-sections for charge transfer to various levels of cadmium ions

7.7(6*G*):8.2(6*F*):8.2(9*S*):7.7(9*P*):5.1(8*D*) (arbitrary units) [20]

$He^*(2\,^3S_1) + Cd \rightarrow He + Cd^+ + e$	50±10	–	[27]
	65±6	–	[14, 20]
	45±2	–	[19, 28]
$He^*(2\,^3S_1) + Cd \rightarrow He + Cd^+(4^2D^*_{5/2}) + e$	3±2	–	[29]

A.3 (*continued*)

Process	Cross-section (10^{-16} cm^2)	Rate constant[a] (cm^3 s^{-1}; cm^6 s^{-1})	References[b]
Relative cross-sections for Penning ionization by He*(2 $^3S_1$) to various levels of cadmium ions			
28(5 $^2S_{1/2}$):1.4(5 $^2P_{1/2}$):1.4(5 $^2P_{3/2}$): 5.4(4 $^2D^*_{5/2}$):5.4(4 $^2D^*_{3/2}$):2.7(6 $^2S_{1/2}$) [arbitrary units]			[30]
13(4 $^2D^*_{3/2}$):22(4 $^2D^*_{5/2}$):40(5 $^2S_{1/2}$) [arbitrary units]			[27]
5(4 $^2D^*_{3/2,5/2}$):2(5 $^2P_{1/2,3/2}$):10(5 $^2S_{1/2}$) [arbitrary units]			[31]
He(2 $^1S_0$) + Cd → He + Cd^{+*} + e	100; 94±38		[23, 32]
He(2 $^1P_0$) + Cd → He + Cd$^+$ + e	175±63	–	[23]
Ne(3 $^3P_2$) + Cd → Ne + Cd$^+$ + e	46.1	–	[19]
Te (tellurium, Z = 52)			
He$^+$ + Te → He + Te^{+*}	40; 10	–	[18, 33]
I (iodine, Z = 53)			
He$^+$ + I → He + I^{+*}	3	–	[34]
He*(2 $^3S_1$) + I → He + I^{+*} + e	1.5	–	[34]
Cs (caesium, Z = 55)			
He$^+$ + Cs → He + Cs^{+*}	100±15	–	[14]
Ne$^+$ + Cs → Ne + Cs^{+*}	0.4+0.3	–	[14]
He*(2 ^3S) + Cs → He + Cs$^+$ + e	6.5; 54±13		[3, 14]
He*(2 ^1S) + Cs → He + Cs$^+$ + e	22; 350±60		[3]
	185±45	–	[25]
He*(2 ^1P) + Cs → He + Cs$^+$ + e	550±150		[25]
Ne*(3 $^3P_2$) + Cs → Ne + Cs$^+$ + e	36±10	–	[14]
Ne*(3 $^3P_0$) + Cs → Ne + Cs$^+$ + e	35±10	–	[14]
Ba (barium, Z = 56)			
He$^+$ + Ba → He + Ba^{2+} + e	30; 16	–	[15, 35]
Ne$^+$ + Ba → Ne + Ba^{2+} + e	80	–	[15]
He*(2 $^3S_1$) + Ba → He + Ba$^+$ + e	55	–	[36]
He*(2 $^1S_0$) + Ba → He + Ba$^+$ + e	105	–	[36]
He*(2 $^3S_1$) + Ba → He + Ba^{2+} + 2e	0.55±0.27		[36]
He*(2 $^1S_0$) + Ba → He + Ba^{2+} + 2e	1.85±1	–	[36]
Hg (mercury, Z = 80)			
He$^+$ + Hg → He + Hg^{+*}	110–140		[14, 18, 37, 38]
	12	–	[39]

(*continued overleaf*)

A.3 (*continued*)

Process	Cross-section (10^{-16} cm^2)	Rate constant[a] (cm^3 s^{-1}; cm^6 s^{-1})	References[b]
Partial cross-sections for charge transfer to various levels of mercury ions			
$20(7\,^2P_{3/2})$; $2.4(6D_{5/2})$; $3.6(7\,^2P_{1/2})$ (absolute units (10^{-16} cm^2))			[40]
$3(5d^{10}n'l')$:$7(5d^96sn'l')$ (arbitrary units)			[40]
$26(7\,^2P_{3/2})$;$3.6(7\,^2P_{1/2})$ (absolute units (10^{-16} cm^2))			[41]
to the level $7\,^2P_{3/2}$	0.15–0.25 (absolute units (10^{-16} cm^2))		[14]
$78(7\,^2P_{3/2})$:$12(7\,^2P_{1/2})$:$4.2(6\,^2P_{3/2})$:$5.8(6\,^2P_{5/2})$ (arbitrary units)			[42]
$He^*(2^3S_1) + Hg \rightarrow He + Hg^{+*} + e$	90±15	–	[38]
	140; 400	–	[3]
	65±11	–	[14]
Relative cross-sections for Penning ionization by $He^*(2^3S_1)$ to various levels of mercury ions			
$1(6\,^2S_{1/2})$:$0.38(5\,^2D^*_{5/2})$: $0.34(5\,^2D^*_{3/2} + 6\,^2P_{1/2})$:$0.15(6\,^2P_{3/2})$ (arbitrary units)			[43]
$1(6\,^2S_{1/2})$:$0.52(5\,^2D^*_{5/2})$: $0.4(5\,^2D^*_{3/2} + 6\,^2P_{1/2})$:$0.32(6\,^2P_{3/2})$ (arbitrary units)			[31]
$He^*(2\,^1S_0) + Hg \rightarrow He + Hg^{+*} + e$	145±55	–	[23]
Relative cross-sections of Penning ionization with $He^*(2\,^3S_0)$ to various levels of mercury ions			
$1.05(6\,^2S_{1/2})$:$1.27(5\,^2D^*_{5/2})$: $0.98(5\,^2D^*_{3/2}+6\,^2P_{1/2})$:$0.29(6\,^2P_{3/2})$ (arbitrary units)			[43]
$1.05(6\,^2S_{1/2})$:$1.23(5\,^2D^*_{5/2})$: $0.94(5\,^2D^*_{3/2}+6\,^2P_{1/2})$:$0.29(6\,^2P_{3/2})$ (arbitrary units)			[31]
$He^*(2\,^3S_1) + Hg(6\,^3P_2) \rightarrow He^+ + Hg^{+*}(7\,^2P_{1/2}) + e$	100	–	[3]
$Ne^*(3\,^3P_2) + Hg \rightarrow Ne + Hg^+ + e$	37±5	–	[44]
Tl (thallium, Z = 81)			
$He^+ + Tl \rightarrow He + Tl^{+*}$	15±5	–	[45]
$Ne^+ + Tl \rightarrow Ne + Tl^{+*}$	30±5	–	[45]
Pb (lead, Z = 82)			
$He^+ + Pb \rightarrow He + Pb^{2+} + e$	4	–	[46]
$He^*(2\,^3S_1) + Pb \rightarrow He + Pb^+ + e$	14.8	–	[46]

[a] Electron temperature, T_e and gas temperature, T are in electron volts.
[b] Besides original papers, books and surveys are cited.
[c] An approximation based on the experimental data of a number of authors given in [8].

A.4 LASER TRANSITIONS OF METAL VAPOUR ION LASERS

The regimes of laser operation and types of discharges referred to in Table A4.1 below are as follows:
PPC, QCWPC, CWPC— pulsed, quasi-CW and CW oscillation in a positive-column (PC) longitudinal discharge.
PHC, QCWHC, CWHC— pulsed, quasi-CW and CW oscillation in a hollow-cathode (HC) discharge.
PTD— pulsed transverse discharge.
TRFD— transverse RF discharge.
LRFD— longitudinal RF discharge.
LPP— laser-produced plasma.
SPE— segmented plasma excitation.
EBP— electron-beam pumped.
NP—nuclear-pumped.
PP—photoionization-pumped.
MWP— microwave-pumped.
As well as original papers, books and surveys are cited.

Table A4.1 Ion laser lines listed according to element.

Spectrum	λ(nm)	Transition	Buffer Gas	Regimes, types of discharge, references
Be (beryllium, Z = 4)				
Be IV	253	5g → 4f	He	LPP [47]
Be II	436.2	$4D \rightarrow 3P$	He	LPP [48]
Be II	467.5	$4F \rightarrow 3D$	He	PPC [11], LPP [49]
Be II	527.2	$4S \rightarrow 3P$	He	PPC [11]
Be II	1012	$5G \rightarrow 4F$	He	LPP [48, 49]
Be II	1209.6	$3P \rightarrow 3S$	Ne	PPC [11]
B (boron, Z = 5)				
B II	345.13	$2p^2\,{}^1D_2 \rightarrow ({}^2S)2p\ {}^1P^o{}_1$	BCl_3 [a]	PPC [50]
C (carbon, Z = 6)				
C IV	154.82	$2s\ {}^2S_{1/2} \rightarrow 2p\ {}^2P^o{}_{3/2}$	CO_2 [a]	PPC [51]
C IV	155.08	$2s\ {}^2S_{1/2} \rightarrow 2p\ {}^2P^o{}_{1/2}$	CO_2	PPC [51]
C III	464.74	$({}^2S)3p\ {}^3P^o{}_2 \rightarrow ({}^2S)3s\ {}^3S_1$	He, Ne, CO_2	PPC [52]
C III	465.02	$({}^2S)3p\ {}^3P^o{}_1 \rightarrow ({}^2S)3s\ {}^3S_1$		PPC [52]
C II	495.41	$3d\ {}^2P_{1/2} \rightarrow 3p\ {}^2P^o{}_{1/2}$	CO_2	PPC [53]
C II	514.57	$3p\ {}^4D_{7/2} \rightarrow 3s\ {}^4P^o{}_{5/2}$	CO_2	PPC [54]
C II	657.8	$3p\ {}^2P^o{}_{3/2} \rightarrow 3s\ {}^2S_{1/2}$	CO_2	PPC [54]
C II	678.38	$3p\ {}^4P_{5/2} \rightarrow 3s\ {}^4P^o{}_{5/2}$	CO_2	PPC [54]
Mg (magnesium, Z = 12)				
Mg II	921.8	$4\ {}^2P^o{}_{3/2} \rightarrow 4\ {}^2S_{1/2}$	Ne, He	PPC [13], LPP [55], CWPC [56]

(continued overleaf)

Table A4.1 (*continued*)

Spectrum	λ(nm)	Transition	Buffer Gas	Regimes, types of discharge, references
Mg II	924.4	$4\ ^2P^o_{1/2} \rightarrow 4\ ^2S_{1/2}$	Ne, He	PPC [13], LPP [55], CWPC [56]
Mg II	1091.4	$4\ ^2P^o_{3/2} \rightarrow 4\ ^2D_{5/2}$	Ne, He	CWPC [56], LPP [55], LPP [49], SPE [57]
Mg II	1095.1	$4\ ^2P^o_{1/2} \rightarrow 4\ ^2D_{3/2}$	Ne, He	CWPC [56]
Mg II	2404.15	–	Ne, He	PC [58]
Mg II	2412.45	–	Ne, He	PC [58]
Al (aluminium, Z = 13)				
Al II	358.744	$4s\ ^3F^o_2 \rightarrow 4d\ ^3D_1$	Ne	QCWHC [59, 60]
Al III	569.6	$4p\ ^2P^o_{3/2} \rightarrow 4s\ ^2S_{1/2}$	He	SPE [61]
Al III	572.2	$4p\ ^2P^o_{1/2} \rightarrow 4s\ ^2S_{1/2}$	He	SPE [61]
Al II	622.8	$4\ ^3D_1 \rightarrow 4\ ^3P^o_0$	He, Ne	PPC [62]
Al II	623.3	$4\ ^3D_2 \rightarrow 4\ ^3P^o_1$	He, Ne	PPC [62]
Al II	624.3	$4\ ^3D_3 \rightarrow 4\ ^3P^o_2$	He, Ne	PPC [62]
Al II	683.0	$5\ ^3S_1 \rightarrow 4\ ^3P^o_2$	He, Ne	PPC [62]
Al II	691.99	$5s\ ^1S_0 \rightarrow 4p\ ^1P^o_1$	Ne	QCWHC [59, 60]
Al II	704.2	$4\ ^3P^o_2 \rightarrow 4\ ^3S_1$	He, Ne	QCWHC [59, 60] SPE [61], LPP [49]
Al II	705.66	$4p\ ^3P^o_1 \rightarrow 4s\ ^3S_1$	Ne	QCWHC [59, 60]
Al II	747.1	$4\ ^1F^o_3 \rightarrow 3\ ^1D_2$	He, Ne	PPC [62], SPE [61], LPP [49], QCWHC [59, 60]
Al II	751.5		He	SPE [61]
Al II	835.4	$5\ ^3F^o_4 \rightarrow 4\ ^3D_3$	He, Ne	PPC [62]
Al II	836.1	$5\ ^3F^o_3 \rightarrow 4\ ^3D_2$	He, Ne	PPC [62]
Al II	836.5	$5\ ^3F^o_2 \rightarrow 4\ ^3D_1$	He, Ne	PPC [62]
Al II	1008	$4\ ^3P^o_2 \rightarrow 3\ ^3D_1$	He	SPE [61]
Al II	1012	$4\ ^3P^o_0 \rightarrow 3\ ^3D_1$	He	SPE [61]
Al II	1063.3	$5\ ^3D_1 \rightarrow 4\ ^3F^o_2$	He, Ne	PPC [62]
Al II	1063.6	$5\ ^3D_2 \rightarrow 4\ ^3F^o_3$	He, Ne	PPC [62]
Al II	1063.9	$5\ ^3D_3 \rightarrow 4\ ^3F^o_4$	He, Ne	PPC [62]
Al II	1246	$5\ ^1D_2 \rightarrow 5\ ^1P^o_2$	He	SPE [61]
Al II	1406	$7\ ^1P^o_1 \rightarrow 6\ ^1S_0$	He	SPE [61]
Si (silicon, Z = 14)				
Si IV	408.89	$4p\ ^2P^o_{3/2} \rightarrow 4s\ ^2S_{1/2}$	Ar	PPC [53]
Si III	455.26	$4p\ ^3P^o_2 \rightarrow 4s\ ^3S_1$	SF_6, PF_6 [e]	PPC [63]
Si III	456.78	$4p\ ^3P^o_1 \rightarrow 4s\ ^3S_1$	SF_6, PF_6	PPC [63]
Si II	634.71	$4p\ ^2P^o_{3/2} \rightarrow 4s\ ^2S_{1/2}$	SF_6, PF_6	PPC [63]
Si II	637.15	$4p\ ^2P^o_{1/2} \rightarrow 4s\ ^2S_{1/2}$	SF_6, PF_6	PPC [63]
Si II	667.19	$4p\ ^4D_{7/2} \rightarrow 4s\ ^4P^o_{5/2}$	SF_6, PF_6	PPC [63]
P (phosphorus, Z = 15)				
P IV	334.77	$4p\ ^3P^o_2 \rightarrow 4s\ ^3S_1$	no buffer [b]	PPC [63]
P III	442.2	$4p\ ^2P^o_{3/2} \rightarrow 4s\ ^2S_{1/2}$	no buffer [b]	PPC [63]

Table A4.1 (*continued*)

Spectrum	λ(nm)	Transition	Buffer Gas	Regimes, types of discharge, references
P II	602.42	4p $^3D_2 \rightarrow$ 4s $^3P^o_1$	no buffer [b] He, Ne	PPC [64], CWPC [65]
P II	603.42	4p $^3D_1 \rightarrow$ 4s $^3P^o_0$	no buffer [b] He,Ne	PPC [63], CWPC [65]
P II	604.325	4p $^3D_3 \rightarrow$ 4s $^3P^o_2$	no buffer [b] He, Ne	PPC [64], CWPC [65]
P II	608.79	4p $^3D_1 \rightarrow$ 4s $^3P^o_1$	no buffer	PPC [63]
P II	616.58	4p $^3D_2 \rightarrow$ 4s $^3P^o_2$	no buffer	PPC [63]
P II	784.56	4p $^1P_1 \rightarrow$ 4s $^1P^o_1$	no buffer	PPC [64]
S (sulphur, Z = 16)				
S III	332.487	4p $^3P_2 \rightarrow$ 3d $^3P^o_2$	He, Ne, Ar, air	PPC [66]
S III	349.734	–	He, Ne, Ar, air	PPC [66]
S III	363.202	4p $^3D_3 \rightarrow$ 3d $^3P^o_2$	He, Ne, Ar, air	PPC [66]
S III	370.937	4p $^3D_2 \rightarrow$ 3d $^3P^o_1$	He, Ne, Ar, air	PPC [66]
S II	532.07	4p′ $^2F^o_{7/2} \rightarrow$ 4s′ $^2D_{5/2}$	He, Ne, Ar, air	PPC [64]
S II	534.57	4p′ $^2F^o_{5/2} \rightarrow$ 4s′ $^2D_{3/2}$	He, Ne, Ar, air	PPC [64]
S II	542.87	4p $^4D_{3/2} \rightarrow$ 4s′ $^4P_{3/2}$	He, Ne, Ar, air	PPC [64]
S II	543.28	4p $^4D_{7/2} \rightarrow$ 4s $^4P_{5/2}$	He, Ne, Ar, air	PPC [64]
S II	545.38	4p $^4D^o_{1/2} \rightarrow$ 4s $^4P_{1/2}$	He, Ne, Ar, air	PPC [64]
S II	547.36	4p $^2D_{5/2} \rightarrow$ 4s $^4P_{3/2}$	He, Ne, Ar, air	PPC [64]
S II	564.0	4p $^2D^o_{3/2} \rightarrow$ 4s $^2P_{1/2}$	He, Ne, Ar, air	PPC [64]
S II	564.70	4p $^2D^o_{3/2} \rightarrow$ 4s $^2P_{1/2}$	He, Ne, Ar, air	PPC [64]
Ca (calcium, Z = 20)				
Ca II	370.6	5s $^2S_{1/2} \rightarrow$ 4p $^2P^o_{1/2}$	He, He+H_2, Ne+H_2	PPC [13]
Ca II	373.7	5s $^2S_{1/2} \rightarrow$ 4p $^2P^o_{3/2}$	He, He+H_2, Ne+H_2	PPC [13], PTD [67]
Ca II	854.2	4p $^2P^o_{3/2} \rightarrow$ 3d $^2D_{5/2}$	He, Ne	PPC [68]
Ca II	866.2	4p $^2P^o_{1/2} \rightarrow$ 3d $^2D_{3/2}$	He, Ne	PPC [68]
Ca II	994.0	6s $^2S_{1/2} \rightarrow$ 5p $^2P^o_{3/2}$	H_2, He, Ne	PPC [13]
Ca II	1183.9	5p $^2P^o_{3/2} \rightarrow$ 5s $^2S_{1/2}$	Kr	PHC [16], SPE [61]
Ca II	1195.0	5p $^2P^o_{1/2} \rightarrow$ 5s $^2S_{1/2}$	Kr	PHC [16], SPE [61]
Cr (chromium, Z = 24)				
Cr ?	844.7	–	He	QCWHC [69]
Cr ?	859.4	–	He	QCWHC [69]
Cr ?	862.94	–	He	QCWHC [69]
Cr II	868.4	$z\ ^4D^o_{3/2} \rightarrow b\ ^2D_{3/2}$	He	QCWHC [69]
Cr ?	870.36	–	He	QCWHC [69]
Cr ?	871.18	–	He	QCWHC [69]
Ni (nickel, Z = 28)				
Ni II	796.2	–	–	QCWHC [70]

(*continued overleaf*)

Table A4.1 (*continued*)

Spectrum	λ(nm)	Transition	Buffer Gas	Regimes, types of discharge, references
Ni II	797.5	–	–	QCWHC [70]
Cu (copper, Z = 29) [c]				
Cu II	248.58	$3d^9\ 5s\ ^3D_1 \rightarrow 3d^94p\ ^3F^o_2$	Ne	QCWHC [71]
Cu II	250.63	$5s\ ^3D_2 \rightarrow {}^3F^o_3$	Ne	QCWHC [71]
Cu II	252.93	$5s\ ^1D_2 \rightarrow {}^3D^o_3$	Ne	QCWHC [72]
Cu II	257.18	$5s\ ^1D_2 \rightarrow {}^3D^o_2$	Ne	QCWHC [73]
Cu II	259.05	$5s\ ^3D_1 \rightarrow {}^3D^o_2$	Ne	CWHC [71]
Cu II	259.88	$5s\ ^3D_2 \rightarrow {}^3F^o_2$	Ne	CWHC [71]
Cu II	260.03	$5s\ ^1D_2 \rightarrow {}^1F^o_3$	Ne	CWHC [72]
Cu II	270.32	$5s\ ^3D_1 \rightarrow {}^3D^o_1$	Ne	CWHC [72]
Cu II	271.88	$5s\ ^1D_2 \rightarrow {}^1P^o_1$	Ne	QCWHC [73]
Cu II	272.2	$5s\ ^3D_1 \rightarrow 4p\ ^1D^o_2$	Ne	QCWHC [74]
Cu II	274.0	$5s\ ^3D_1 \rightarrow 4p\ ^1P^o_1$	Ne	QCWHC [73]
Cu II	450.60	$4s^2\ ^3P_1 \rightarrow 4p\ ^3P^o_2$	He (+Ar)	QCWHC [75]
Cu II	455.59	$4s^2\ ^3P_2 \rightarrow {}^3P^o_2$	He (+Ar)	QCWHC [75]
Cu II	467.35	$4f\ ^3P^o_0 \rightarrow 4d\ ^3S_1$	He (+Ar)	QCWHC [75], TRFD [76]
Cu II	468.20	$4f\ ^3P^o_1 \rightarrow 4d\ ^3S_1$	He (+Ar)	QCWHC [75]
Cu II	485.49	$4f\ ^3D^o_5 \rightarrow 4d\ ^3G_5$	He (+Ar)	QCWHC [75]
Cu II	490.97	$4f\ ^3H^o_6 \rightarrow 4f\ ^3G_5$	He (+Ar)	QCWHC [75]
Cu II	493.16	$4f\ ^3H^o_5 \rightarrow 4d\ ^3D_3$	He (+Ar)	QCWHC [75]
Cu II	501.26	$4f\ ^3G^o_4 \rightarrow 4d\ ^3F_3$	He (+Ar)	QCWHC [75]
Cu II	502.129	$4f\ ^3D^o_3 \rightarrow 4d\ ^3D_3$	He (+Ar)	QCWHC [75]
Cu II	505.178	$4f\ ^3G^o_5 \rightarrow 4d\ ^3F_4$	He (+Ar)	QCWHC [75], TRFD [76]
Cu II	506.06	$4s^2\ ^3P_1 \rightarrow 4p\ ^3P^o_0$	He (+Ar)	QCWHC [75]
Cu II	627.66	$6s\ ^3D_3 \rightarrow 4p''\ ^3F^o_4$	He	QCWHC [73]
Cu II	725.58	$6s\ ^1D_2 \rightarrow 5p\ ^3D^o_2$	Ne	QCWHC [77]
Cu II	739.99	$5p\ ^3D_3 \rightarrow 5s\ ^3D_3$	Ne	QCWHC [77]
Cu II	740.43	$6s\ ^3D_3 \rightarrow 5p\ ^3P^o_2$	He (+Ar)	CWHC [75]
Cu II	743.82	$6s\ ^3D_1 \rightarrow 5p\ ^3P^o_1$	He	QCWHC [77]
Cu II	766.47	$6s\ ^3D_2 \rightarrow 5p\ ^3F^o_3$	He (+Ar)	CWHC [75]
Cu II	773.868	$6s\ ^3D_2 \rightarrow 5p\ ^3F^o_2$	He (+Ar)	CWHC [75]
Cu II	777.874	$6s\ ^3D_2 \rightarrow 4p''\ ^1D^o_2$	He (+Ar)	CWHC [75]
Cu II	780.52	$6s\ ^3D_2 \rightarrow 5p\ ^3P^o_1$	He (+Ar)	CWHC [75]
Cu II	780.77	$6s\ ^3D_3 \rightarrow 5p\ ^3F^o_4$	He (+Ar)	CWHC [75, 78], PHC [79], SPE [61], TRFD [76]
Cu II	782.57	$5p\ ^3F^o_4 \rightarrow 5s\ ^3D_3$	He (+Ar)	CWHC [75]
Cu II	784.50	$6s\ ^1D_2 \rightarrow 5p\ ^1F^o_3$	He (+Ar)	CWHC [75]
Cu II	789.58	$5p\ ^3F^o_2 \rightarrow 5s\ ^3D_1$	He (+Ar)	CWHC [75]
Cu II	790.25	$5p\ ^1F^o_3 \rightarrow 5s\ ^1D_3$	He (+Ar)	CWHC [75]
Cu II	794.44	$5p\ ^3P^o_1 \rightarrow 5s\ ^3D_2$	He (+Ar)	CWHC [75]
Cu II	798.81	$6s\ ^3D_3 \rightarrow 4p''\ ^1F^o_3$	He (+Ar)	CWHC [75]

Table A4.1 (*continued*)

Spectrum	λ(nm)	Transition	Buffer Gas	Regimes, types of discharge, references
Cu II	808.85	6s $^3D_1 \rightarrow$ 5p $^3D^o{}_1$	He	CWHC [80]
Cu II	809.6	–	He	QCWHC [80]
Cu II	819.22	6s $^3D_2 \rightarrow$ 5p $^3D_3$ (s $^1D_2 \rightarrow$ 5p $^1D^o{}_2$)	He	CWHC [80]
Cu II	827.7	5p $^3P^o{}_2 \rightarrow$ 5s $^3D_3$	He	QCWHC [81]
Cu II	828.321	6s $^3D_3 \rightarrow$ 5p $^3D_3$	He	CWHC [80]
Cu II	851.10	6s $^3D_2 \rightarrow$ 5p $^3D_2$	He	QCWHC [80]
Cu II	1743.8	–	He	QCWHC [81]
Cu II	1770.8	5p $^3P^o{}_0 \rightarrow$ 4d $^3P^o$	He	QCWHC [81]
Cu II	1787	–	He	SPE [61]
Cu II	1800.4	–	He	QCWHC [81]
Cu II	1844.8	–	He	QCWHC [81]
Cu II	1915.4	5p $^3F^o{}_4 \rightarrow$ 4d $^3G_5$	He, Ne	QCWHC [81]
Cu II	1932.8	–	He	QCWHC [81]
Cu II	1948.0	5p $^1F^o{}_3 \rightarrow$ 4d $^1G_4$	Ne	QCWHC [81]
Cu II	1971.0	5p $^1D^o{}_2 \rightarrow$ 4d $^1F_3$	He	QCWHC [81]
Cu II	2000.6	5p $^3F_2 \rightarrow$ 4d $^3G_3$	Ne	QCWHC [81]
Zn (zinc, Z = 30)				
Zn II	491.16	4f $^2F^o{}_{5/2} \rightarrow$ 4d $^2D_{3/2}$	He, Ne	PPC [32], PHC [32], QCWHC [82], CWHC [70], TRFD [83], EBP [84]
Zn II	492.40	4f $^2F^o{}_{7/2} \rightarrow$ 4d $^2D_{5/2}$	He, Ne	PPC [85, 32], PHC [32], CWPC [86, 87], QCWHC [82, 88], CWHC [70], TRFD [83], LPP [49], SPE [57]
Zn II	589.44	4s^2 $^2D^*{}_{3/2} \rightarrow$ 4p $^2P^o{}_{1/2}$	He, Ne [b] Ar, no buffer [b]	PPC [82], PHC [32] CWPC [65, 82, 86, 87], CWHC [89], TRFD [83]
Zn II	602.13	5d $^2D_{3/2} \rightarrow$ 5p $^2P^o{}_{1/2}$	He	PPC [32], PHC [32], TRFD [90]
Zn II	610.25	5d $^2D_{5/2} \rightarrow$ 5p $^2P^o{}_{3/2}$	He	PPC [32, 91], PHC [32], CWHC [70], TRFD [90], EBP [84]
Zn II	621.4	4s^2 $^2D^*{}_{3/2} \rightarrow$ 4p $^2P^o{}_{3/2}$	He	CWHC [92]
Zn II	747.88	4s^2 $^2D^*{}_{5/2} \rightarrow$ 4p $^2P^o{}_{3/2}$	He, Ne Ar, no buffer [b]	PPC [32, 91], PHC [32], CWPC [65, 82, 86, 93], CWHC [94], TRFD [90], PP [95]

(continued overleaf)

Table A4.1 (*continued*)

Spectrum	λ(nm)	Transition	Buffer Gas	Regimes, types of discharge, references
Zn II	758.85	5p $^2P^o_{3/2} \rightarrow$ 5s $^2S_{1/2}$	He	PPC [32, 91], PHC [32], QCWWHC [82, 88], CWHC [94], CWPC [65, 86, 93], EBP [84], TRFD [90]
Zn II	761.29	6s $^2S_{1/2} \rightarrow$ 5p $^2P^o_{1/2}$	He	PPC [91], PHC [32]
Zn II	773.25	5p $^2P^o_{1/2} \rightarrow$ 5s $^2S_{1/2}$	He	PHC [32]
Zn II	775.79	6s $^2S_{1/2} \rightarrow$ 5p $^2P^o_{3/2}$	He	PPC [85], PHC [32], SPE [57]
Zn II	947.9	–	He+Xe	HC [96]
Zn II	1104	–	He	LPP [55]
Zn II	1476	–	He	LPP [49, 55]
Zn II	1685	7s $^2S_{1/2} \rightarrow$ 6p $^2P^o_{3/2}$	He	SPE [57]
Zn II	1830.8	6p $^2P^o_{3/2} \rightarrow$ 6s $^2S_{1/2}$	He	PTD [97]
Zn II	2149.50	5p $^2P^o_{3/2} \rightarrow$ 4d $^2D_{5/2}$	He	PHC [98]
Zn II	2243.50	5p $^2P^o_{1/2} \rightarrow$ 4d $^2D_{3/2}$	He	PHC [98]
Zn II	5084.8	6p $^2P^o_{3/2} \rightarrow$ 5d $^2D_{5/2}$	He	PTD [97]
Ga (gallium, Z = 31)				
Ga II	633.4	5p $^3P_2 \rightarrow$ 5s $^3S_1$	Ne	PPC [99], PHC [79, 100]
Ga II	719.8	5p $^1P_1 \rightarrow$ 5s $^1S_0$	Ne	PPC [99], PHC [79, 100]
Ge (germanium, Z = 32)				
Ge II	513.17	4f $^2F^o_{5/2} \rightarrow$ 4d $^2D_{3/2}$	He, Ne	PPC [91], QCWHC [101]
Ge II	517.9	4f $^2F^o_{7/2} \rightarrow$ 4d $^2D_{5/2}$	He, Ne	PPC [91], QCWHC [101]
Ge II	589.3	5p $^2P^o_{3/2} \rightarrow$ 5s $^2S_{1/2}$	He, Ar	QCWHC [101]
As (arsenic, Z = 33)				
As II	538.52	6s $^1P_1 \rightarrow$ 5p $^3P_1$	He	CWHC [102]
As II	549.69	6s $^3P_2 \rightarrow$ 5p $^3D_3$	He	CWHC [102]
As II	549.8	5p $^3D_1 \rightarrow$ 5s $^3P^o_0$	Ne, He	PPC [103], CWPC [65], CWHC [104]
As II	555.81	5p $^3D_2 \rightarrow$ 5s $^3P^o_1$	Ne, He	PPC [103], CWPC [65], CWHC [102]
As II	565.2	5p $^3D_3 \rightarrow$ 5s $^3P^o_2$	Ne	PPC [103], CWPC [65]
As II	583.79	6s $^3P_2 \rightarrow$ 5p $^3D_2$	He	CWHC [102]
As II	617.03	5p $^1P_1 \rightarrow$ 5s $^3P^o_1$	Ne, He	PPC [103], CWPC [65], CWHC [102]
As II	651.17	6s $^1P_1 \rightarrow$ 5p $^1D_2$	He	CWHC [102], EBP [84]
As II	710.27	5p $^3P_1 \rightarrow$ 4p $^{3\,3}P^o_2$	He	CWHC [102]
Se (selenium, Z = 34)				
Se II	446.760	5p $^2P_{1/2} \rightarrow$ 5s $^2P_{1/2}$	He	CWPC [105]

Table A4.1 (*continued*)

Spectrum	λ(nm)	Transition	Buffer Gas	Regimes, types of discharge, references
Se II	460.434	5p $^2D_{5/2} \rightarrow$ 5s $^4P_{1/2}$	He	CWPC [106], CWHC [94]
Se II	461.877	5p $^4P_{5/2} \rightarrow$ 5s $^4P_{3/2}$	He	CWPC [105],
Se II	464.844	5p $^4P_{3/2} \rightarrow$ 5s $^4P_{1/2}$	He	CWPC [106]
Se II	471.823	5p $^4S_{3/2} \rightarrow$ 3	He	CWPC [105]
Se II	474.097	5p $^2P_{3/2} \rightarrow$ 4sp^4 $^2P_{3/2}$	He	CWPC [105]
Se II	476.365	5p $^2D_{3/2} \rightarrow$ 5s $^4P_{3/2}$	He	CWPC [106]
Se II	476.552	5p $^2P_{1/2} \rightarrow$ 5s $^2P_{3/2}$	He	CW PC [105]
Se II	484.063	5p $^2S_{1/2} \rightarrow$ 5s $^4P_{3/2}$	He	CWPC [106]
Se II	484.496	5p $^4S_{3/2} \rightarrow$ 5s $^4P_{5/2}$	He	CWPC [65, 106], CWHC [94]
Se II	497.566	5p $^2D_{5/2} \rightarrow$ 4s4p^4 $^2P_{3/2}$	He	CWPC [65, 106], CWHC [94], EBP [84]
Se II	499.275	5p $^4P_{3/2} \rightarrow$ 5s $4P_{3/2}$	He	CWPC [65, 106], CWHC [94], EBP [84]
Se II	506.865	5p $^4P_{5/2} \rightarrow$ 5s $^4P_{5/2}$	He	CWPC [65, 106], CWHC [94], TRFD [83], EBP [84]
Se II	509.650	5p $^4D_{7/2} \rightarrow$ 4d $^4P_{9/2}$	He	PPC [103], CWPC [65, 107]
Se II	514.214	5p $^4D_{3/2} \rightarrow$ 5s $^4P_{1/2}$	He	CWPC [65, 106], EBP [84]
Se II	517.598	5p $^4D_{5/2} \rightarrow$ 5s $^4P_{3/2}$	He	PPC [108], CWPC [65, 107], CWHC [94], EBP [84], TRFD [83]
Se II	522.751	5p $^4D_{5/2} \rightarrow$ 5s $^4P_{3/2}$	He	PPC [103], CWPC [65, 106], TRFD [83], CWHC [94], EBP [84]
Se II	525.307	5p $^2D_{5/2} \rightarrow$ 5s $^2P_{3/2}$	He	PPC [108], CWPC [65, 106], CWHC [94], EBP [84]
Se II	525.363	5p $^4D_{3/2} \rightarrow$ 5s $^4P_{1/2}$	He	PPC [108], CWPC [65, 106], CWHC [94], EBP [84]
Se II	527.111	5p $^4D_{5/2} \rightarrow$ 4d $^4F_{7/2}$	He	PPC [108], CWPC [65, 106]
Se II	530.535	5p $^2D_{3/2} \rightarrow$ 5s $^2P_{1/2}$	He	PPC [108], CWPC [65, 106], EBP [84], TRFD [83]
Se II	552.242	5p $^4P_{3/2} \rightarrow$ 5s $^4P_{5/2}$ (or 5p $^4P_{5/2} \rightarrow$ 4s4p^4 $^2P_{3/2}$)	He	CWPC [65, 106], CWHC [94], EBP [84]

(continued overleaf)

Table A4.1 (*continued*)

Spectrum	λ(nm)	Transition	Buffer Gas	Regimes, types of discharge, references
Se II	556.693	5p $^4D_{3/2} \rightarrow$ 5s $^4P_{3/2}$	He	CWPC [65, 105]
Se II	559.116	5p $^4P_{3/2} \rightarrow$ 5s $^2P_{1/2}$	He	CWPC [106], EBP [84]
Se II	562.313	5p $^4P_{1/2} \rightarrow$ 5s $^4P_{1/2}$	He	CWPC [105]
Se II	569.788	5p $^4D_{1/2} \rightarrow$ 5s $^4P_{3/2}$	He	CWPC [106]
Se II	574.762	5p $^4D_{5/2} \rightarrow$ 5s $^4P_{5/2}$	He	CWPC [106]
Se II	584.268	5p $^2S_{1/2} \rightarrow$ 5s $^2P_{3/2}$	He	CWPC [105]
Se II	586.627	5p $^4P_{5/2} \rightarrow$ 5s $^2P_{3/2}$	He	CWPC [105]
Se II	605.2		He	CWPC [65]
Se II	605.596	5p $^2P_{3/2} \rightarrow 7$	He	CWPC [65, 106], CWHC [94], EBP [84]
Se II	606.583	5p $^4P_{3/2} \rightarrow$ 5s $^2P_{3/2}$	He	CWPC [105]
Se II	610.196	5p $^2D_{3/2} \rightarrow$ 5s $^2P_{3/2}$	He	CWPC [105]
Se II	644.0	–	He	CWPC [65]
Se II	644.425	5p $^2P_{1/2} \rightarrow 7$	He	CWPC [65, 106], CWHC [94], EBP [84]
Se II	648.8	–	He	CWPC [65]
Se II	649.048	5p $^4D_{1/2} \rightarrow$ 5s $^2P_{1/2}$	He	CWPC [65, 106], CWHC [94], EBP [84]
Se II	653.495	5p $^2P_{1/2} \rightarrow 9$	He	CWPC [65, 106]
Se II	706.389	5p $^4P_{1/2} \rightarrow$ 5s $^2P_{1/2}$	He	CWPC [105]
Se II	739.199	5p $^4P_{5/2} \rightarrow 7$	He	CWPC [105]
Se II	767.482	5p $^2P3/2 \rightarrow$ 5s′ $^2D_{5/2}$	He	CWPC [105]
Se II	772.404	5p $^4D_{1/2} \rightarrow$ 5s $^2P_{3/2}$	He	CWPC [105]
Se II	779.615	5p $^2P_{1/2} \rightarrow$ 5s′ $^2D_{3/2}$	He	CWPC [105]
Se II	783.881	5p $^4P_{1/2} \rightarrow$ 4s4p^4 $^2P_{3/2}$	He	CWPC [105]
Se II	830.952	5p $^2D_{3/2} \rightarrow$ 5s′ $^2D_{5/2}$	He	CWPC [105], EBP [84]
Se II	924.93		He	CWPC [105]
Se II	995.515	5p $^4P_{5/2} \rightarrow$ 5s′$^2D_{5/2}$	He	CWPC [105]
Se II	1040.881	5p $^4D_{1/2} \rightarrow 6$	He	CWPC [105]
Se II	1258.678	5p $^4P_{3/2} \rightarrow 10$	He	CWPC [105]
Br (bromine, Z = 35)				
Br II	474.266	5p″ $^3D_3 \rightarrow$ 5s″ $^3P^o{}_2$	no buffer [b]	PPC [109]
Br II	494.14	5p′ $^3F_2 \rightarrow$ 5s′ $^3D^o{}_3$	HBr [a]	PPC [110]
Br II	505.463	5p′ $^3F_3 \rightarrow$ 4d $^3D^o{}_2$	no buffer	PPC [109]
Br II	518.238	5p $^3P_2 \rightarrow$ 5s $^3S^o{}_1$	no buffer	PPC [109]
Br II	523.826	5p $^3P_1 \rightarrow$ 5s $^3S^o{}_1$	no buffer	PPC [109]
Br II	533.20	5p′$^1F_3 \rightarrow$ 5s$^1D^o{}_2$	no buffer	PPC [109]
Br II	583.07	5p′$^3D_1 \rightarrow$ 4d$^3D^o{}_2$	HBr [a]	PPC [110]
Br II	611.756	5p $^5P_2 \rightarrow$ 5s $^3S^o{}_1$	no buffer	PPC [109]
Br II	616.878		no buffer	PPC [109]
Sr (strontium, Z = 38)				
Sr II	416.2	6s $^2S_{1/2} \rightarrow$ 5p $^2P^o{}_{1/2}$	He + H_2	PPC [111, 13]

Table A4.1 (*continued*)

Spectrum	λ(nm)	Transition	Buffer Gas	Regimes, types of discharge, references
Sr II	430.5	6s $^2S_{1/2} \rightarrow$ 5p $^2P^o{}_{3/2}$	He, He+H_2	PPC [111, 13], PTD [112, 113, 114], PHC [115], EBP [116, 115], MWP [117]
Sr II	870	6d $^2D_{5/2} \rightarrow$ 6p $^2P^o{}_{3/2}$	He, Ne, H_2	PPC [111, 13, 118]
Sr II	1033.0	5p $^2P^o{}_{3/2} \rightarrow$ 4d $^2D_{5/2}$	He	PPC [119], MWP [98]
Sr II	1087.6	7s $^2S_{1/2} \rightarrow$ 6p $^2P^o{}_{1/2}$	Kr, H_2	PPC [62, 118], PHC [16], CWHC [120]
Sr II	1091.8	5p $^2P^o{}_{1/2} \rightarrow$ 4d $^2D_{3/2}$	He	PPC [119], MWP [98]
Sr II	1123.0	7s $^2S_{1/2} \rightarrow$ 6p $^2P^o{}_{3/2}$	He, Ne, Kr, H_2	PPC [13, 62, 118], PHC [16], CWHC [120]
Sr II	1201.7	6p $^2P^o{}_{3/2} \rightarrow$ 6s $^2S_{1/2}$	Kr	PPC [62], PHC [16], CWHC [120]
Sr II	1245.0	6p $^2P^o{}_{1/2} \rightarrow$ 6s $^2S_{1/2}$	Kr	PHC [16]
Ag (silver, Z = 47)				
Ag II	224.34	5d $^1S_0 \rightarrow$ 5p $^1P^o{}_1$	He [c]	CWHC [121]
Ag II	227.74	5d $^3D_2 \rightarrow$ 5p $^3P^o{}_1$	He	CWHC [121]
Ag II	318.07	5s^2 $^1G_4 \rightarrow$ 5p $^3F^o{}_3$	Ne	CWHC [121]
Ag II	408.59	5s^2 $^1G_4 \rightarrow$ 5p $1F_3$	Ne	CWHC [122]
Ag II	462.1	5s $^1D_2 \rightarrow$ 5p $^1F^o{}_3$	Ne	CWHC [73]
Ag II	478.84	5s^2 $^1D_2 \rightarrow$ 5p $^1P^o{}_1$	Ne	CWHC [122]
Ag II	502.73	5s^2 $^1D_2 \rightarrow$ 5p $^1D^o{}_2$	Ne	QCWHC [122]
Ag II	640.27	6p $^3F^o{}_2 \rightarrow$ 6s $^3D_2$	He	QCWHC [123]
Ag II	800.54	6p $^3D^o{}_3 \rightarrow$ 6s $^3D_3$	He	PPC [26], QCWHC [123]
Ag II	825.47	6p $^3D^o{}_3 \rightarrow$ 6s $^3D_2$	He	CWHC [123]
Ag II	826.3	6p $^3D^o{}_1 \rightarrow$ 6s $^3D_1$	He	QCWHC [81]
Ag II	832.44	6p $^3D^o{}_2 \rightarrow$ 6s $^3D_2$	He	CWHC [123]
Ag II	837.95	6p $^3P^o{}_1 \rightarrow$ 6s $^1D_2$	He	CWHC [80]
Ag II	838.0	6p $^3P^o{}_1 \rightarrow$ 6s $^3D_2$	He	QCWHC [81]
Ag II	840.32	6p $^3F^o{}_4 \rightarrow$ 6s $^3D_3$	He	PPC [80], CWHC [123]
Ag II	874.76	6p $^1P^o{}_1 \rightarrow$ 6s $^1D_2$	He	CWHC [80]
Ag II	877.2	6p $^3F^o{}_2 \rightarrow$ 6s $^3D_1$	He	CWHC [81]
Ag II	1375.9	6p $^3P^o{}_1 \rightarrow$ 5d $^3S_1$	He	QCWHC [81]
Ag II	1598.0	$^3D^o{}_3 \rightarrow {}^3P_2$	He	QCWHC [81]
Ag II	1646.2	$^3P^o{}_1 \rightarrow {}^3P_1$	He	QCWHC [81], SPE [57]
Ag II	1665.6		He	QCWHC [81]
Ag II	1720.3	6p $^3D^o{}_3 \rightarrow$ 5d $^3D_3$	He	CWHC [81], SPE [57]
Ag II	1734.5	6p $^3F^o{}_4 \rightarrow {}^3G_5$	He	CWHC [81], SPE [57], LPP [49, 55]
Ag II	1747.8	6p $^3D_1 \rightarrow {}^3D_1$	He	QCWHC [81]

(continued overleaf)

Table A4.1 (*continued*)

Spectrum	λ(nm)	Transition	Buffer Gas	Regimes, types of discharge, references
Ag II	1766.4		He	QCWHC [81]
Ag II	1840.8	6p $^3D_2 \rightarrow$ 5d $^3F_3$	He	CWHC [81]
Ag II	1846.3	6p $^3D^o_3 \rightarrow {}^3F_4$	He	CWHC [81], LPP [49, 55]
Ag II	1872.5	6p $^3F_3 \rightarrow {}^3G_4$	He	QCWHC [81]
Ag II	1879.5	6p $^3P_1 \rightarrow {}^3D_2$	He	CWHC [81]
Ag II	1897.9	6p $^3F^o_2 \rightarrow {}^3G_3$	He	CWHC [81]
Ag II	1971.4	$^3D_1 \rightarrow {}^3F_2$	He	QCWHC [81]
Ag II	1982.3	$^1P^o_1 \rightarrow {}^1D_2$	He	QCWHC [81]
Ag II	2079.6	$^3F_3 \rightarrow {}^3D_3$	He	QCWHC [81]
Cd (cadmium, Z = 48)				
Cd II	325.0	5s^2 $^2D^*_{3/2} \rightarrow$ 5p $^2P_{1/2}$	He	PHC [32], CWPC [82, 124], CWHC [94], TRFD [83], PP [125]
Cd II	441.563	5s^2 $^2D^*_{5/2} \rightarrow$ 5p $^2P_{3/2}$	He, Ne, Ar, no buffer [b]	PPC [32, 126, 91], PHC [32], CWPC [124, 127, 82, 128], CWHC [89], LRFD [129], TRFD [83], PTD [130], PP [125], NP [131], EBP [84]
Cd II	488.2	5p′ $^4F^o_{5/2} \rightarrow$ 5d $^2D_{3/2}$	He	CWPC [87]
Cd II	502.59	5p′ $^4F^o_{7/2} \rightarrow$ 5d $^2D_{5/2}$	He	CWPC [87]
Cd II	533.749	4f $^2F^o_{5/2} \rightarrow$ 5d $^2D_{3/2}$	He, Ne, no buffer	PPC [32, 126, 85] PHC [32], QCWHC [82], CWHC [132, 89], TRFD [83], SPE [133], PTD [130], NP [134], EBP [84]
Cd II	537.804	4f $^2F^o_{7/2} \rightarrow$ 5d $^2D_{5/2}$	He, Ne, no buffer	PPC [32, 126, 85] PHC [32], QCWHC [88, 82], CWHC [132, 89], TRFD [83], LPP [49, 135], SPE [133], PTD [130], NP [134], EBP [84]
Cd II	635.48	6g $^2G_{7/2} \rightarrow$ 4f $^2F^o_{5/2}$	He	CWHC [132], TRFD [83], EBP [84]
Cd II	636.01	6g $^2G_{9/2} \rightarrow$ 4f $^2F^o_{7/2}$	He	PHC [32], CWHC [132], TRFD [83], EBP [84]
Cd II	646.5	6d $^2D_{3/2} \rightarrow$ 6p $^2P^o_{1/2}$	He	TRFD [136]
Cd II	672.6	6d $^2D_{5/2} \rightarrow$ 6p $^2P^o_{3/2}$	He	TRFD [136]

Table A4.1 (*continued*)

Spectrum	λ(nm)	Transition	Buffer Gas	Regimes, types of discharge, references
Cd II	723.69	6f $^2F^o{}_{5/2} \rightarrow$ 6d $^2D_{3/2}$	He	PHC [32], CWHC [132]
Cd II	728.4	6f $^2F^o{}_{7/2} \rightarrow$ 6d $^2D_{5/2}$	He	PHC [32], CWHC [132]
Cd II	806.7	6p $^2P^o{}_{3/2} \rightarrow$ 6s $^2S_{1/2}$	He	PHC [32], QCWHC [82, 88], CWHC [94], CWPC [65], TRFD [83], LPP [49], SPE [133], EBP [84]
Cd II	838.78	7s $^2S_{1/2} \rightarrow$ 6p $^2P^o{}_{3/2}$	He	PPC [137], PHC [32]
Cd II	853.0	6p $^2P^o{}_{1/2} \rightarrow$ 6s $^2S_{1/2}$	He	PHC [32], CWHC [94], SPE [133]
Cd II	865.2	9s $^2S_{1/2} \rightarrow$ 5p′ $4P_{1/2}$	He	QCWHC [74]
Cd II	887.8	9s $^2S_{1/2} \rightarrow$ 7p $^2P^o{}_{3/2}$	He	CWHC [94]
Cd II	967.15	5g $^2G_{9/2} \rightarrow$ 4f $^2F^o{}_{7/2}$	He	PHC [32]
Cd II	1039	–	He	LPP [55]
Cd II	1194.5	5f $^2F^o{}_{5/2} \rightarrow$ 6d $^2D_{3/2}$	He	EBP [84]
Cd II	1219.9	5f $^2F^o{}_{7/2} \rightarrow$ 6d $^2D_{5/2}$	He	SPE [133], EBP [84]
Cd II	1425	–	He	LPP [55]
Cd II	1752.26	6g $^2G_{9/2} \rightarrow$ 5f $^2F^o{}_{7/2}$	He	EBP [84]
Cd II	1783.96	6g $^2G_{7/2} \rightarrow$ 5f $^2F^o{}_{5/2}$	He	EBP [84]
Cd II	1805.1	6p $^2P^o{}_{3/2} \rightarrow$ 5d $^2D_{5/2}$	He	PHC [32], SPE [133], EBP [84]
Cd II	1926.96	7p $^2P^o{}_{3/2} \rightarrow$ 7s $^2S_{1/2}$	He	EBP [84]
Cd II	1991.6	6p $^2P^o{}_{1/2} \rightarrow$ 5d $^2D_{3/2}$	He	HC [32], EBP [84]
Cd II	3288.2	9s $^2S_{1/2} \rightarrow$ 8p $^2P^o{}_{3/2}$	He	TD [97]
In (indium, Z = 49)				
In III	298.3	4f $^2F^o{}_{5/2} \rightarrow$ 5d $^2D_{3/2}$	He	SPE [133]
In III	300.8	4f $^2F^o{}_{7/2} \rightarrow$ 5d $^2D_{5/2}$	He	SPE [133]
In II	468.0	4f $^3F^o{}_4 \rightarrow$ 5d $^3D_3$	He, Ne	PPC [91], SPE [133]
In III	524.9	6p $^2P^o{}_{3/2} \rightarrow$ 6s $^2S_{1/2}$	He	SPE [133]
In III	582.0	6p $^2P^o{}_{3/2} \rightarrow$ 5d $^2D_{5/2}$	He	SPE [133]
In II	689.1	6p $^3P^o{}_2 \rightarrow$ 6s $^3S_1$	He	SPE [133]
In III	704.9	5f $^2F^o{}_{7/2} \rightarrow$ 6d $^2D_{5/2}$	He	SPE [133]
In II	718.2	6p $^3P^o{}_1 \rightarrow$ 6s $^3S_1$	He	SPE [133]
In II	727.5	6p $^3P^o{}_0 \rightarrow$ 6s $^3S_1$	He	SPE [133]
In II	735.3	7s $^3S_1 \rightarrow$ 6p $^3P^o{}_1$	He	SPE [57]
In II	768.5	7s $^3S_1 \rightarrow$ 6p $^3P^o{}_2$	He	SPE [57]
Sn (tin, Z = 50)				
Sn II	558.9	4f $^2F^o{}_{5/2} \rightarrow$ 5d $^2D_{3/2}$	He, Ne	PPC [138]
Sn II	579.9	4f $^2F^o{}_{7/2} \rightarrow$ 5d $^2D_{5/2}$	He, Ne	PPC [138, 91], PHC [139], LPP [49], SPE [57]

(continued overleaf)

Table A4.1 (*continued*)

Spectrum	λ(nm)	Transition	Buffer Gas	Regimes, types of discharge, references
Sn II	645.35	6p $^2P^o{}_{3/2} \rightarrow$ 6s $^2S_{1/2}$	He, Ne	PPC [91], CWHC [139], CWPC [93]
Sn II	684.4	6p $^2P^o{}_{1/2} \rightarrow$ 6s $^2S_{1/2}$	He, Ne	PPC [91], CWPC [93]
Sn II	719.1	7s $^2S_{1/2} \rightarrow$ 6p $^2P^o{}_{3/2}$	He, Ne	PHC [139]
Sn II	1062	5f $^2F^o{}_{5/2} \rightarrow$ 6d $^2D_{3/2}$	He, Ne	PPC [138], PHC [139], SPE [57], LPP [55]
Sn II	1074	5f $^2F^o{}_{7/2} \rightarrow$ 6d $^2D_{5/2}$	He, Ne	PPC [138], PHC [139], SPE [57], LPP [49, 55]
Sn II	1239	7d $^2D_{5/2} \rightarrow$ 7p $^2P^o{}_{3/2}$	He	SPE [57]
Sn II	1363	–	He	SPE [57]
Sb (antimony, Z = 51)				
Sb II	441.54	7s $^1P^o{}_1 \rightarrow$ 6p $^3D_1$	He, Ne	PPC [140]
Sb II	612.99	6p $^3D_3 \rightarrow$ 6s $^3P^o{}_2$	Ne, He	PPC [141], CWPC [142]
Sb II	699.64	6p $^3D_2 \rightarrow$ 6s $^3P^o{}_1$	He, Ne	PPC [140], CWPC [142]
Te (tellurium, Z = 52)				
Te II	484.29	$122'_{5/2} \rightarrow 102_{3/2}$	He	CWPC [143], QCWHC [101]
Te II	502.04	–	He	CWPC [143], QCWHC [101]
Te II	525.64	–	He	CWPC [143]
Te II	544.98	$103_{3/2} \rightarrow 85_{5/2}$	Ne	CWPC [143], QCWHC [101]
Te II	545.4	–	Ne, He	PPC [144], CWPC [65]
Te II	547.91	$105_{3/2} \rightarrow 86_{3/2}$	Ne	CWPC [143], QCWHC [101]
Te II	548.79	$100_{5/2} \rightarrow 81_{3/2}$	Ne	QCWHC [101]
Te II	557.62	–	Ne	PPC [141]
Te II	557.635	$112_{7/2} \rightarrow 94_{5/2}$	He	PPC [144], CWPC [65]
Te II	564.05	–	Ne, He	PPC [144], CWPC [65]
Te II	564.92	$96_{3/2} \rightarrow 78_{1/2}$	Ne	QCWHC [101]
Te II	566.62	$101_{3/2} \rightarrow 83_{1/2}$	He	CWPC [143]
Te II	570.81	$103_{7/2} \rightarrow 85'_{5/2}$	He, Ne	PPC [141], CWPC [65], QCWHC [101]
Te II	574.16	$112_{3/2} \rightarrow 94_{5/2}$	Ne	CWPC [143]
Te II	575.59	$100_{5/2} \rightarrow 82_{3/2}$	He, Ne	CWPC [65], QCWHC [101]
Te II	576.53	$112_{5/2} \rightarrow 95_{3/2}$	He	CWPC [143]
Te II	585.11	$111_{5/2} \rightarrow 94_{5/2}$	He, Ne	CWPC [143]
Te II	593.61	$99_{3/2} \rightarrow 82_{3/2}$	Ne	PPC [144], CWPC [65], QCWHC [101]
Te II	597.26	$111_{5/2} \rightarrow 95_{3/2}$	Ne, He	CWPC [143]
Te II	597.47	$102_{5/2} \rightarrow 85'_{5/2}$	He	CWPC [65]

Table A4.1 (*continued*)

Spectrum	λ(nm)	Transition	Buffer Gas	Regimes, types of discharge, references
Te II	601.45	$105_{5/2} \rightarrow 88_{3/2}$	Ne	CWPC [143]
Te II	608.23	–	He	CWPC [143]
Te II	623.07	$105_{3/2} \rightarrow 88_{3/2}$	Ne	CWPC [143], QCWHC [101]
Te II	624.54	$99_{3/2} \rightarrow 83_{1/2}$	He, Ne	PPC [144], CWPC [65], QCWHC [101]
Te II	629.36	$97^0{}_{1/2} \rightarrow 81_{3/2}$	Ne	QCWHC [101]
Te II	634.97	–	He	PPC [141], CWPC [65]
Te II	658.51	–	He	CWPC [143]
Te II	664.86	$97_{1/2} \rightarrow 82_{3/2}$	Ne	CWPC [143], QCWHC [101]
Te II	667.61	$103_{3/2} \rightarrow 88_{3/2}$	Ne	CWPC [143]
Te II	688.51	$100_{5/2} \rightarrow 85_{5/2}$	Ne	CWPC [143]
Te II	703.91	$97_{1/2} \rightarrow 83_{1/2}$	Ne	PPC [144], CWPC [65], QCWHC [101]
Te II	780.17	$105_{3/2} \rightarrow 92_{3/2}$	Ne	CWPC [143]
Te II	792.17	$97_{1/2} \rightarrow 85_{3/2}$	Ne	CWPC [143]
Te II	860.46	–	He	CWPC [143]
Te II	873.38	–	He	CWPC [143]
Te II	897.21	$103_{3/2} \rightarrow 92_{5/2}$	Ne	CWPC [143]
Te II	899.82	–	Ne	CWPC [143]
Te II	937.85	$99_{3/2} \rightarrow 88_{5/2}$	Ne	CWPC [143], QCWHC [101]
I (iodine, Z = 53)				
I II	261.45	7p' $^3D_2 \rightarrow$ 5d' $^3F^o{}_3$	HI[a]	PPC [110]
I II	288.22		HI[a]	PPC [110]
I ?	290.63		HI[a]	PPC [110]
I II	293.17	7d $^5D_4 \rightarrow$ 6p $^5P_3$	HI[a]	PPC [110]
I ?	407.08		HI[a]	PPC [110]
I II	448.855	6p' $^3D_1 \rightarrow$ 5d $^3D^o{}_1$	He	QCWHC [145]
I II	453.379		He	PPC [146]
I II	453.40		HI[a]	PPC [110]
I II	467.42		HI[a]	PPC [110]
I II	467.440		He	PPC [146]
I II	467.553	6p' $^1D_2 \rightarrow$ 6s' $^1D^o{}_2$	He	QCWHC [145]
I II	493.467		He	PPC [146]
I II	493.47		HI[a]	PPC [110]
I II	498.692	6p' $^3D_2 \rightarrow$ 5d $^3D^o1$	He	PPC [147], CWHC [148]
I II	521.408	6p' $^3D_2 \rightarrow$ 5d $^3D^o{}_3$	He	QCWHC [145]
I II	521.627	6p' $^3F_2 \rightarrow$ 5d $^3D^o{}_1$	He	PPC [147], CWHC [148]
I II	533.82	6p' $^3D_3 \rightarrow$ 6s' $^3D^o{}_2$	HI	PPC [110]

(*continued overleaf*)

Table A4.1 (*continued*)

Spectrum	λ(nm)	Transition	Buffer Gas	Regimes, types of discharge, references
I II	534.50		HI[a]	PPC [110]
I II	540.736	6p' $^3D_2 \rightarrow$ 6s' $^3D^o_2$	He	PPC [147], CWPC [65], CWHC [149]
I II	541.9	.	He	PPC [147]
I II	559.312	6p' $^3D_1 \rightarrow$ 5d $^3D_1$	He	QCWHC [145]
I II	562.569	6p $^3P_2 \rightarrow$ 6s $^3S^o_1$	Ne	PPC [147]
I II	567.808	6p' $^3F_2 \rightarrow$ 6s' $^3D^o_2$	He	PPC [147], CWPC [65], CWHC [149]
I II	576.072	6p' $^3D_2 \rightarrow$ 6s' $^3D^o_1$	He	PPC [150], CWPC [65], CWHC [149], EBP [84]
I II	595.03	6p $^3P_1 \rightarrow$ 6s $^3S_1$	HI[a]	PPC [110]
I II	606.893	6p' $^3F_2 \rightarrow$ 6s' $^3D_1$	He	PPC [151], QCWHC [145]
I II	612.749	6p' $^3D_1 \rightarrow$ 6s' $^3D^o_2$	He	PPC [150], CWPC [65], CWHC [149], EBP [84]
I II	620.486	6p' $^1P_1 \rightarrow$ 6s' $^1D^o_2$	He	QCWHC [145]
I II	633.997	6p' $^3F_3 \rightarrow$ 6s' $^3D^o_3$	He	CWHC [145]
I II	651.618	$^3F_2 \rightarrow$ 5p^5 $^1P^o_1$	He	PPC [146], CWHC [149]
I II	658.521	$^3D_1 \rightarrow$ 6s' $^3D^o_1$	He	PPC [147], CWPC [65], CWHC [149]
I II	662.235	$^3D_2 \rightarrow$ 5d' $^3D^o_2$	He	QCWHC [145]
I II	682.523	$^3F_2 \rightarrow$ 6s' $^3D^o_3$	He	PPC [151], QCWHC [145]
I II	690.48	6p' $^3D_2 \rightarrow$ 6s' $^1D^o_2$	HI	PPC [110]
I II	690.777	6p' $^3D_2 \rightarrow$ 6s' $^1D^o_2$	Ne, Kr	PPC [147]
I II	703.299	$^3F_2 \rightarrow$ 5d' $^3D^o_2$	He	PPC [147], CWPC [152], CWHC [149]
I II	713.897	$^3D_2 \rightarrow$ 5d' $^3D^o_3$	He	PPC [151], CWHC [149]
I II	761.850	6p' $^3F_2 \rightarrow$ 5d' $^3D^o_3$	He	CWPC [152], CWHC [148]
I II	773.578	$^3D_1 \rightarrow$ 5d' $^3D^o_2$	He	CWHC [148]
I II	817.007	$^3D_2 \rightarrow$ 5d' $^3F^o_3$	He	CWPC [152], CWHC [148]
I II	825.384	$^3D_1 \rightarrow$ 5d' $^3P^o_0$	He	PPC [147], CWHC [149]
I II	880.423	$^3F_2 \rightarrow$ 5d' $^3F^o_3$	He	PPC [147], CWPC [152], CWHC [149]
I II	887.761	$^3F_3 \rightarrow$ 5d' $^3G^o_4$	He	CWHC [148]
I ?	980	–	He	PPC [147]
I ?	1010	–	He	PPC [147]
I ?	1030	–	He	PPC [147]
I ?	1060	–	He	PPC [147]
Ba (barium, Z = 56)				
Ba II	614.17	6p $^2P^o_{3/2} \rightarrow$ 5d$^2D_{5/2}$	He, Ne, Xe	PPC [153], PHC [154]
Ba II	649.69	6p $^2P^o_{1/2} \rightarrow$ 5d$^2D_{3/2}$	He, Ne, Xe	PPC [153], PHC [154]
Ba II	677.1	5g $^2G_{7/2} \rightarrow$ 4s $^2F^o_{5/2}$	He, Ne	PPC [62]
Ba II	687.6	5g $^2G_{9/2} \rightarrow$ 4s $^2F^o_{7/2}$	He, Ne	PPC [62]
Ba II	1109.2	9d $^2D_{5/2} \rightarrow$ 8p $^2P^o_{3/2}$	He, Ne	PPC [62]

Table A4.1 (*continued*)

Spectrum	λ(nm)	Transition	Buffer Gas	Regimes, types of discharge, references
Ba II	1158.1	8s $^2S_{1/2} \rightarrow$ 7p $^2P^o{}_{1/2}$	He, Ne	PPC [62]
Ba II	1193.5	10s $^2S_{1/2} \rightarrow$ 8p $^2P^o{}_{3/2}$	He, Ne	PPC [62]
Ba II	1247.8	8s $^2S_{1/2} \rightarrow$ 7p $^2P^o{}_{3/2}$	He, Ne	PPC [13]
Ba II	2592.4	7p $^2P^o{}_{3/2} \rightarrow$ 6d $^2D_{5/2}$	He, Ne, Xe	PPC [155]
Ba II	2905.7	7p $^2P^o{}_{1/2} \rightarrow$ 6d $^2D_{3/2}$	He, Ne, Xe	PPC [155]
Eu (europium, Z = 63)				
Eu II	664.5	6p $^9P_5 \rightarrow$ 5d $^9D^o{}_6$	He ^{d}	PPC [156], QCWPC [156]
Eu II	989.8	6p $^7P_3 \rightarrow$ 5d $^7D^o{}_4$	He	QCWPC [156]
Eu II	1001.9	6p $^7P_4 \rightarrow$ 5d $^7D^o{}_5$	He	QCWPC [157, 156]
Eu II	1016.8	6p $^7P_4 \rightarrow$ 5d $^7D^o{}_4$	He	QCWPC [157, 156]
Eu II	1361.1	6p $^9P_4 \rightarrow$ 5d $^7D^o{}_5$	He ^{d}	PPC [157, 156], QCWPC [156]
Eu II	1477	6p $^9P_3 \rightarrow$ 5d $^7D^o{}_4$	He ^{d}	PPC [156], QCWPC [156]
Yb (ytterbium, Z = 70)				
Yb II	1271.4	–	He, Ne, Ar	PPC [158]
Yb II	1345.3	5p $^2P^o{}_{3/2} \rightarrow$ 5d $^2D_{3/2}$	He, Ne, Ar	PPC [158]
Yb II	1649.8	5p $^2P^o{}_{3/2} \rightarrow$ 5d $^2D_{5/2}$	He, Ne, Ar	PPC [158]
Yb II	1805.7	–	He, Ne, Ar	PPC [158]
Yb II	2148.0	–	He, Ne, Ar	PPC [158]
Yb II	2437.7	5p $^2P^o{}_{1/2} \rightarrow$ 5d $^2D_{3/2}$	He, Ne, Ar	PPC [158, 159]
Au (gold, Z = 79)				
Au II	226.4	7s $^3D_1 \rightarrow$?	He	QCWHC [74]
Au II	253.37	7s $^1D_2 \rightarrow$?	He	QCWHC [160]
Au II	261.65	7s $^3D_1 \rightarrow$?	He	QCWHC [160]
Au II	282.25	7s $^1D_2 \rightarrow$?	He	CWHC [160]
Au II	284.70	7s $^3D_1 \rightarrow$?	He	CWHC [160]
Au II	286.3	–	He	QCWHC [160]
Au II	288.82	–	He	QCWHC [160]
Au II	289.33	7s $^1D_2 \rightarrow$?	He	QCWHC [160]
Au II	291.82	7s $^3D_1 \rightarrow$?	He	QCWHC [160]
Au II	295.94	–	He	QCWHC [160]
Au II	551.63		He	QCWHC [160]
Au II	552.21		He	QCWHC [160]
Au II	621.23		He	QCWHC [160]
Au II	670.14		He	QCWHC [160]
Au II	690.29		He	QCWHC [160]
Au II	694.03		He	QCWHC [160]
Au II	755.58	? $\rightarrow$ 7s $^3D_2$	He	QCWHC [160]
Au II	759.29		He	QCWHC [160]
Au II	760.05	? $\rightarrow$ 7s $^3D_2$	He	QCWHC [160]

(continued overleaf)

Table A4.1 (*continued*)

Spectrum	λ(nm)	Transition	Buffer Gas	Regimes, types of discharge, references
Au II	760.67		He	QCWHC [160]
Au II	763.51		He	QCWHC [160]
Au II	827.29	? → 7s $^3D_3$	He	QCWHC [80]
Au II	859.9	? → 7s $^3D_2$	He	QCWHC [74]
Au II	886.76	? → 7s $^3D_3$	He	QCWHC [80]
Au II	1695		He	LPP [49]
Hg (mercury, Z = 80)				
Hg II	398.4	6p $^2P^o{}_{3/2}$ → 6s^2 $^2D^*{}_{5/2}$	He	PPC [161]
Hg III	479.701	5d^{8}6s^2(J=4) → 5d^{9}6p(J=3)	He, no buffer [b]	PPC [53], QCWPC [162]
Hg III	521.0	5d^{8}6s^2(J=1) → 5d^{9}6p(J=2)	no buffer	QCWPC [162]
HgII	567.73	5f $^2F^o{}_{7/2}$ → 6p $^2D_{5/2}$	He, Ne, Ar, no buffer [b]	PPC [53, 126, 163–5]
Hg II	614.950	7p $^2P^o{}_{3/2}$ → 7s $^2S_{1/2}$	He	PPC [53, 163–6, 165], PHC [167], QCWHC [82], CWHC [168], CWPC [41, 169], NP [170], TRFD [83], PTD [130], EBP [84]
Hg III	650.15	5d^{8}6s^2(J=2) → 5d^{9}6p(J=2)	no buffer	QCWPC [162]
Hg II	734.66	7d $^2D_{5/2}$ → 7p $^2P^o{}_{3/2}$	He	PPC [163, 164]
Hg II	741.81	5d^{9}6s$_{1/2}$ 6p$_{3/2}$ $^2P^o{}_{3/2}$ (109189) → 7s $^2S_{1/2}$	He	PC [171]
Hg II	794.47	7p $^2P^o{}_{1/2}$ → 7s $^2S_{1/2}$	He	PHC [167], CWPC [169], CWHC [168], TRFD [83], PPC [171]
Hg II	854.98	5g $^2G_{7/2}$ → 'C' $^2F^o{}_{5/2}$	He	PPC [164]
Hg II	862.2	8p $^2P^o{}_{3/2}$ → $^4D^2{}_{5/2}$ (116200)	He	PPC [164]
Hg II	867.7	–	He	PPC [164]
Hg II	939.68	10s $^2S_{1/2}$ → 8p $^2P^o{}_{3/2}$	He	PPC [164]
Hg II	1058.3	8s $^2S_{1/2}$ → 7p $^2P^o{}_{3/2}$	He	PPC [163, 164]
Hg II	1117.9	7g → 6f $^2F^o{}_{5/2}$	He	PPC [163, 164]
Hg II	1555.5	7p $^2P^o{}_{3/2}$ → 6d $^2D_{5/2}$	He	PPC [163, 164]
Tl (thallium, Z = 81)				
Tl III	468±2	–	He	PPC [172]
Tl II	473.7	5f $^1F_3$ → 6d $^1D_2$	He	PPC [170], PHC [99]
Tl III	482±2	–	He	PPC [172]
Tl II	498.1	5f $^3F_2$ → 6d $^3D_1$	He	PPC [173], PHC [99]
Tl II	507.8	5f $^3F_3$ → 6d $^3D_2$	He	PPC [173], PHC [99]
Tl II	515.214	5f $^3F_4$ → 6d $^3D_3$	He	PPC [173], TRFD [83], PHC [99]
Tl II	582.9	7p $^1P_1$ → 7s $^3S_1$	Ne	PHC [83]

Table A4.1 (*continued*)

Spectrum	λ(nm)	Transition	Buffer Gas	Regimes, types of discharge, references
Tl II	595.0	7p $^3P_2 \rightarrow$ 7s $^3S_1$	Ne	PPC [173], CWPC [99], PHC [99], QCWHC [174], TRFD [83]
Tl II	695.0	7p $^1P_1 \rightarrow$ 7s $^1S_0$	Ne	PPC [173], PHC [99], QCWHC [175], TRFD [83]
Tl II	707.0	7p $^3P_0 \rightarrow$ 7s $^3S_1$	Ne	CWHC [99]
Tl III	770±5	–	Ne	PPC [172]
Tl III	806±2	5d^{10}6p $^2P^o_{3/2} \rightarrow$5d^{9}6s^2 $^2D_{5/2}$	Ne	PPC [172]
Tl II	913.0	5 $^3G_4 \rightarrow$ 5 $^3F_3$	He	PHC [99]
Tl II	922.7	5 $^3G_5 \rightarrow$ 5 $^3F_4$	He	PHC [99]
Tl II	935.0	–	Ne	PPC [173]
Tl II	1135.0	–	He	PPC [173], PHC [99]
Tl II	1175.0	6 $^3F_4 \rightarrow$ 7 $^3D_3$	He	PPC [173], PHC [99]
Pb (lead, Z = 82)				
Pb II	537.2	5f $^2F^o_{7/2} \rightarrow$ 6p^2 $^4P_{5/2}$	He, Ne+Kr	PPC [111, 91], CWHC [176], SPE [57], PHC [177]
Pb II	560.88	7p $^2P^o_{3/2} \rightarrow$ 7s $^2S_{1/2}$	He, Ne	PPC [91]
Pb II	666.0	7p $^2P^o_{1/2} \rightarrow$ 7s $^2S_{1/2}$	Ar, He, Ne	CWHC [176, 177], PPC [91]
Pb II	839.6	8s $^2S_{1/2} \rightarrow$ 7p $^2P^o_{3/2}$	He	CWHC [176], SPE [57]
Pb II	1159	6s6p^2 $\rightarrow$ 6s^{2}7p	Ne	PPC [178]
Pb II	1162.8	6f $^2F^o_{5/2} \rightarrow$ 7d $^2D_{3/2}$	Ar, Ne	PPC [111], SPE [57], PHC [177]
Pb II	1322.1	6f $^2F^o_{7/2} \rightarrow$ 7d $^2D_{5/2}$	He, Ne, Kr	PPC [111], CWHC [176, 177], LPP [49, 55], SPE [57]
Bi (bismuth, Z = 83)				
Bi III	456.084	7p $^2P^o_{1/2} \rightarrow$ 7s $^2S_{1/2}$	He, Ne	PPC [140]
Bi II	571.92	7p $^3P^o \rightarrow$ 7s $^3P^o_1$	He, Ne	PPC [140], CWPC [142], CWHC [179]
Bi II	660.0	7p $^3D_1 \rightarrow$ 7s $^3P_0$	He+Xe, Ar	CWHC [179]
Bi II	680.9	7p $^3D_1 \rightarrow$ 7s $^3P_1$	He+Xe, Ar	CWHC [179]
Bi III	759.9	6 $^2F^o_{5/2} \rightarrow$ 7 $^2D_{3/2}$	He, Ne	PPC [140], SPE [57]
Bi III	806.89	6 $^2F^o_{7/2} \rightarrow$ 7 $^2D_{5/2}$	He, Ne	PPC [140], SPE [57]

[a] The vapour of the element is produced by dissociation of the given chemical compound in the discharge.
[b] Electric discharge without a buffer gas ('no buffer').
[c] When the metal vapour is produced by cathode sputtering, heavier gases (argon, krypton, etc.) are added to helium (see Chapter 6).
[d] At low buffer-gas pressures, the transition is 'self-terminating'; at high buffer-gas pressures, quasi-CW laser oscillation is observed.
[e] Vapour (atoms) of the element are removed from the wall of a quartz tube by electric discharges in SF_6 and PF_6.

Table A4.2 Ion laser lines.

224.34	Ag II	441.563	Cd II	498.692	I II
226.4	Au II	442.2	P III	499.275	Se II
277.4	Ag II	446.760	Se II	501.26	Cu II
248.58	Cu II	448.855	I II	502.04	Te II
250.63	Cu II	450.60	Cu II	502.129	Cu II
252.93	Cu II	453.379	I ?	502.59	Cd II
253	Be IV	453.40	I II	502.73	Ag II
253.37	Au II	455.26	Si III	505.178	Cu II
257.18	Cu II	455.59	Cu II	505.46	Br II
259.05	Cu II	456.084	Bi III	506.06	Cu II
259.88	Cu II	456.78	Si III	506.865	Se II
260.03	Cu II	460.434	Se II	507.8	Tl II
261.45	I II	461.877	Se II	509.650	Se II
261.65	Au II	462.1	Ag II	513.17	Ge II
270.32	Cu II	464.74	C III	514.214	Se II
271.88	Cu II	464.844	Se II	514.57	C II
272.2	Cu II	465.02	C III	515.2	Tl II
274.00	Cu II	467.35	Cu II	517.598	Se II
282.25	Au II	467.42	I II	517.9	Ge II
284.70	Au II	467.440	I ?	518.238	Br II
286.33	Au II	467.5	Be II	521.0	Hg III
288.2	I ?	467.553	I II	521.408	I II
288.82	Au II	468.0	In II	521.627	I II
289.33	Au II	468.20	Cu II	522.751	Se II
290.63	I	468	Tl III	523.826	Br II
291.82	Au II	471.823	Se II	524.9	In III
293.17	I II	473.7	Tl II	525.307	Se II
295.94	Au II	474.097	Se II	525.363	Se II
298.3	In III	474.266	Br II	525.64	Te II
300.8	In III	476.365	Se II	527.111	Se II
318.07	Ag II	476.552	Se II	527.2	Be II
325.0	Cd II	478.84	Ag II	530.535	Se II
332.487	S III	479.701	Hg III	532.07	S II
334.77	P IV	482	Tl III	533.20	Br II
345.13	B II	484.063	Se II	533.749	Cd II
349.734	S III	484.29	Te II	533.82	I II
359.744	Al II	484.496	Se II	534.50	I II
363.202	S III	485.49	Cu II	534.57	S II
370.6	Ca II	488.2	Cd II	537.2	Pb II
370.937	S III	490.97	Cu II	537.804	Cd II
373.7	Ca II	491.16	Zn II	538.52	As II
398.4	Hg II	492.40	Zn II	540.736	I II
407.08	I ?	493.16	Cu II	541.9	I II
408.59	Ag II	493.467	I ?	542.87	S II
408.89	Si IV	493.47	I II	543.28	S II
416.2	Sr II	494.14	Br II	544.98	Te II
430.5	Sr II	495.41	C II	545.38	S II
436.2	Be II	497.566	Se II	545.4	Te II
441.54	Sb II	498.1	Tl II	547.36	S II

[45] V. V. Vainer, S. P. Zinchenko, I. G. Ivanov *et al., Zh. Prikl. Spektrosk.,* **31**(5), 905–7 (1979).
[46] L. A. Cross and M. C. Gokay, *J. Appl. Phys.,* **49**(5), 2639–50 (1978).
[47] B. A. Bryunetkin, V. I. Derzhiev, V. M. Dyakin, S. A. Maiorov and S. I. Yakovlenko, *Sov. Tech. Phys. Lett.,* **12**(5), 252–3 (1986).
[48] V. A. Boiko, B. A. Bryunetkin, F. V. Bunkin *et al., Sov. J. Quantum Electron.,* **13**(5), 565–6 (1983); **15**(11), 1543–4 (1985).
[49] O. R. Wood II, J. J. Macklin and W. T. Silfvast, *Appl. Phys. Lett.,* **44**(12), 1123–5 (1984).
[50] H. G. Cooper and P. K. Cheo, *IEEE J. Quantum Electron.,* **2**(12), 785 (1966).
[51] R. W. Waynant, *Appl. Phys. Lett.,* **22**(8), 419–20 (1973).
[52] R. A. McFarlane, *Appl. Phys. Lett.,* **5**(5), 91–3 (1964).
[53] W. B. Bridges and A. N. Chester, *IEEE J. Quantum Electron.,* **1**(2), 66–84 (1965).
[54] A. B. Peterson and M. Birnbaum, *IEEE J. Quantum Electron.,* **10**(4), 468 (1974).
[55] C. Momma, M. Hube, A. Tünnermann, K. Mossavi and B. Wellegehausen, *Appl. Phys. B,* **54**(3), 234–8 (1992).
[56] D. T. Hodges, *Appl. Phys. Lett.,* **18**(10), 454–6 (1971).
[57] W. T. Silfvast and O. R. Wood II, *Opt. Lett.,* **7**(1), 34–6 (1982).
[58] Ph. Cahuzac, *IEEE J. Quantum Electron.,* **8**(6), 500 (1972).
[59] D. C. Gerstenberger, R. D. Reid and G. J. Collins, *Appl. Phys. Lett.,* **30**(9), 466–8 (1977).
[60] W. K. Schuebel, *Appl. Phys. Lett.,* **30**(10), 516–9 (1977).
[61] J. J. Macklin, O. R. Wood II and W. T. Silfvast, *IEEE J. Quantum Electron.,* **18**(11), 1832–5 (1982).
[62] V. V. Zhukov, E. L. Latush and M. F. Sem, *Izv. Ser.-Kav. Nauch. Sentr. Vissh. Shkoli. Ser. est. Nauki.,* **11,** 38–42 (1977) (in Russian).
[63] P. K. Cheo and H. G. Cooper, *Appl. Phys. Lett.,* **7**(7), 202–4 (1965).
[64] G. R. Fowles, W. T. Silfvast and R. C. Jensen, *IEEE J. Quantum Electron.,* **1**(4), 183–4 (1965).
[65] M. F. Sem, V. F. Keidan and V. C. Mikhalevskii, *Elektron. Tekh., Ser. 1 Elekronika SVCh,* No. 1, 69–76 (1971) (in Russian).
[66] M. S. Chiu and C. C. Chou, *IEEE J. Quantum Electron.,* **17**(9), 1592–3 (1981).
[67] M. S. Butler and J. A. Piper, *Appl. Phys. Lett.,* **42**(12), 1008–10 (1983); **43**(9), 823–5 (1983).
[68] W. T. Walter, N. Solimene, M. Piltch and G. Gould, *IEEE J. Quantum Electron.,* **2**(9), 474–9 (1966).
[69] K. Jain, *Appl. Phys. Lett.,* **37**(4), 362–4 (1980).
[70] K. Jain, *Appl. Phys. Lett.,* **34**(12), 845–6 (1979).
[71] J. R. McNeil, G. J. Collins, K. B. Persson and D. L. Franzen, *Appl. Phys. Lett.,* **28**(4), 207–9 (1976).
[72] J. R. McNeil and G. J. Collins, *IEEE J. Quantum Electron.,* **12**(6), 371–2 (1976).
[73] R. Solanki, W. M. Fairbank Jr and G. J. Collins, *IEEE J. Quantum Electron.,* **16**(12), 1292–4 (1980).
[74] K. Jain, *IEEE J. Quantum Electron.,* **16**(4), 387–8 (1980).
[75] J. R. McNeil, G. J. Collins, K. B. Persson and D. L. Franzen, *Appl. Phys. Lett.,* **27**(11), 595–8 (1975).
[76] V. S. Mikhalevskii, M. F. Sem, G. N. Tolmachev and V. Ya Khasilev, *J. Appl. Spectrosc.,* **32**(4), 321–3 (1980).

[77] R. Solanki, E. L. Latush, D. C. Gerstenberger, W. M. Fairbank Jr and G. J. Collins, *Appl. Phys. Lett.,* **35**(4), 317–19 (1979).
[78] L. Csillag, M. Jánossy, K. Rósza and T. Salamon, *Phys. Lett. A,* **50**(1), 13–14 (1974).
[79] S. P. Zinchenko, I. G. Ivanov and M. F. Sem, *Summaries of Papers. Population Inversion and Oscillation on Atomic and Molecular Transitions,* Part 1, ed. O. V. Ravodina and I. I. Murav'ev, p. 146 Tomsk: Institute of Atmospheric Optics (Sept. 1986) (in Russian).
[80] K. Jain, *Appl. Phys. Lett.,* **34**(6), 398–9 (1979).
[81] R. Solanki, E. L. Latush, W. M. Fairbank Jr and G. J. Collins, *Appl. Phys. Lett.,* **34**(9), 568–70 (1979).
[82] É. K. Karabut, V. F. Kaydan, V. F. Kravchenko *et al., Elektron. Tekh., Ser. 1 Elektronika SVCh,* No. 9, 103–8 (1969) (in Russian).
[83] E. L. Latush, V. S. Mikhalevskii, M. F. Sem, G. N. Tolmachev and V. Ya Khasilev, *JETP Lett.,* **24**(1), 69–71 (1976).
[84] J. J. Rocca and G. J. Collins, *Optroelectronics, Instrumentation and Data Processing,* No. 1, 1–16 (1984).
[85] G. R. Fowles and W. T. Silfvast, *IEEE J. Quantum Electron.,* **1**(3), 131 (1965).
[86] R. C. Jensen, G. J. Collins and W. R. Bennett Jr, *Phys. Rev. Lett.,* **23**(7), 363–7 (1969).
[87] A. L. Bloom and J. P. Goldsborough, *IEEE J. Quantum Electron.,* **6**(3), 164 (1970).
[88] É K. Karabut, V. C. Michalevskii, V. F. Papakin and M. F. Sem, *Sov. Phys. – Tech. Phys.,* **14**(10), 1447–8 (1970).
[89] Y. Sugawara, V. Tokiwa and T. Iijima, *Jpn. J. Appl. Phys.,* **9**(12), 1537 (1970); **9**(5), 588–9 (1970).
[90] N. Sabotinov and P. Telbizov, *Bulg. J. Phys.,* **11**(5), 559–64 (1984) (in Russian); *Opt. Commun.,* **59**(4), 290–2 (1986).
[91] W. T. Silfvast, G. R. Fowles and B. D. Hopkins, *Appl. Phys. Lett.,* **8**(12), 318–19 (1966).
[92] J. A. Piper and P. Gill, *J. Phys. D,* **8**(1), 127–34 (1975).
[93] W. T. Silfvast, *Appl. Phys. Lett.,* **15**(1), 23–5 (1969).
[94] J. A. Piper and C. E. Webb, *J. Phys. D,* **6**(4), 400–7 (1973).
[95] H. Lundberg, J. J. Macklin, W. T. Silfvast and O. R. Wood II, *Appl. Phys. Lett.,* **45**(4), 335–7 (1984).
[96] T. Iijima and Y. Sugawara, *Phys. Lett.,* **54A**(1), 93–5 (1975).
[97] M. S. Chou and T. A. Cool, *J. Appl. Phys.,* **48**(4), 1551–5 (1977).
[98] V. F. Kravchenko, V. S. Mikhalevskii, S. P. Chubar' and A. P. Shelepo, *Sov. J. Quantum Electron.,* **14**(6), 725–7 (1984).
[99] I. G. Ivanov and M. F. Sem, *Elektron. Tekh., Ser. 4 Elektrovakuumiye i Gazorazryadniye Pribori,* No. 2, 12–16 (1974) (in Russian).
[100] L. Jian-bang, *Appl. Phys. B,* **29**(4), 251–3 (1982).
[101] L. Jian-bang, *Appl. Phys. B,* **32**(4), 211–15 (1983).
[102] J. A. Piper and C. E. Webb, *J. Phys. B,* **6**(5), L116–20 (1973).
[103] W. E. Bell, A. L. Bloom and J. P. Goldsborough, *IEEE J. Quantum Electron.,* **1**(9), 400 (1965).
[104] J. A. Piper, *Opt. Commun.,* **31**(3), 374–6 (1979).
[105] M. B. Klein and W. T. Silfvast, *Appl. Phys. Lett.,* **18**(11), 482–5 (1971).
[106] W. T. Silfvast and M. B. Klein, *Appl. Phys. Lett.,* **17**(9), 400–3 (1970).
[107] L. Lis, *Acta Phys. Pol. A,* **44**(2), 173–6 (1973).
[108] V. F. Keidan and V. S. Mikhalevskii, *Zh. Prikl. Spektrosk.,* **15**(2), 331–2 (1971)

(in Russian).
[109] W. M. Keeffe and W. J. Graham, *Appl. Phys. Lett.*, **7**(10), 263–4 (1965); *Phys. Lett.*, **20**(6), 643 (1966).
[110] Z. Zhou, M. Shao, M. Qiu and F. Qiu, *Opt. Commun.*, **47**(3), 218–22 (1983).
[111] V. V. Zhukov, V. S. Kucherov, E. L. Latush and M. F. Sem, *Sov. J. Quantum Electron.*, **7**(6), 708–14 (1977).
[112] M. S. Butler and J. A. Piper, *Appl. Phys. Lett.*, **42**(12), 1008–10 (1983); **43**(9), 823–5 (1983).
[113] M. Brandt, *Appl. Phys. Lett.*, **42**(2), 127–9 (1983); *IEEE J. Quantum Electron.*, **20**(9), 1006–7 (1984).
[114] A. M. Bogus, V. L. Dzhikiya and A. A. Chernov, *Sov. J. Quantum Electron.*, **8**(2), 259–60 (1978).
[115] Yu V. Koptev, E. L. Latush and M. F. Sem, *Sov. J. Quantum Electron.*, **20**(4), 346–8 (1990).
[116] P. A. Bokhan and A. R. Sorokin, *Opt. Quantum Electron.*, **23**(4), S523–38 (1991).
[117] J. W. Bethel, A. Maitland, P. M. Beecham, T. C. Bell, S. Webb and C. E. Little, *Opt. Quantum Electron.*, **26**(12) 1079–87 (1994).
[118] É K. Karabut, V. F. Kravchenko and V. S. Mikhalevskii and A. P. Shelepo, **5**(11), 1371–2 (1975).
[119] J. S. Deech and J. H. Sanders, *IEEE J. Quantum Electron.*, **4**(7), 474 (1968).
[120] E. L. Latush, R. Solanki and G. J. Collins, *Phys. Lett. A,* **73**(5,6), 387–8 (1979).
[121] J. R. McNeil, W. L. Johnson, G. J. Collins and K. B. Persson, *Appl. Phys. Lett.*, **29**(3), 172–4 (1976).
[122] W. L. Johnson, J. R. McNeil and G. J. Collins, *Appl. Phys. Lett.*, **29**(2), 101–2 (1976).
[123] R. D. Reid, W. L. Johnson, J. R. McNeil and G. J. Collins, *IEEE J. Quantum Electron.*, **12**(12), 778–9 (1976).
[124] J. P. Goldsborough, *Appl. Phys. Lett.*, **15**(6), 159–61 (1969).
[125] W. T. Silfvast, J. J. Macklin and O. R. Wood II, *Opt. Lett.*, **8**(11), 551–3 (1983).
[126] V. V. Zhukov, I. G. Ivanov and M. F. Sem, *Zh. Prikl. Spektrosk.*, **26**(3), 544–7 (1977) (in Russian).
[127] W. T. Silfvast, *Appl. Phys. Lett.*, **13**(5), 169–71 (1968).
[128] G. R. Fowles and B. D. Hopkins, *IEEE J. Quantum Electron.*, **3**(10), 419 (1967).
[129] E. A. Verni, L. D. Mash and B. M. Rabkin, *Zh. Prikl. Spektrosk.*, **16**(1), 156–7 (1972) (in Russian).
[130] V. G. Il'yushko, V. F. Kravchenko, V. S. Mikhalevskii and V. A. Polunin, *J. Appl. Spectrosc.*, **28**(1), 106–7 (1978).
[131] A. I. Mis'kevich, V. S. Il'yashenko, B. S. Salamakha, A. A. Sipailo, V. A. Stepanov and E. M. Gorodkov, *Sov. Phys. - Tech. Phys.*, **27**(2), 260–1 (1982).
[132] W. K. Schuebel, *Appl. Phys. Lett.*, **16**(11), 470–2 (1970); *IEEE J. Quantum Electron.*, **6**(9), 574–5 (1970).
[133] W. T. Silfvast, L. H. Szeto and O. R. Wood II, *Appl. Phys. Lett.*, **36**(8), 615–7 (1980); **39**(3), 212–14 (1981).
[134] A. I. Mis'kevich, A. B. Dmitriev, V. S. Il'yashchenko, B. S. Salamakha, V. A. Stepanov and E. M. Gorodkov, *Sov. Tech. Phys. Lett.*, **6**(7), 352–3 (1980).
[135] K. Toyoda, M. Kobiyama and S. Namba, *Jpn. J. Appl. Phys.*, **15**(10), 2033–4 (1976).
[136] A. A. Aleksandrov, V. V. Elagin and A. E. Fotiadi, *Pis'ma Zh. Tekh. Fiz.*, **6**(3), 160–1 (1980) (in Russian).

[137] A. S. Tibilov, *Opt. Spectrosc.*, **19,** 463–4 (1965).
[138] V. V. Zhukov, E. L. Latush, V. S. Mikhalevskii and M. F. Sem, *Sov. J. Quantum Electron.*, **5**(4) 468–9 (1975).
[139] K. Gnädig and L. Fucheng, *Opt. Commun.*, **34**(2), 218–20 (1980).
[140] V. F. Keidan and V. S. Mikhalevskii, *J. Appl. Spectrosc.*, **9**(4), 1154 (1968).
[141] W. E. Bell, A. L. Bloom and J. P. Goldsborough, *IEEE J. Quantum Electron.*, **2**(6), 154 (1966).
[142] I. G. Ivanov and M. F. Sem, *Elektron. Tekh., Ser. 4 Elektrovakuumiye i Gazorazryadniye Pribori,* No. 9, 102–3 (1972) (in Russian).
[143] W. T. Silfvast and M. B. Klein, *Appl. Phys. Lett.*, **20**(12), 501–4 (1972).
[144] C. E. Webb, *IEEE J. Quantum Electron.*, **4**(6), 426–7 (1968).
[145] J. A. Piper and C. E. Webb, *IEEE J. Quantum Electron.*, **12**(1), 21–5 (1976).
[146] C. S. Willet and O. S. Heavens, *Opt. Acta,* **13**(3), 271–3 (1966).
[147] R. C. Jensen and G. R. Fowles, *Proc. IEEE,* **52**(11), 1350 (1964).
[148] J. A. Piper, G. J. Collins and C. E. Webb, *Appl. Phys. Lett.*, **21**(5), 203–5 (1972).
[149] J. A. Piper, *J. Phys. D,* **7**(2), 323–8 (1974).
[150] C. S. Willett, *IEEE J. Quantum Electron.*, **3**(1), 33 (1967).
[151] G. R. Fowles and R. C. Jensen, *Proc IEEE,* **52**(7), 851–2 (1964).
[152] T. Goto, H. Kano, S. Yoshino, J. K. Mizeraczyk and S. Hattori, *J. Phys. E,* **10**(3), 292–5 (1977).
[153] A. A. Isaev, M. A. Kazaryan and G. G. Petrash, *Sov. J. Quantum Electron.*, **3**(4), 358–9 (1974).
[154] K. U. Baron and B. Stadler, *IEEE J. Quantum Electron.*, **11**(10), 852–3 (1975).
[155] Ph Cahuzac, *Phys. Lett. A,* **32**(3) 150–1 (1970).
[156] P. A. Bokhan, V. M. Klimkin and V. E. Prokop'ev, *Sov. J. Quantum Electron.*, **4**(6), 752-4 (1974).
[157] P. A. Bokhan, V. M. Klimkin and V. E. Prokop'ev, *JETP Lett.,* **18**(2), 44–5 (1973).
[158] V. M. Klimkin, *Sov. J. Quantum Electron.,* **5**(3), 326–9 (1975).
[159] Ph Cahuzac, *Phys. Lett. A,* **31**(10), 541–2 (1970).
[160] R. D. Reid, J. R. McNeil and G. J. Collins, *Appl. Phys. Lett.,* **29**(10), 666–8 (1976).
[161] A. A. Isaev and G. G. Petrash, *Zh. Prikl. Spektrosk.,* **12**(6), 1118–20 (1970) (in Russian).
[162] P. Burkhard, H. R. Lüthi and W. Seelig, *Opt. Commun.,* **18**(4), 485–7 (1976).
[163] W. E. Bell, *Appl. Phys. Lett.,* **4**(2), 34–5 (1964).
[164] A. L. Bloom, W. E. Bell and F. O. Lopez, *Phys. Rev.,* **135**(3A), 578–9 (1964).
[165] G. Convert, M. Armand and P. Martinot-Lagarde, *C. R. Acad. Sci.,* **258**(12), 3259–60 (1964).
[166] N. Suzuki, *Jpn. J. Appl. Phys.,* **4**(6), 452–7 (1965).
[167] R. L. Byer, W. E. Bell, E. Hodges and A. L. Bloom, *J. Opt. Soc. Am.,* **55**(12), 1598–602 (1965).
[168] W. K. Schuebel, *IEEE J. Quantum Electron.,* **7**(1), 39–40 (1971).
[169] A. Ferrario, *Opt. Commun.,* **7**(4), 376–8 (1973).
[170] M. A. Akerman, G. H. Miley and D. A. McArthur, *Appl. Phys. Lett.,* **30**(8), 409–12 (1977).
[171] J. P. Goldsborough and A. L. Bloom, *IEEE J. Quantum Electron.,* **5**(9), 459–60 (1969).
[172] E. L. Latush, M. F. Sem and G. D. Chebotarev, *Summaries of Papers. Population Inversion and Oscillation on Atomic and Molecular Transitions,*

Part 1, ed. O. V. Ravodina and I. I. Murav'ev, p. 145, Tomsk: Institute of Atmospheric Optics (Sept. 1986) (in Russian).

[173] I. G. Ivanov and M. F. Sem, *Zh. Prikl. Spektrosk.*, **19**(2), 358–60 (1973) (in Russian).

[174] M. G. Grozeva, N. V. Sabotinov and N. K. Vuchkov, *Opt. Commun.*, **29**(3), 339–40 (1979).

[175] M. G. Grozeva, N. V. Sabotinov, P. K. Telbizov and N. K. Vuchkov, *Opt. Commun.*, **31**(2) 211–13 (1979).

[176] K. Gnädig, *Opt. Commun.*, **35**(3), 427–9 (1980).

[177] Lin Fucheng, *Opt. Commun.*, **36**(5), 387–90 (1981).

[178] L. H. Szeto, W. T. Silfvast and O. R. Wood II, *IEEE J. Quantum Electron.*, **15**(12) 1332–4 (1979).

[179] Zhu Xu-hui and Liu Jian-bang, *Appl. Phys. B*, **29**(4), 291–2 (1982)

Index

acoustic oscillations in discharge 53
afterglow 48–53
  recombination pumping in 13–21
  spectra 109–11
alkaline earth metal lasants 122, 130, 151, 159
ambipolar diffusion 28–30, 41, 83, 106–7, 177
  coefficient 30
  in afterglow 48–9
applications of MVILs
  active optical system 227
  colour displays 233
  dye laser pumping 156, 228
  fluorescence spectra 194
  holography 157, 194, 228, 233
  impurity and pollutant detection 194
  information processing 227, 233
  isotope separation 194
  medicine 157, 194
  microtechnology 157
  multicolour generation 233–41
  photo-lithography 228
  photoresist exposure 157, 194
  projection television 157
  quality control 228
  recording on photosensitive materials 193
  scattering (Raman, Brillouin, Rayleigh) 194
  semiconductor substrate cleaning 157
  solid state laser pumping 156–7
Ar ion lasers 156–7, 178, 193–4, 197–9, 234
axial electric field intensity 28-34

Bessel
  current profile 35
  electron density 29
Bethe approximation 6
Beutler levels 39–40, 88–97, 108–11, 113, 177, 181, 189–90, 205, 207, 218
Blanc's law 36
Blumlein-type circuit 81–2, 144
Boltzmann distribution 15–16
Born approximation 5
buffer gas 1, 7–9, 11–12, 15
  efficiency of excitation and ionization 74
  electrophoresis of 43–5
  metastables
    density in positive column 36–8
    density in hollow cathode 54, 62–3
    excitation of (table) 30
    in hollow cathode 63–5
    in pulsed lasers 108–9, 111
  role in noise generation in cataphoretic lasers 191
  role in recombination lasers 129–33
burst-mode excitation 71, 220

cataphoresis 1, 3, 28, 40–3, 216
cataphoretic lasers 167–94, 227
  above–optimum discharge current 186–8
  applications 192–4
  dependence of gain on
    buffer gas pressure 184–5
    current 181–3
    metal vapour pressure 183–4
    tube diameter 185–6
  multi-colour output 233–4, 236–40
  multiple-metal lasers 236–40
  output power and efficiency 188–90
  parameters of commercial devices 193
cathode 80–1
  helical hollow 67, 204–5, 211–12
  sputtering 1, 3, 55, 67–8, 167, 197, 209–12, 219
cathode dark space (CDS) 26–7, 53, 68–9, 73–4, 204, 208, 215–16, 224–5
charge transfer 8–12
  criteria for population inversion by 12
  efficiency 71–4
  in cataphoretic lasers 177–88

in hollow cathode 64–5, 199, 205–7, 234
from molecular helium ions 38
in pulsed lasers 88, 97–103, 108, 112–3, 116, 118, 124, 141–3
in transverse RF lasers 219
charged particle densities
contribution to metal ionization 38–40
in afterglow 48
measurement 34–6
CIE chromatic diagram 234
CO_2 lasers 160–2
colour triangle 233–4, 241
copper vapour lasers (self–terminating) 2, 157, 212
Coulomb logarithm 14, 50
cross sections
charge transfer 11–13
electron impact 5, 6
heavy body processes 9, 10
ionization 30
optical pumping 21
Penning ionization 10
tabulated 248–52
current–voltage characteristic 34, 69
CW laser oscillation
advantages 4
due to favourable lifetime ratio 22

deexcitation
electron collisional 4, 6–7, 15–21, 65, 100, 103–6, 111, 113, 212
of helium metastables (table) 30
heavy body collisional 12–13
probability of 7
radiative 4, 7, 16, 22
Detailed Balance, Principle of 6
diffusion
metal ions, atoms 1, 2, 167
diffusion limited discharge 28–
discharge heating, *see* self-heating
double–pulse excitation 84, 133
Duffendack reactions, *see* charge transfer

E/*p* ratio 5, 32, 48, 75, 85
efficiency of excitation 71–5
cataphoretic lasers 188–90
hollow cathode lasers 212
transversely excited lasers 221–3
electric field
longitudinal 32–48, 172–3, 192, 203–4
radial 40, 216
electrical excitation 81–2, 151, 158
clipping thyratron 149
peaking capacitor 81–2
storage capacitor 47, 81–2, 144
transversely excited lasers 226–8
electron beam pumped lasers 75, 109, 156, 222–5
electron cooling 15
electron density
measurement 35–6
effect of metal atoms 48, 53
in recombination lasers 127–9
electron energy distribution function 5
in hollow cathode 26, 56–64, 204
in pulse hollow cathode 68–71, 215
in pulsed positive column 48
in stationary positive column 28–34
electron impact excitation
direct 4–6, 82
contribution to metal ionization 38–40
in cataphoretic lasers 176–7
lasers 82–7, 97–103
probability 7
stepwise 4–6, 9, 82
electron temperature 1, 5–6
and buffer gas in afterglow 129–33
and recombination 14–15, 20
dependence on key parameters 31–4
during current pulse 48
effects of additives 3
effect of metal atom density 31–2, 48
in afterglow 48–53
influence on laser ion spectra 106–8
in recombination lasers 127–9
electrophoresis 28, 43–5, 172
energy level diagrams
Al II 139
Ba II 134
Be II 138
Ca II 121
Cd II 11, 90
Eu II 85
Hg II 91
Mg II 136
Sr II 120
Zn II 89
excitation mechanisms, *see* electron impact excitation, ‘Penning ionization’, ‘charge transfer’, etc.

excitation potentials of metastable levels 244–6

gain coefficient
  cataphoretic lasers 180–1
  determination of 174
  hollow cathode lasers 205
  influence of isotopic and hyperfine splitting 22–3
gas discharges 26–75
  hollow–cathode 6–7, 26–7, 112, 167, 197–218
    pulsed 68–71, 151, 155, 215
    longitudinal 203–4
    steady state 53–68
      electron energy and concentration 56–62
      excitation processes 63–7
      helium metastables 62–3
      metal atom density 64
    transverse 203–4
  microwave 156
  'open' 151, 155–6
  positive column 26–53
    axial electric field 28–34
    electron temperature 28–34
    longitudinal 26–53
    pulsed 26, 46–53
      afterglow 48–53
      breakdown 46–7
      electron density 46
      high current stage 47–8
      lasers 79–113
    steady-state 26–7
      lasers 167–94
    transverse 26–7, 109, 150–6, 235
  RF
    longitudinal 28
    transverse 6, 53–68, 73–4, 197–8, 219
    types 26–7
  segmented arc 158–60
gas return path 172–3, 180
gold vapour lasers (self–terminating) 2, 157

He-Ne laser 3, 193, 227, 236
  efficiency 190
'heat-pipe' effect 202–3
heavy–body collisions
  efficiency of excitation 71–4
  excitation 4, 7–13, 87–113
  rate constants tabulated 248–52
helium
  molecular 92, 133
  sorption of 171–2
hollow-cathode lasers 6, 7, 26–7, 112, 167, 197–218, 224–5
  applications 227–8
  efficiency 221–3
  longitudinal discharge 203–4
  multi-colour output 234–9, 241
  mixed-metal 241
  multiple–metal 236–9
  pulsed 151, 155, 212–18
  stability 208
  transverse discharge 203–4, 207
hyperfine splitting
  influence on lasing 22–3

ion drift velocity 41–3
ion lasers
  aluminium 112, 117, 137, 139, 159, 198, 209, 212, 254
  antimony 113, 170, 175, 180, 238
  arsenic 27, 112, 170, 180, 198, 225, 258
  barium 82–3, 117, 134–5, 159, 198, 266–7
  beryllium 112, 117, 124, 137–8, 161–2, 198, 212, 253
  bismuth 170, 175, 180, 198, 238, 269
  boron 253
  bromine 27, 87, 260
  cadmium 7, 10, 11, 21–2, 27–8, 31–2, 35–45, 49, 63, 87–100, 103, 105–13, 116–19, 122, 124, 142, 159–62, 167–83, 185–94, 197–9, 204–9, 212–13, 217–22, 225–6, 234–7, 239–41, 262–3
  calcium 7, 13, 49, 82–4, 112, 117, 119–33, 141–59, 198, 212–13, 255
  carbon 159, 253
  chlorine 87
  chromium 197–8, 198, 255
  copper 3, 112–3, 159, 167, 197–8, 209–12, 219, 256–7
  europium 82–3, 85–7, 170, 267
  gallium 112, 198, 212, 258
  germanium 198, 258
  gold 3, 112, 197–8, 209–12, 267–8
  indium 159, 263

iodine 112, 170, 175, 179, 198, 206–7, 225, 238, 241, 265–6
lead 82–3, 112–13, 117, 140, 159, 162, 198, 269
lithium 159
magnesium 112–13, 117, 135–7, 159, 162, 170, 253–4
mercury 82–3, 87–100, 103–8, 112, 116–19, 122, 124, 141–2, 170, 175, 178–9, 197–9, 206, 212–15, 217–18, 221–2, 225–6, 234, 236, 238, 241, 268
nickel 198, 209, 255–6
phosphorus 170, 180, 254–5
selenium 27–8, 33, 44, 112, 170, 175, 178–9, 191–4, 198, 225, 233, 236–40, 258–60
silicon 254
silver 3, 112, 159, 167, 197–8, 209–12, 221, 225, 261–2
sulphur 255
strontium 7, 13, 49, 51–2, 82–5, 112, 117, 119–33, 141–58, 198, 212–13, 260–1
tellurium 112, 170, 175, 179–80, 198, 238, 264–5
thallium 112, 117, 138–40, 170, 175, 198, 204, 212–13, 217–18, 268–9
tin 112–13, 117, 140, 159, 162, 170, 198, 263–4
ytterbium 82–3, 267
zinc 7, 10, 11, 21–2, 87–100, 103–13, 116–19, 122, 124, 142, 159, 162, 170, 175, 177–8, 189–90, 197–9, 204–7, 209, 212–13, 219, 225, 233, 236–7, 239–40, 257–8

ion laser transitions
by element 253–69
by wavelength 270–3

ionization
of metal atoms 1, 30, 38–41, 71–4
role in noise generation 190–2

ionization potentials 244–6
ionization regime 4
isoelectronic scaling 159
isotopes
influence on lasing 22–3, 92
single 171, 174, 179, 188, 192, 217

kinetic modelling
afterglow 101–3
afterglow spectra 109–11
inversion on Sr II 430.5 nm transition 126–7

KrF lasers 162
Kr ion lasers 157, 197–9, 204, 233

laser–induced plasma 116
recombination lasers 160–2

laser tube
construction
cataphoretic lasers 167–73, 192–4
hollow–cathode lasers 199–203, 209–12
mixed–metal lasers 239–40
multiple–metal lasers 237
pulsed positive column lasers 79–82
recombination lasers 144–51
SPER lasers 158–9
transverse RF 218–19, 226–7
rectangular bore 148
temperature 2

metal halides 3
metal seeding 2
by cataphoresis 167–73
by cathode sputtering 67–8, 167, 197, 209–12
by molecular donor 2, 167–8, 197, 212, 241

metal vapour pressure 1
metastable ions 83–4
microwave diagnostics 35
multi–colour laser output 220–1, 228, 233–41
multiple and mixed metal ion lasers 236–41

Nd:YAG lasers 160–2
negative glow 26–7, 53–, 73–4, 204–7
noise on laser output 173
in cataphoretic lasers 190–2
low in hollow cathode lasers 207, 227
low in transverse RF lasers 221, 227

optical cavity
losses 174
Sr and Ca ion lasers 144

optical pumping 21–22
output power
saturation with current 7

oven, *see* reservoir

peaking capacitor 81–2
Penning ionization 8–11
  criteria for population inversion by 10
  contribution to metal ionization 38–40
  efficiency 71–4
  in cataphoretic lasers 176–88, 191
  in hollow cathode 64–5, 199, 205, 207, 215, 234
  in pulsed lasers 88, 97–103, 105, 108–13, 117, 119, 141
  in transverse RF lasers 219
population inversion 3
  by radiative decay 22
  in pulsed regime 4
  in recombining plasma 15–21
  role of electron deexcitation 7
  in cataphoretic lasers 180–8
  in Sr and Ca recombination lasers 124–7
positive–column (cw) excited lasers, *see* cataphoretic lasers
probe technique 34, 36, 51, 99, 128–9
pulsed lasing advantages 4
pulsed power 80–1

radial profiles
  in hollow cathode
    electron density 57–9
    metal atom density 68, 170
    spectral line intensities 57–8
radiation trapping 12, 20, 111
  absence of 14
Raoult's law for alloys 238
rate constants 6
  electron impact deexcitation 6–7, 248–52
  electron impact excitation 6
  electron impact for atoms and ions compared 7
  excitation and deexcitation of helium metastables (table) 30
  heavy body collisions 8, 248–52
  recombination 14, 124, 161, 248–52
  tabulated 248–52
recombination pumped lasers 1, 8, 13–21, 48–53, 88, 116–62
  advantages 13–14
  alternatives to recombination pumping 141–3
  applications 152–8
  short and long pulse oscillation 132–3, 145
  pulse recurrence frequency 133
recombination regime 4, 13–21
reservoir (metal source) 2, 41–3, 80, 167–71, 202, 204

scaling of laser power
  recombination lasers 144–5, 150–1, 160
  transversely excited lasers 226
second-kind collisions, *see* heavy–body collisions
secondary emission coefficient (g) 55
self-heating 2, 79–80, 143–4, 167
self-terminating lasers 2, 6, 47, 82–, 150, 212
side-arm oven, *see* reservoir
similarity
  in cataphoretic lasers 189
  in gas discharges 32, 34
SPER (segmented arc) lasers 116, 158–60
storage capacitor 47, 81–2, 144

temperature stabilization 173
temperature
  laser operation 247
thermal transfer 145–9
transverse RF discharge lasers 73–4, 197–8, 207, 218–25
  applications 227–8
  efficiency 221–3
  noise 221
  multi-colour output 221–1, 234, 237–9

'white' light lasers 208, 233–41

X-ray lasers 162